Quantum Mechanics
- A Philosophical Perspective

DON HAINESWORTH

AuthorHouse™
1663 Liberty Drive
Bloomington, IN 47403
www.authorhouse.com
Phone: 1 (800) 839-8640

This book is printed on acid-free paper.

ISBN: 978-1-7283-2557-6 (sc)
ISBN: 978-1-7283-2558-3 (e)

Print information available on the last page.

Published by AuthorHouse 09/17/2019

authorHOUSE®

Contents

An Overview of Quantum Mechanics v
Preface - Quantum Mechanics Interpretation ix
Ch 1 The Nature of Light 1
Ch 2 Light Quanta 16
Ch 3 Early Quantum Physics 23
Ch 4 Quantum Physics 47
Ch 5 Particle Physics 102
Ch 6 The Structure of Particle Physics – The Standard Model 145
Ch 7 Philosophies and Paradoxes in Physics 156
Ch 8 Quantum Entanglement (Supplemental) 208
Ch 9 Delayed Choice Quantum Eraser Experiment 211
Ch 10 Extra Dimensions 216
Ch 11 The Fabric of Space-Time 226
Ch 12 Parallel Worlds 243
Ch 13 In Search of the Multiverse 247
Ch 14 The Philosophy of Neuroscience 257
Ch 15 Altered States of Consciousness 259
Ch 16 The Mystery of Consciousness 268
Ch 17 Quantum Mind Theories 269
Ch 18 Quantum Mechanics Mathematical Examples 275
Epilogue 323
Glossary 324
References 349

An Overview of Quantum Mechanics

What is quantum mechanics?

Quantum Mechanics attempts to explain the behavior of subatomic particles at the *nanoscopic* level.

Through the work or Planck and Einstein, we came to the realization that energy is quantized. And that light exhibits wave particle duality. Then De Broglie extended this duality to include matter as well. Meaning that all matter has a wavelength from a tiny electron to your whole body to a massive star. However, an object's wavelength is inversely proportional to its mass. So objects larger than a molecule have a wavelength that is so small that it is completely negligible. But an electron is incredibly small, so small that it's wavelength is indeed relevant. Being around the size of an atom. So we must view electrons as both particles and waves. Therefore, we must discuss the wave nature of the electron. So what kind of wave might this be? We can regard an electron in an atom as a standing wave. Just like the kind in classical physics. Except that rather than something like a plucked guitar string, and electron is a circular standing wave surrounding the nucleus. If we understand this, it becomes immediately apparent why quantization of energy applies to the electron. Because any circular standing wave must have an integer (a positive whole number) number of wavelengths in order to exist. Given that an increasing number of wavelengths means more energy carried by the wave, we can see the Bohr model for the hydrogen atom begin to emerge. As we imagine a standing wave with one wavelength and then two, three, four and so forth. This is the reason that an electron in an atom can only inhabit a discrete set of energy levels. A circular standing wave that represents the electron can only have an integer number of wavelengths. When an electron is struck by a photon of light of a particular energy, this energy is absorbed promoting the electron to a higher energy state which increases the number of wavelengths contained within the standing wave. This is why the electron goes to inhabit a higher energy level. And this is what is fundamentally occurring during electron excitation.

Furthermore, it is the constructive interference of these standing waves that explains how orbital overlap results in covalent bonding. So we can enjoy a little more clarity in our understanding of chemistry thanks to modern physics. Once it was realized that electrons exhibit wave behavior the physics community set out to find a mathematical model that could describe this behavior. Erin Schrodinger achieved this goal in 1925 when the developed his Schrodinger equation which incorporated the de Broglie relation. The equation is a differential equation in which utilizes concepts in mathematics that are beyond the scope of this introductory discussion. However, we can certainly discuss the conceptual implications of the equation. Essentially, just as **F=ma** applies to Newtonian systems, the Schrodinger equation applies to quantum systems by describing the system's three-dimensional wave function represented by the Greek letter Psi. In this equation, this term is called the Hamilton operator. Which is a set of mathematical operations that describes all the interactions that affect the state of the system in which can be interpreted as the total energy of a particle. But while the Schrodinger equation can calculate the wave function of a system, it doe not specifically reveal what the wave function is. Max Born proposed that we interpret the wave function as a probability-amplitude where the square of the magnitude of the wave function describes the probability of an electron existing in a particular location. Looking back at the double-slit experiment we understand the diffraction pattern as a wave of probabilities. The pattern is not the electron itself. It is the probability that an electron will arrive at each location on the screen. We can't predict where one electron will go, only the probability that it will arrive at a particular location. If many electrons arrive at the screen it becomes apparent how their distribution obeys the wave function. So the Schrodinger equation does compute the wave function deterministically. But what the wave function tells us is probabilistic in nature. This idea that nature is probabilistic on the most fundamental level was a lot for the scientific committee to handle at the time. In addition, quantum mechanics' corresponding equations are not mathematical artifacts. They truly describe the inner workings of nature at the atomic and subatomic level. So, just the way sound waves are mechanical waves, and light waves are oscillations in an electromagnetic field. An electron can be a cloud of probability density. There are many such *interpretations* of quantum mechanics in which involve different ways of viewing the relationship between the wave function, experimental results and the nature of reality. But still there is no firm consensus as to which view is correct, be it the Copenhagen Interpretation, the many worlds interpretation or a number of others.

The following list contains the leading interpretations of quantum mechanics :

Copenhagen Interpretation

Many-worlds Interpretation

Quantum DE coherence

Bohmian mechanics

A word about the Many-worlds Interpretation

Why the MWI?

The reason for adopting the MWI is that it avoids the *collapse* of the quantum wave. (Other non-collapse theories are not better than MWI for various reasons, e.g., nonlocality of Bohmian mechanics; and the disadvantage of all of them is that they have some additional structure.) The advantage of the MWI is that it allows us to view quantum mechanics as a complete and consistent physical theory which agrees with all experimental results obtained to date.

Quantum Mechanics Terminology

Quantum Leap – In a Quantum Leap, it is theorized that at the subatomic level, energy can only be released and absorbed in discrete, divisible units called quanta. This means that electrons have fixed orbits around the nucleus of the atom as their energy comes in discrete amounts. When the electron gets excited or de-excited they will absorb or admit specific quanta of energy respectively, which will mean they leap from one orbit to another without inhabiting the space in between. This is called the quantum leap. In essence, there are places within the atom that the electron will be likely to be and other places where they won't where energy is being absorbed and released in discrete units.

Particles behave like waves - There is a famous experiment in quantum physics called the double-slit experiment, which reveals that particles display both particle like and wave like behavior. In a version of the experiment on a larger scale, we have a gun that shoots tennis balls one by one, at a detector, which will register where the tennis balls land. In between thee gun and the detector, we will place a barrier with two slits, which will leave openings for any tennis ball to go through. Over a period of time and after many tennis balls, a pattern emerges on the detector indicating where the tennis balls have landed. The results show balls, which have not been blocked, have landed directly behind the slits in the barrier. If we replicate this experiment but on the subatomic scale, and use electrons instead of tennis balls we expect similar results. But this is not the case. Scientists found that when the gun shoots electrons one by one toward the detector and past the double-slit barrier the pattern that emerges from the detector look like an interference pattern. They land in narrow strips across the length of the detector. A significant number of them even landed directly straight behind the middle of the double-slit barrier. The pattern that emerged is an interference pattern and is associated with the behavior of waves. Imagine that we have two waves that interact. The peaks of the two waves will combine or interact to form a single higher peak and the troughs will combine to form a single deeper trough. And when a peak and trough meet they will cancel each other out. So if we imagine that our electrons are less tennis ball like and more wave like, then what happens? An electron goes through a double slit and it's wave is split in two. And these waves then interact. When the peaks meet, they reinforce each other creating higher peaks. And when the toughs meet, the reinforce each other to create a lower or deeper trough. And when a peak and a trough meet, they cancel each other out.

The interaction between the waves results in an interference pattern at the detector screen. Where the waves are most intense we find more of the electrons on the detector screen. And where they cancel each other out, there are no electrons on the detector screen.

Schrodinger's equation - Erin Schrodinger came up with an equation for the electron's wave function. And using this equation, we can find out the probability of the electron being in a particular location. Think of the wave as a bundle of probabilities. And the size of the wave in any location predicts the likelihood that the electron will be found there if it is looked for. That is why on the detector screen you observe most of the electrons landing in the places where the electron wave is at its most intense. It seems that the electron is not in a fixed position. But has many probabilities of being in many different places at once.

The act of Measurement - To observe the electron's wave function going both slits at the same time, detectors were placed next to the slits to capture this activity. But when this was done something strange happened. The electrons stop behaving like waves and went through one or the other of the slits and landed on the detector screen to form the two stripe pattern rather than the interference pattern. It seemed as if the act of measuring did something to collapse the wave function.

The Superposition Principle - the superposition principle states that while we do no measure the electron for it's position, it is in all the possible positions it could be in at the same time. And when we observe it the superposition collapses. So in this thought experiment, when our detector is off, our electron is in all of the possible positions or states it could be in simultaneously. But when we switch our detector on the superposition collapses and the electron gives up all of its states save one. This is why we are not able to observe the electron's wave function gong through the double slits. The very act of attempting to observe it made the wave function's collapse. The electron gave up it's superposition and choose just the one state to be in. ie, the electron actually reversed its wave particle-like behavior. And that is why instead of going through both of its slits, the electrons that landed on the back of the detector chose to go through either one or the other of the slits.

Schrodinger's cat - Erin Schrodinger described the thought experiment to further illustrate what quantum mechanics is saying about the electron and the superposition. He described a scenario where a cat was placed in a covered box with radioactive sample that has a fifty percent chance of decaying and killing the cat. While the box is covered, we have no idea if the cat is dead or alive. Only once we open the box will we know if the cat survived or not. So if the cat were similar to an electron using the superposition principle we would be saying that while the box were covered and the cat was not being observed the cat was both alive and dead at the same time in order for it to be in all the states that it could possibly be in. Only when we lifted the cover to observe the cat did the superposition collapse for it to be either dead or alive. Obviously, this feels intuitively wrong. And the reason why this thought experiment was employed was to illustrate how odd the laws of quantum mechanics are when it comes to describing the behavior of particles. However, has this experiment corroborated the results predicted by quantum mechanics? It does indeed seem that particles have a wave function and that particles have in all the states it could possibly be in simultaneously until it is observed. But why do we see evidence of wave-like behavior from electrons and not cats? After all, cats are made up of particles. Well the reason is, the bigger the object the smaller its wavelength. And at the size of a cat, the wavelength is simply too small to be detected. Why do we accept Quantum Mechanics - The predictions of Quantum Mechanics have proved so reliable that we cannot ignore the experimental evidence that corroborates the theory. The entire electronics industry is built on using quantum theory. Those principles have led to the inventions of lasers, transistors and the integrated circuit and other solid state devices.

Preface - Quantum Mechanics Interpretation

Quantum Mechanics (Interpretation)

The foundation of Quantum Mechanics is explained by the following experiment. Suppose we have a barrier with two holes. Now, suppose we shoot marbles at it one marble at a time. Each time the marble hits a cloth located behind the barrier it marks a spot on the cloth where it landed. As marbles hit the same spot more than once we mark the spot, which is arbitrarily colored red. Some marbles make it through the holes by bouncing off by an angle. But most marbles that make it through the holes enter in through a straight line.

After awhile, there will be many red marks on the wall. The darkest red marks will be directly behind the two holes. Now suppose the two holes are very narrow. Suppose the marble is very small. Now the result is very different. A striped pattern is produced. The marbles never hit the area of the wall between two adjacent stripes. All particles in the universe produce this striped pattern provided that they and the holes are small enough. No matter how many times we repeat this experiment, and no matter what type of object we use to replace the marbles, the result is always the same. Only one known phenomena can explain this result. The phenomena can only be waves. When a wave passes through a hole it spreads out on the other side. If there are two holes, two waves are produced. When you have two waves, they interact with one another. In some areas, they strengthen each other and in other areas they cancel each other out.

This creates a striped pattern. This is the exact same pattern that was described earlier with the small marble. This means that all objects really behave like waves. But if all objects behave like waves, then why don't we see a striped pattern for the large marbles? Large objects have much more energy than small objects. Waves have more energy by having a higher frequency. When waves with higher frequencies interact with each other the pattern is different. Large objects have more energy and therefore behave like high frequency waves.

This is why large objects do not produce a striped pattern but small objects do. But, there is still a problem. For a wave to produce a striped pattern, each wave must simultaneously pass through both holes so that there will be two new waves that interact with one another. But we are shooting marbles at the wall only one marble at a time. This means that each marble must somehow simultaneously pass through both holes in order to create the striped pattern. Now, lets see if this is what actually happens by blocking one of the holes.

The striped pattern disappears. Most of the marks are directly behind the one open hole. Now lets block the other hole instead. Again, the darkest lines are directly behind the one open hole. But when we unblock both holes, the striped pattern returns. Areas that were hit many times when one of the holes was blocked are now never hit when both holes are open.

This means that each marble really does have to simultaneously pass through both holes to produce a striped pattern. Lets test this by putting a detector in front of each hole. We should expect that both detectors should simultaneously indicate that the marble passes through it. However, this is not what happens. Each marble only passes through one detector or the other but never both. 08:01

Also, once we placed detectors in front of the holes, the striped pattern disappears. Now the darkest lines are directly behind the two holes. Just as when we blocked one hole at a time.

Lets put a detector in front of only one of the two holes. It turns out that just having only one detector has the same effect as having two detectors, and causes the striped pattern to disappear. Any attempt to discover which of the two holes the marble passes through, forces the marble to pass through one hole or the other and not both. One detector has the same effect as two because once we know if the marble passed through one hole then we also automatically know whether or not it passed through the other one.

A marble goes through both holes only when we are not trying to find out which hole it went through. But when we do try to find out the marble only goes through only one hole or the other. So what if we placed detectors in front of both holes and just closed our eyes and not look? We don't know for sure what is happening when we are not looking, but we do know what the mathematics describing the waves tells us. When waves pass through a detector, the waves are altered so that they no longer interact with one another. This means that the striped pattern will disappear even when we are not watching it. The detectors will cause this to happen on their own. However, the mathematics also says that each wave still simultaneously passes through both holes even when the detector is present but when we open our eyes and look, we always see the detector indicating that the marble passed through only one hole or the other but never both. Each wave still simultaneously passes through both holes even with the detectors present. But when we open our eyes and look we always see the detector indicating that the marble passed through only one hole or the other but not both.

This means that the marble must be more than just a wave. The wave only describes the probability of where we will see the marble when we will look at it. The probability of the marble being at a particular location is given by the wave's amplitude. The higher the amplitude of the wave at a particular location, the higher the probability is that we will see the marble there when we look. This means that we can never simultaneously know both the position and the momentum. Before the wave hits the detector we know exactly the direction the momentum is in. However, we know nothing about the object's position. Immediately after we see the marble hit the detector, we know exactly where its position is, but now we know nothing about the direction of the momentum. So are we not able to simultaneously measure the position and momentum of the object, the object doesn't even have an exact position or momentum until we observe it. If the marble always had a specific position then the marble would not be able to go through both holes simultaneously, which is necessary to produce the striped pattern. But if all objects are just a wave of probability until we observe them then this means that the detectors and all the objects the marbles interact with are just a wave of probability too.

Suppose we place an object behind each of the two holes. The marble will knock down one of the two objects depending on which hole it passes through. If we close our eyes and don't look, then the wave of probability passes through both holes. And each object being knocked down also becomes a wave of probability. Just as each marble simultaneously passes through both holes, each object is both simultaneously both standing up and knocked down. No matter how long we wait after the marbles have hit the objects, each object will continue to have a probability of still being in the standing position and each object will also continue to have a probability of being in the knocked down position. According to the mathematics describing the probability waves neither outcome is certain. Its only when we open our eyes and look that we see only one outcome or the other. Its not just we ourselves do not know the outcome until we look, it seems the universe itself does not know which object is standing up and which object is knocked down until we open our eyes and observe the results. To explain why this is the case, and what this means about the fundamental nature of our universe, let us talk about spin.

The concept of SPIN

The direction of spin (an electronic dipole), of a particle can be described by a imaginary arrow. Particle spinning in opposite directions will have their arrows pointing in opposite directions. This can only be imagined with classical particles spinning, no one can imagine what the spin of a quantum particle looks like. Particles are too small to see the spin directly with our eyes but we can built detectors which can tell us the direction the particle's arrow is pointing at, for instance at a red plate as opposed to a blue plate just parallel to the red plate.

Suppose we think we know the spin of a particle ahead of time because we measured it previously, and we line-up the detector with this direction of arrow-spin. The detector will always give us the same result we measured previously. But if the spin we measured previously is not in the same direction as the detector then the act of measuring the spin ends up changing it (the direction of the arrow-spinning particle). It is not possible to simultaneously measure the spin of a particle in more than one direction at a time. If we want to know the direction of the spin in the horizontal direction, then we need to rotate the detector (by 90 degrees). But why does it mean that the universe does not know what an object is doing until we observe it?

The answer lies in the fact that we can produce two pairs of particles that always spin in opposite directions. If the two detectors are aligned in the same direction then when the spin of one particle was measured to be towards the red plate the spin of its partner is always measured to be towards the blue plate.

This is true 100% of the time. This is still true the vast majority of the time even when we offset the detectors by 45 degrees. (Vast majority means 85% of the time). We know this based on experiment. If we offset the detectors by 45 degrees, then when the spin of one particle moves towards the red plate, the spin of its partner will move towards the blue plate the vast majority of the time.

Suppose the two detectors are perfectly aligned with each other and they are both in the diagonal position, does the universe know ahead of time that the first particle will be measured moving towards the red plate and the spin of the second particle therefore be towards the blue plate? If the universe knows ahead of time that the spin of the second particle will be towards the blue plate than this means that the universe must know that the spin of the first particle will probability be towards the red plate, even if we rotate the red detector by 45 degrees even in the vertical position or even in the horizontal position.

Therefore, if the universe knows the results of the diagonal observation ahead of time then the universe would also know that one detector will be rotated in the vertical position and the other detector will be rotated into the horizontal position then the two particles will probability be spinning in opposite directions. But when we actually do the experiment, this not what happens. When the two detectors are offset by 90 degrees, there is no correlation between the measured spins of the two particles. When the two detectors are offset by 90 degrees the spins of two particles are just as likely to be read in the same direction as they are to be read in the opposite directions. Therefore, if we start out by assuming that the universe knows ahead of time what the measurements will be this leads to a contradiction. Assuming that the universe knows the answers ahead of time implies that most of the time the measured vertical spin of one particle must be in the opposite direction of the measured horizontal spin of the second particle. But we know that this is not the case.

The apparent implication is that the universe cannot know ahead of time what the measurement of the spin will be, and the universe makes up its mind only when the spin is actually observed. The apparent implication is also that when the spin of one particle is measured it sends an instantaneous message to its partner to spin in the opposite direction. If each particle makes up its mind about what direction its spinning only when its observed, then we need this instantaneous message from one partner to the other in order to guarantee that the two particles will always decide to spin in opposite directions when its detectors are aligned. This is true no matter how far apart the two particles have traveled away from each other, even if we wait until the two particles are on opposite sides of the universe before we make our observation. This instantaneous message still seems to occur.

Until we observe the particles, their spins are nothing more than probabilities. But we need to observe only one of the two particles for both of them to simultaneously decide in what direction to spin. If everything in the universe is made out of these particles including the detectors themselves, then the detectors are nothing more than a probability until they are observed. According to the mathematics describing the probabilities, the particles passing through the detector is not what causes the particle-spin to decide to spin in one direction or the other.

Passing through the detector entangles the detector to the particle in the same way that the two particles are entangled to each other. The moment we look at the detector it seems to send the instantaneous message to the particle so that the detectors' measurement will agree with the spin of the particle. This is the same way in which the two particles send instantaneous messages to each other.

There are many possible explanations as to what is actually happening, and how to interpret these results. And this is a matter of considerable debate. But if we really do have these types of instantaneous messages, which are faster than the speed of light, than this creates an interesting situation for Einstein's theory of relativity. According to Einstein's theory of Relativity, different observers will disagree about which two events will happen first. And no one observer is more correct than the other. From one observer's point of view, the right particle was observed first, and caused the left particle to change its spin. From another observer's point of view, the left particle was observed first and caused the right particle to change its spin. Therefore, we can't know which event particle is the cause and which is the effect. Since both points are equally valid. In fact, according to quantum mechanics, we can't even know which particle is which.

Suppose we have two particles in a container. Each particle does not have its own separate probability wave. There is only one probability wave in which describes the probability of measuring the two particles in every possible combination. The probability that particle one will be in one position and that particle two will be in another position is exactly equal to the probability that the two particles will be in the swapped position.

Therefore, we cannot know if a particle we are observing is the same particle we measured earlier. If we think of our container as the entire universe, than this would imply that the universe consists of just one probability wave, governing the probability of all the particles in existence. But if we ourselves are made out of the exact same particles, then why is the act of observing something so fundamentally different from everything else in the universe? This is one of the greatest unsolved scientific and philosophical mysteries of all time.

The Heisenberg's Uncertainty Principle (Revisited)

The Heisenberg's Uncertainty Principle, one of the crowning achievements of Quantum Mechanics, relates measurements to things such as particles at the atomic and subatomic level.

Heisenberg's uncertainty principle is usually written in the following form:

$$Delta(x) * Delta(p) > \; or = h - bar/2$$

Delta(*x*) refers to the *uncertainty* (not a small change) in position. For instance, if you want to know where an electron is, then you actually want to know how precise you can know its true position. Delta(*p*) refers to the *uncertainty* of an electron's momentum, which is it's velocity and direction. And the idea in uncertainty in position multiplied by the uncertainty in momentum must be greater than or equal to *h*-bar over two.

What results is that you cannot precisely define position and momentum with exact precision at the same instant. For example, lets say you have an electron and that electron has a certain position (*x*) and it also has a certain momentum (*p*). And we cannot measure both of them precisely at the same time using natural light a measuring probe. The reason is for this is that electrons and even atoms are very small elements in nature compared with anything you are going to use to measure them with.

For example, lets say that we use light to look at an electron. The light wave will go right over and past the electron due to the large displacement of its wavelength pattern. A light-wave has a wavelength of approximately 5x10 to the minus **7** meters. An atom has the diameter of approximately 10 to the minus **10** meters. A proton has a diameter of 10 to the minus **15** meters. And an electron has a diameter of 10 to the minus **18** meters. So in other words, even an atom is over a thousand times smaller then the wavelength of light.

A proton would be 100 million times smaller than the wavelength of the light. So any visible light would go straight past any of the fore-mentioned particles.

If you want to actually "see" or measure something of the size of an atom or proton, you have to use some form of radiation (electromagnetic radiation) who's wavelength is comparable to the size of the particle you are trying to "see" (measure). But therein lies the problem. Because the smaller the wavelength the larger the momentum.

It is given by a formula that says:

$P = h/lambda$,

where P is the momentum of a particle

So as lambda gets smaller, *P* gets larger.

This formula comes from the famous formula of Einstein,

$E = m * c^2$

And the momentum is classically given by:

$P = mv$,

where m = mass and v = velocity

Now when we are talking about electromagnetic radiation like light photons, which are the constituents or parts of light, do not have a mass. But we can get away by saying that the mass of a photon is equivalent to

$m = E/ c^2$

So the P (momentum) is the mass equaled to E(energy) over c^2,

where c is the speed of light

$P = E/c^2 * C$ (Velocity) $= E/c$

The energy of an electronic wave is:

$E = hf$,

where h = Planck's constant * the frequency of the radiation

This is known as the photon packet-of-energy or quantized energy.

So now we can say that p (momentum) = E (hf) / c

and c/f = lambda, so p = hf/c = h/lambda

So we have derived the original equation P = h/*lambda.*

So if we have an electron, and we send in a very high wavelength (high frequency) electromagnetic radiation wave in order to be able to "see" the electron, it will have such a high momentum, that although we may now be able to detect the electron, the wave will give the electron such a strong kick that it will send it scurrying away in some unknown velocity and direction. So as a result you will not be able to tell what the momentum of the electron is.

We can perhaps explain this by analogy of a *single*-slit Experiment. This is where you take a very narrow single slit and you pass ordinary light through it. So what happens is as the light beam goes through the slit it spreads out onto a back screen. Now what you find out is that the intensity of the light on the screen creates a bell curve. At the far left and right ends of the bell curve there is no light due to a pattern interference of opposite or opposing light waves that are completely out of phase with respect each other. This means that they completely cancel each other out. At the top or center of the bell curve is where the most intensity of light is due to patterns of light waves that are in perfect phase with each other. This means that they sum up to a stronger intensity. This produces a uniform pattern, not a fringe pattern. If it were a *double-slit* apparatus, then it would produce a fringe pattern.

The spread of the light waves is at an angle theta. This is due to the many waves of the light beam going in various angular directions past the slit on to the back screen.

Performing some basic geometry, one can see that the superimposed waves of opposite phase, which is due to their perpendicular angles toward each other creating a cancelation of wave light intensity, results in no light at the fringes of the bell curve light wave pattern on the back screen. See the definition of diffraction.

This is all due to the distance between the theta angles, called lambda. So for every light photon that is lambda/2 apart from another photon, will produce a cancellation.

Therefore, Lambda = *d*(sin)-theta. This is the width of the slit.

Given the above apparatus and process, we can perform the same procedure using electrons.

So using electron beams, we can attempt to measure the initial position and momentum of an incoming electron particle. Now, since lambda– *d*(sin)-theta, and since we know that lambda is fixed (does not change)., that means that if you make (*d*) smaller then the angle theta will increase. So as the slit gets smaller, the angle will get larger. And as a result, the wave will spread out more.

(*d*) is in a sense the measure in the y-direction of the position of the electron. We know that the electron must be somewhere between the two points on either side of the slit opening otherwise it would not get through the gap.

So (*d*) = delta-*x*, it is the uncertainty in the y-position of the electron. Now, what about the momentum? Well, the *p* has a component part call delta-*p*. It is the uncertainty of the momentum in the *y*-direction at the back screen.

And

delta-*p* = *p*(sin)theta

And *p* = *h*/lambda or lambda = *h/p*

So, h/p = d(sin)theta, which follows that h = (d)(p)(sin)theta, which follows that

h = delta(x)delta(p).

This is the derived Heisenberg Uncertainty Principle equation.

So, the take away point is not an issue with any measuring device or measuring instrument, it is not because we cannot measure things precisely with our instruments, but instead it's a fundamental inability to measure the two aspects of both position and momentum with 100 percent accuracy simultaneously. This is a fundamental aspect of nature.

Ch 1 The Nature of Light

Huygens

1.1 Light Wave / Particle Phenomena

Light was a motif of the age: the symbolic enlightenment of freedom of thought and religion and light as an object of scientific inquiry, as in Snell's study of refraction, Leeuwenhoek's invention of the microscope and Huygens' own wave theory of light.

Isaac Newton admired Christian Huygens and thought him *'the most elegant mathematician'* of their time, and the truest follower of the mathematical tradition of the ancient Greeks, then, as now, a great compliment. Newton believed, in part because shadows had sharp edges, that light behaved as if it were a stream of tiny "particles." He thought that red light was composed of the largest particles and violet the smallest. Huygens argued that instead light behaved as if it were a "wave" propagating in a vacuum, as an ocean wave does in the sea, which is why we talk about the wavelength and frequency of light. Many properties of light, including diffraction, are naturally explained by the wave theory, and in subsequent years Huygens' view carried the day.

Newton

In Italy, Galileo had announced other worlds, and Giordano Bruno had speculated on other life forms. For this they had been made to suffer brutally. But in Holland, the Dutch physicist and astronomer Christiaan Huygens, who believed in both, was showered with honors. His father was Consantijn Huygens, a master diplomat of the age, a litterateur, poet, composer, musician, close friend and translator of the English poet John Donne, and the head of an archetypical great family. Consantijn admired the painter Rubens, and "discovered" a young artist named Rembrandt van Rijn, in several of whose works he subsequently appears. After their first meeting,

Descartes wrote of Christiaan: "I could not believe that a single mind could occupy itself with so many things, and equip itself so well in all of them." The Huygens home was filled with goods from all over the world. Distinguished thinkers from other nations were frequent guests. Growing up in this environment, the young Christiaan Huygens became simultaneously adept in languages, drawing, law, science, engineering, mathematics and music. His interests and allegiances were broad.

"The world is my country," he said, "science my religion." The microscope and telescope both developed in early 17th-century Holland, represent the extension of human vision to the realms of the very small and the very large. Our observations of atoms and galaxies were launched in this time and place. Christiaan Huygens loved to grind and polish lenses for astronomical telescopes and constructed one five meters long.

His discoveries with the telescope would by themselves have ensured his place in the history of human accomplishments. In the footsteps of Eratosthenes, he was the first person to measure the size of another planet. He was also the first to speculate that Venus is completely covered with cloud; the first to draw a surface feature on the planet Mars. And he was the first to recognize that Saturn was surrounded by a system of rings which nowhere touches the planet.

Huygens did much more. A key problem for marine navigation in this age was the determination of longitude. Latitude could easily be determined by the stars, the farther south you were the more southern constellations you could see. But longitude required precise timekeeping. An accurate shipboard clock would tell the time in your home port; the rising and setting of the Sun and stars would specify the local shipboard time; and the difference between the two would yield your longitude. Huygens invented the pendulum clock (its principle had been discovered earlier by Galileo), which was then employed, although not fully successfully, to calculate position in the midst of the great oceans.

Huygens was delighted that the Copernican view of the Earth as a planet in motion around the Sun was widely accepted even by the ordinary people in Holland. Indeed, he said, Copernicus was acknowledged by all astronomers except those who "were a bit slow-witted or under the superstitions imposed by merely human authority."

In the middle ages, Christian philosophers were fond of arguing that since the heavens circle the Earth once every day, they can hardly be infinite in extent; and therefore an infinite number of worlds, or even a large number of them, is impossible. The discovery that the Earth is turning rather than the sky moving had important implications for the uniqueness of the Earth and the possibility of life elsewhere.

Copernicus held that not just the solar system but the entire Universe was *heliocentric*, and Kepler denied that the stars have planetary systems.

The first person to make explicit the idea of a large – indeed, an infinite number of other worlds in orbit about other suns seems to have been Giordano Bruno. But others thought that the plurality of worlds followed immediately from the ideas of Copernicus and Kepler. Huygens was, of course, a citizen of his time. He claimed science as his religion and then argued that the planets must be inhabited because otherwise God had made worlds for nothing.

1.2 Waves vs. Particles

With his physics of particles such a success, it is hardly surprising that when Newton tried to explain the behavior of light he did so in terms of particles. After all, light rays are observed to travel in straight lines, and the way light bounces off a mirror is very much like the way a ball bounces off a hard wall. Newton built the first reflecting telescope, explained "*white light*" as a superposition of all the colors of the rainbow, and did much more with optics, but always his theories rested upon the assumption that light consisted of a stream of tiny particles, called corpuscles. Light rays bend as they cross the barrier between a lighter and a denser substance, such as from air to water or glass. But even in Newton's day, there was alternative ways of explaining all of this. Huygens was a contemporary of Newton, although thirteen years older, having been born in 1629. He developed the idea that light is not a stream of particles but a wave, rather like the waves moving across the surface of a sea or lake, but propagating through an invisible substance called the "*luminiferous ether*". Like ripples produced by a pebble dropped into a pond, light waves in the ether were imagined to spread out in all directions from a source of light. The wave theory explained reflection and refraction just as well as the corpuscular theory. So, the two theories conflicted with respect to the observations in their predictions.

Three hundred years ago, the evidence clearly favored the corpuscular theory, and the wave theory, although not forgotten, was discarded. By the early 19th century, however, the status of the two theories had been almost completely reversed. In the 18th century, very few people took the wave theory of light seriously. One of the few who not only took it seriously but wrote in support of it was the Swiss Leonard Euler, the leading mathematician of his time, who made major contributions to the development of geometry, calculus and trigonometry. Modern mathematics and physics are described in arithmetical terms, by equations; the techniques on which that arithmetical description rests on were largely developed by Euler. And the only other prominent contemporary of Euler who did share those views was Benjamin Franklin; but physicists found it easy to ignore them until crucial new experiments were performed by the Englishman Thomas Young just at the beginning of the 19th century, and by the Frenchman Augustin Fresnel soon after.

Young used his knowledge of how waves move across the surface of a pond to design an experiment that would test whether or not light propagates in the same way. We all know what a water wave looks like, although it is important to think of a ripple, rather than a large breaker, to make the analogy accurate. The distinctive feature of a wave is that it raises the water level up slightly, then depresses it, as the wave passes; the height of the crest of the wave above the undisturbed water level is its amplitude, and for a perfect wave this is the same as the amount by which the water level is pushed down as the wave passes. A series of ripples, like the ones from a stone dropped into the pond follow one another with a regular spacing, called the wavelength which is measured from one crest to the next. Around the point where our pebble drops into the water, the waves spread out in circles. The observed propagating waves on the surface of a pond look almost flat although technically they are three dimensional, however light travels in a spherical geometric shape from its source and not in a flat dispersion. The number of wave crests passing by some fixed point, like a rock, in each second tells us the frequency of the wave. The frequency is the number of wavelengths passing each second, so the velocity of the wave, the speed with which each crest advances, is the wavelength multiplied by the frequency.

Young

Now if we have two ripples spreading out across the water, this produces a more complicated pattern of ripples on the surface of the water. Where both waves are lifting the water surface upward, we get a more pronounced crest; where one wave is trying to create a crest and the other is trying to create a trough the two cancel each other out and the water level is undisturbed. The effects are called constructive and destructive interference, and are easy to see. So, if light is a wave, then an equivalent experiment should be able to produce similar interference among light waves, and that is exactly what Young discovered.

Born in 1773, Young was the eldest of ten children. Young had been a child prodigy. He was reading fluently by the age of two and had read the entire Bible twice by six

A master of more than a dozen languages, Young went on to make important contributions towards the deciphering of Egyptian hieroglyphics. He studied medicine at the universities of Edinburgh and Gottingen, where he graduated in 1796. He practiced medicine all of his life but was not a very good doctor because of his poor bedside manner. As trained physician, he could indulge his myriad intellectual pursuits after a bequest from an uncle left him financially secure. He was more interested in science than in medicine and didn't pay enough attention to his patients.

While in medical school, Young discovered how the lens of the eye changes shape when focusing at different distances. He later discovered that astigmatism was due to imperfections in the curvature of the cornea. From studying the eye, Young moved to the nature of light.

He showed that light formed an interference pattern, the signature of a wave. From his calculations, Young found out that the value of the wavelength of light was much smaller then thought. The longest wavelength in the visible spectrum is the one for red light, which is less than one-thousandth of a millimeter. That's why light casts sharp shadows and doesn't appear to bend around corners. You need tiny objects, like Young's pinholes, to detect the bending.

His interest in the nature of light led Young to examine the similarities and differences between light and sound, and ultimately to 'one or two differences in the Newtonian system.' Convinced that light was a wave he devised an experiment that was to prove the beginning of the end for Newton's particle theory. Young shone *monochromatic* light onto a screen

with a single slit. From this slit a beam of light spread out to strike a second screen with two very narrow and parallel slits close together. Like a car's headlights, these two slits acted as new sources of light, or as Young wrote, 'as centers of divergence, from whence the light diffracted in every direction.' What Young found on another screen placed some distance behind the two slits was a central bright band surrounded on each side by a pattern of alternating dark and bright bands.

1.3 Demonstration that light is a wave

A century after Newton in 1802 Thomas Young improved Grimaldi's experiments. As mentioned earlier, Young passed a beam of light through a pinhole that he punched on a screen. This light spread out form a hole and passé through a set of two pinholes that he had punched side-by-side on a second screen. He used a third screen to observe the pattern of dark and bright regions that he had made. Young knew very well that what he was seeing was telling him that light was a wave. The light behaved like water ripples that run into each other as it made similar patterns.

1.4 Coherent beams of light

Young's interference experiment, as we call it today, was very clever. It turns out that you can't get these interference patterns with a regular source of light. The reason is that the light beams from a regular source don't vibrate in lockstep, and when they run into each other, they don't form these patterns. You need two beams that vibrate in step; this is what is called coherent beams. The light beams from the two holes in the second screen had the same origin; they both came from the light passing through the first hole. Because Young placed the two holes in the second screen at equal distances from the first hole, the light beams that came out of the holes in the second screen were coherent. Young did more experiments, replacing the two pinholes in the second screen with slits.

The pattern of bright and dark regions became parallel bands. Using simple geometry, he calculated the wavelength of the light he used from the distances between the lines.

1.5 Resistance to Young's pinhole experiment

Young's experiment is what we call a landmark experiment. His experiment showed, without any doubt, that light is a wave. The interference pattern that Young saw with his setup is the mark of a wave. You can't get that pattern with particles bouncing off the edges of the pinhole. The particles would have to have a coordinated motion to be able to form such a symmetric pattern. One would think that with this irrefutable proof, the wave nature of light would become well-established right away. But that did not happen. Young's experiment went against the teachings of the great Isaac Newton, and the English physicists were not going to have any of it.

To explain the appearance of these bright and dark 'fringes', Young used an analogy. Two stones are dropped simultaneously and close together into a still lake. Each stone produces waves hat spread out across the lake. As they do so, the ripples originating from one stone encounter those from the other. At each point where two wave troughs or two wave crests meet, they coalesce to produce a new single trough or crest. This was constructive interference. But where a trough meets a crest or vice versa, they cancel each other out, leaving the water undisturbed at that point. This is what is described as destructive interference. In Young's experiment, light waves originating from the two slits similarly interfere with each other before striking the screen. The bright fringes indicate constructive interference while the dark fringes are a product of destructive interference. Young recognized that only if light is a wave phenomenon could these results be explained. Newton's particles would simply produce two bright images of the slits with nothing but darkness in between. An interference pattern of bright and dark fringes was simply impossible.

When he first put forward the idea of interference and reported his early results in 1801, Young was viciously attacked in print for challenging Newton. He tried to defend himself by writing a pamphlet in which he let everyone know his feelings about Newton: 'But, much as I venerate the name of Newton, I am not therefore obliged to believe that he was infallible. I see, not with exultation, but with regret, that he was liable to err, and that his authority has, perhaps, sometimes even retarded the progress of science.' Only a single copy was sold.

Feeling the need to convince the doubters, Young tried to tell people that Newton himself was really not against the wave nature of light. Newton accepted the idea that the different colors of light had different wavelengths. But sometimes when people have strong beliefs, arguments are not enough to persuade them. It wasn't until 1818, when two French physicists came up with a complete wave theory based on mathematics, that the wave theory was finally accepted. It hasn't been challenged since.

It was a French civil engineer who followed Young in stepping out of Newton's shadow.

Augustin Fresnel, fifteen years his junior, independently rediscovered interference and much else of what Young, unknown to him, had already done. However, compared to the Englishman, Fresnel's elegantly designed experiments were more extensive, with the presentation of results and accompanying mathematical analysis so impeccably thorough that the wave theory started to gain distinguished converts by the 1820s. Fresnel convinced them that the wave theory could better explain an array of optical phenomena than Newton's particle theory. He also answered the long-standing objection to the wave theory: light cannot travel around corners. He insisted that it does. However, since light waves are millions of times smaller than sound waves, the bending of a beam of light forming a straight path is very small and therefore extremely difficult to detect. A wave bends only around an obstacle not much longer than itself. Sound waves are much longer and can easily move around most barriers that they encounter.

One way to get opponents and skeptics to finally decide between the two rival theories was to find observations for which they predicted different results. Experiments conducted in France in 1850 revealed that the speed of light was slower in a dense medium such as glass of water than the air. This was exactly what the wave of light predicted, while Newton's corpuscles failed to travel as fast as expected. But the question remained: if light was a wave, what were its properties?

Young's experiment didn't exactly set the world of science on fire, especially in Britain. The scientific establishment there regarded opposition to any idea of Newton's as almost heretical, and certainly unpatriotic.

Newton had only died in 1727, and in 1805, less than a hundred years before, Young announced his discoveries. Newton had been the first man to receive a knighthood for his scientific works. It was too soon for the idol to be dethroned in England, so perhaps it was appropriate at that time of the Napoleonic wars that it was a Frenchman, Augustin Fresnel, who took up this unpatriotic idea and eventually established the wave explanation of light. Years later, everybody knew that light was a form of wave motion propagated through the ether. Still, however, it would be nice to know exactly what was "waving" in the beam of light. In the 1860s and 1870s, the theory of light seemed at last to have been completed when the great Scottish physicist James Clerk Maxwell established the existence of waves involving changing electric and magnetic fields. This electromagnetic radiation was predicted by Maxwell to involve patterns of stronger and weaker electric and magnetic fields in the same way that water waves involve crests and troughs in the height of the water.

In 1887, only a hundred and twenty-four years ago, Heinrich Hertz succeeded in transmitting and receiving electromagnetic radiation in the form of radio waves, which are similar to light waves but have much longer wavelengths. At last the wave theory of light was complete but just in time to be overturned by the greatest revolution in scientific thinking since the time of Newton and Galileo. By the end of the 19th century, only a genius or a fool would have suggested that light is "*corpuscular.*" His name was Albert Einstein.

In 1905, Einstein showed that the particle theory of light could explain the photoelectric effect, the ejection of electrons from a metal upon exposure to a beam of light. Modern quantum mechanics combines both ideas, and it is customary today to think of light as behaving in some circumstances as a beam of particles and in others as a wave. This wave-particle dualism may not correspond readily to our common-sense notions, but it is in excellent accord with what experiments have shown light really does. There is something mysterious and stirring in this marriage of opposites, and it is fitting that Newton and Huygens, bachelors both, were the parents of our modern understanding of the nature of light.

1.6 Historical overview of the nature of light

Scientists have been trying to understand the nature of light for centuries. The Greeks thought that light was made up of small particles traveling in straight lines that entered our eyes and stimulated our sense of vision. Isaac Newton also thought that light was made up of little particles. But the English scientist Thomas Young thought that light was a wave and he came up with a clever experiment to prove it. Newton started experiments on what later became his theory of colors. This work was of great importance, what we know today about color started with the experiments he did in that marvelous year. Newton knew that a beam of light passing through a prism broke up into a splash of the colors of the rainbow; violet, blue, green, yellow, orange, and red. This knowledge was commonplace even at the time of Aristotle. But no one knew why light could be broken up into colors until Newton came along.

During his miracle year of 1666, Newton wanted to try the "celebrated phenomena of colors."

He bought a glass prism, brought it home to his mother's farm, went into his room, closed the doors and windows, and made a small hole in the shutters to let a narrow beam of light come into

the room. He placed his prism in the path of the light beam. The beam broke out into colors on the opposite wall. People had seen this happen many times – it was nothing new, but it was spectacular when viewed. Newton noticed something peculiar. He had carefully made a circular hole, but the shape of the spot on the wall was oval, not circular. Before Newton, people thought that a prism changed the color of the light. The theory was that Sunlight passing through the thick end of a prism was darkened more, so it became blue, while light passing through the thin end was darkened less and became red. But the prevailing theory didn't explain why the round hole made an oval shape on the wall. Newton wanted to know why. He made the hole bigger, then smaller. He changed the location of the prism and the place where the beam entered the prism, but the spectrum never changed. He placed a second prism a few yards away so that the light beam would pass through both prisms. Then, he noticed something remarkable. The blue end of the rainbow was bent even farther than the red, but no additional colors appeared. The second prism didn't change the color of light at all: Red was still red and blue was still blue.

Newton was able to give a detailed mathematical account for a host of optical observations, including *reflection* and *refraction*; the bending of light as it passes from a less to a denser medium. However, there were other properties of light the Newton could not explain. For example, when a beam of light hit a glass surface, part of it passed through and the rest was reflected. The question Newton had to address was why some particles of light were reflected and others not? To answer it, he was forced to adapt his theory. Light particles caused wavelike disturbances in the ether. These 'Fits of easy Reflection and easy Transmission', as he called them, were the mechanism by which some of the beam of light was transmitted through the glass and the remainder reflected. He linked the 'bigness' of these disturbances to color. The biggest disturbances, those having the longest wavelength, in the terminology that came later, were responsible for producing red. The smallest, those having the shortest wavelength, produced violet.

Christiaan Huygens argued that there was no Newtonian particle of light. Thirteen years older than Newton, by 1678 Huygens had developed a wave theory of light that explained reflection and refraction. However, his book on the subject, *"Traite de la Lumiere"* was not published until 1690. Huygens believed that light was a wave traveling through the ether. It was akin to the ripples that fanned out across the still surface of a pond from a dropped stone. If light was really made up of particles, Huygens asked, then where was the evidence of collisions that should occur when two beams of light crossed each other? There was none, argued Huygens. Sound waves do not collide; therefore, light must also be wavelike. Although the theories of Newton and Huygens were able to explain reflection and refraction, each predicted different outcomes when it came to certain other phenomena. None could be tested with any degree of precision for decades. However, there was one prediction that could be observed. A beam of light made up of Newton's particles travelling in straight lines should cast sharp shadows when striking objects, whereas Huygens' waves, like water waves bending around an object they encounter, should result in shadows whose outline is slightly blurred. The Italian Jesuit and mathematician Farther Francesco Grimaldi christened this bending of light around the edge of an object, or around the edges of an extremely narrow slit, diffraction. In a book published in 1665, two years after his death, Bruno described how an opaque object placed in a narrow shaft of Sunlight allowed to enter an otherwise room through a very small hole in a window shutter, cast a shadow larger than expected if light consisted of particles travelling in straight lines. He also found that around the shadow were fringes of colored light and fuzziness where there should have been a sharp, well-defined separation between light and dark.

Newton was well aware of Grimaldi's discovery and later conducted his own experiments to investigate diffraction, which seemed more readily explicable in terms of Huygens' wave theory. However, Newton argued that diffraction was the result of forces exerted on light particles and indicative of the nature of light itself. Given his pre-eminence, Newton's particle theory of light, though in truth a strange hybrid of particle and wave, was accepted as the orthodoxy.

It helped that Newton outlived Huygens, who died in 1695, by 32 years. Alexander Pope's famous epitaph bears witness to the awe in which Newton was held in his own day by writing: 'Nature and Nature's Laws lay hid in the Night; God said, Let Newton be! And all was light.'

In the years after his death in 1727, Newton's authority was undiminished and his view on the nature of light barely questioned. At the dawn of the 19th century the English polymath, Thomas Young did challenge it, and in time his work led to a revival of the wave theory of light. As mentioned earlier, the English physicist Thomas Young used his knowledge of how waves move across the surface of a pond to design an experiment that would test whether or not light propagates in the same way. We all know what a water wave looks like, although it is important to think of a ripple, rather than a large breaker, to make the analogy accurate. The distinctive feature of a wave is that it raises the water level up slightly, then depresses it, as the wave passes; the height of the crest of the wave above the undisturbed water

level is its amplitude, and for a perfect wave this is the same as the amount by which the water level is pushed down as the wave passes. A series of ripples, like the ones from a stone dropped into the pond, follow one another with a regular spacing, called the wavelength which is measured from one crest to the next. Around the point where our pebble drops into the water, the waves spread out in circles. The observed propagating waves on the surface of a pond look almost flat, although technically they are three dimensional, however light travels in a spherical geometric shape from its source and not in a flat dispersion. The number of wave crests passing by some fixed point, like a rock, in each second tells us the frequency of the wave. The frequency is the number of wavelengths passing each second, so the velocity of the wave, the speed with which each crest advances, is the wavelength multiplied by the frequency. *See figure 6-1.*

Now if we have two ripples spreading out across the water, this produces a more complicated pattern of ripples on the surface of the water. Where both waves are lifting the water surface upward, we get a more pronounced crest; where one wave is trying to create a crest and the other is trying to create a trough the two cancel out and the water level is undisturbed. The effects are called *"constructive"* and *"destructive"* interference, and are easy to see. So, if light is a wave, then an equivalent experiment should be able to produce similar interference among light waves, and that is exactly what Young discovered.

Young's experiment didn't exactly set the world of science on fire, especially in Britain. The scientific establishment there regarded opposition to any idea of Newton's as almost heretical, and certainly unpatriotic. Newton had only died in 1727, and in 1805, less than a hundred years before Young announced his discoveries. Newton had been the first man to receive a knighthood for his scientific works. It was too soon for the idol to be dethroned in England, so perhaps it was appropriate at that time of the Napoleonic wars that it was a Frenchman, Augustin Fresnel, who took up this unpatriotic idea and eventually established the wave explanation of light. Years later, everybody knew that light was a form of wave motion propagated through the ether. Still, however, it would be nice to know exactly what was "waving" in the beam of light.

In the 1860s and 1870s, the theory of light seemed at last to have been completed when the great Scottish physicist James Clerk Maxwell established the existence of waves involving changing electric and magnetic fields. This electromagnetic radiation was predicted by Maxwell to involve patterns of stronger and weaker electric and magnetic fields in the same way that water waves involve crests and troughs in the height of the water.

In 1887, only a hundred and twenty-four years ago, Heinrich Hertz succeeded in transmitting and receiving electromagnetic radiation in the form of radio waves, which are similar to light waves but have much longer wavelengths. At last the wave theory of light was complete but just in time to be overturned by the greatest revolution in scientific thinking since the time of Newton and Galileo. By the end of the 19th century, only a genius or a fool would have suggested that light is *corpuscular.* His name was Albert Einstein.

In 1905, Einstein showed that the particle theory of light could explain the photoelectric effect, the ejection of electrons from a metal upon exposure to a beam of light. Modern quantum mechanics combines both ideas, and it is customary today to think of light as behaving in some circumstances as a beam of particles and in others as a wave. This "wave-particle" dualism may not correspond readily to our common-sense notions, but it is in excellent accord with what experiments have shown light really does. There is something mysterious and stirring in this marriage of opposites, and it is fitting that Newton and Huygens, bachelors both, were the parents of our modern understanding of the nature of light.

1.7 The true nature of light

We have seen that light, and all electromagnetic energy as well as electric and magnetic fields, travel at a constant speed through the Universe, and that this speed in vacuum is the same as measured from all non-accelerating reference points. But what is electromagnetic energy? How do we explain its propagation? Scientists have developed two models to explain the behavior of electromagnetic propagation, and these theories are commonly called the "particle theory" and the "wave theory." These are self-explanatory terms. Light displays certain properties that make it seem like it is composed of particles. In certain other ways, it behaves like a wave. These two models of electromagnetic energy, both with merit, make it difficult for us to say exactly what light is, just as atomic theory makes it quite a task to explain exactly what matter is.

1.8 The ultraviolet catastrophe of radiation

Around the end of the 19th century and the beginning of the 20th, evidence began to emerge that indicated that describing light as a wave is not sufficient to account for all its observed properties.

Two particular areas of study were central in this. The first area concerns the properties of the heat radiation emitted by hot objects. At reasonably high temperatures, this heat radiation becomes visible and we describe the object as 'red hot' or, at even higher temperatures, giving off a 'white heat'.

We note that red corresponds to the longest wavelength in the optical spectrum, so it appears that light of long wavelength (i.e. at a lower temperature) can be generated more easily than that of shorter wave-length; indeed, heat radiation of longer wavelengths is commonly known as *'infrared'*. Following the emergence of Maxwell's theory of electromagnetic radiation and progress in the understanding of heat (a topic to which Maxwell also made major contributions), physicists tried to understand these properties of heat radiation. It was known by then that temperature is related to energy: the hotter an object is the more heat energy it contains. Also, Maxwell's theory predicted that the energy of an electromagnetic wave should depend only on its amplitude and, in particular, should be independent of its wavelength. One might therefore expect that a hot body would radiate at all wavelengths, the radiation becoming brighter, but not changing color, as the temperature rises. In fact, detailed calculations showed that because the number of possible waves of a given wavelength increases as the wavelength reduces, shorter wavelength heat radiation should actually be brighter than that with long wavelengths, but again this should be the same at all temperatures. If this were true, all objects should appear violet in color, their overall brightness being low temperatures and higher at high temperatures, which of course is not what we observe. This discrepancy between theory and observation was known as the *"ultraviolet catastrophe."*

1.9 The Particle theory of Light

Scientists first developed the idea that light might consist of discrete corpuscles in the time of Isaac Newton. Today, we notice that light behaves as a stream of particles, known as quanta or photons, which represent the smallest possible packet of electromagnetic energy. As such, they are indivisible. You can't have part of a photon. According to the corpuscular theory of light, however, you cannot continue a process of flashing a light beam on a surface giving shorter and shorter durations of flashes (photons) forever. After a certain number of repetitions, you will reach a point where just a few photons, perhaps eight of them or even four photons will hit the wall, and then two, and finally only one. What after that? You cannot have half a photon.

On the next flash of the light, all we can say is that we might get a photon and we might not, and that the probability is 50 percent that we will. On the next flash, the probability is 25 percent that a photon will strike the wall; after that, 12.5 percent, and so on. Of course, there might be extraordinary occurrences where we get two photons that happen to be emitted almost simultaneously, but the point is that we will never, ever get part of a photon. They have also actually observed that a beam of light puts physical pressure on any object it shines upon.

The amount of energy carried by a single photon can vary. A photon of red light carries less energy than a photon of green light, which in turn carries less energy than a photon of violet light. Knowing the amount of energy carried by a photon, we can determine the kind (amount) of energy it contains.

A single photon carries only a minute amount of energy. Mathematically, this energy is on the order of *quintrillionths* of a joule for visible light, *septrillionths* or *octrillionths* of a joule for radio signals, and perhaps a few trillionths of a joule for *X*-rays. But for a given type of electromagnetic energy or frequency, all the photons carry the same amount of energy, and it is not possible to have any quantity of energy not a multiple of this amount. Mathematically, the relation between the mass of a photon and the energy it carries is given by the formula: $E = mc^2$ where E is the energy and m is the mass in joules and grams respectively. The constant in this equation is c, the speed of light.

1.10 Electromagnetic energy as a Wave

Electromagnetic energy shows wave-like behavior under the right conditions. This can be easily demonstrated by passing a beam of light through two narrow slits spaced very close to each other. The resulting interference pattern gives strong evidence to the idea that light is a wave. Light waves like ripples on a pond, tend to bend around sharp corners and diffract through narrow slits. The wave theory of light gives rise to the thought that there must be some kind of conductive medium to carry electromagnetic energy. This theory is not, however, consistent with other observed facts. How do waves travel through a total void? Apparently the simple existence of electric and magnetic forces is sufficient; the idea that there must be some kind of medium to carry these forces is a notion that appeals to our intuition but need not necessarily be a fact.

All waves result from oscillations. *Electromagnetic waves are caused by the vibration of electrically charged particles.* Wave motion displays two properties, frequency represented by (f) and wavelength represented by the Greek letter lambda (λ). These two properties of a wave are related to each other by the constant factor c: $c = \lambda f$, where λ is specified in meters, f in oscillations per second, and c in meters per second.

Radio signals have a wavelength that can be measured in meters; visible light waves must be measured in millionths of meters, or microns; they are often also measured in units of 10^{-10} meters, known as Angstrom units. Electromagnetic radiation can occur at any wavelength. Electromagnetic waves travel more slowly in substances such as water, as compared to their speed in a perfect vacuum. Hence the speed of light c is less in such substances. Moreover, the speed of light in different substances varies.

1.11 Particle / Wave duality

This puzzling effect mandates a further refinement of the fundamental axiom: We must restrict our measurement of the speed of light to the same medium, and the same wavelength if the medium is not a vacuum, in order for c to be constant under all conditions. The wave theory and the particle theory of light are quite different "models" for the same phenomenon, and this might lead you to wonder whether scientists really know what light is. In fact, about all we can say to answer that question is that light is energy.

It displays some properties of both a particle and a wave. Sometimes the particle model provides a better explanation of observed data, and sometimes the wave model seems more descriptive.

One postulate of the particle-wave duality, at least from this author's point of view, is that the particle attribute of light is a much more concentrated form of energy, one of higher density, while the wave attribute of light is a much less concentrated form of energy propagating through space-time at a much higher velocity (the speed of light). Inertia may play a vital role in the different velocities between the particle and wave speed of propagation. But, both forms of energy are really one in the same as Einstein's equation $E=mc^2$ suggests.

1.12 Light as electromagnetic radiation

Another commonly encountered wavelike phenomenon is *'electromagnetic radiation'*, exemplified in the radio waves that bring signals to our radios and televisions and in light. These waves have different frequencies and wavelengths: for example, typical *FM* radio signals have a wavelength of *3m*, whereas the wavelength of light depends

on its color, being about $4 \times 10^{-8}\,m$ for blue light and 7×10^{-8} m for red light: other colors have wavelengths between these values. Light waves are different from water waves and sound waves in that there is nothing corresponding to the vibrating medium (i.e. water, and air) in the examples discussed earlier.

Indeed, light waves are capable of travelling through empty space, as is obvious from the fact that we can see the light emitted by the Sun and stars. This property of light waves presented a major problem to scientists in the 18th and 19th centuries. Some concluded that space is not actually empty, but filled with an otherwise undetectable substance known as *'ether'* which was thought to support the oscillation of light waves. However, this hypothesis ran into trouble when it was realized that the properties required to support the very high frequencies typical of light could not be reconciled with the fact that the ether offers no resistance to the movement of objects (such as the Earth in its orbit) through it.

So, it was James Clerk Maxwell who around 1860 showed that the *ether* postulate was unnecessary. At that time, the physics of electricity and magnetism was being developed and Maxwell was able to show that it was all contained in a set of equations (now known as "Maxwell's equations"). He also showed that one type of solution to these equations corresponds to the existence of waves that consist of oscillating electric and magnetic fields that can travel through empty space without requiring a medium. The speed these electromagnetic waves travel is determined by the fundamental constants of electricity and magnetism, and when this speed was calculated, it was found to be identical to the measured speed of light. This led directly to the idea that light is an electromagnetic wave and it is now known that this model also applies to a range of other phenomena, including radio waves, infrared radiation (heat), micro-waves, *X*-rays, etc.

1.13 Interference

Direct evidence that a phenomenon, such as light, is a wave is obtained from studying "interference". Interference is commonly encountered when two waves of the same wavelength, are added together. When two waves

are in step, the technical term is *"in phase."* They add together to produce a combined wave that has twice the amplitude of either of the originals. If, on the other hand, they are exactly out of step (in *"anti-phase"*) they cancel each other out. In intermediate situations the waves partially cancel and the combined amplitude has a value between these extremes. Interference is crucial evidence for the wave properties of light

and no other classical model can account for this effect. Suppose, for example, that we instead had two streams of classical particles: the total number of particles would always equal the sum of the numbers in the two beams and they would never be able to cancel each other out in the way that waves can. Despite all this, we shall soon see that there is evidence that light does exhibit particle properties in some circumstances and a more comprehensive understanding of the quantum nature of light will introduce us to 'wave-particle' duality.

Note: *Electromagnetic waves are caused by the vibrations of electrically charged particles.*

Ch 2 Light Quanta

2.1 Discovering the Quanta

At about the time that J.J. Thomson was doing his experiments with cathode ray tubes, Max Planck in Germany was trying to resolve another big unexplained problem in physics. This problem had to do with the way hot objects radiate energy. His explanation eventually helped scientists explain why the electron in an atom doesn't collapse into the nucleus. But before it did that Planck's solution needed Einstein's interpretation.

Planck

To understand the problem physicists were grappling with, consider an example close to home. When you turn on an electric stove, you can feel the element getting hot before there's any appreciable change in color. Within a couple minutes, the "element" begins to glow red, and eventually, when it's extremely hot it has an orange glow. This thermal radiation is an electromagnetic wave. You can't see it if the radiation is in the infrared range, for example.

But sometimes the wavelength is in the visible range of the spectrum. And other times, the radiation is in the short wavelengths of the ultraviolet region, where again it can't be seen.

2.2 Max Planck

Planck was born in Kiel, Germany, in 1858, the sixth son of a professor of law at the University of Kiel. Planck came from a long line of academics; his grandfather and great-grandfather had also been professors. When Planck was 9, his father accepted a position at the University of Munich. The high school that Planck attended in Munich had a great math and physics teacher, and Planck became very interested in these subjects.

He was always a top student. Planck entered the University of Munich to study physics but didn't get along with his professor, Philipp von Jolly. Von Jolly told Planck that there wasn't anything new to be discovered in physics.

Unhappy with the university, Planck decided to move to the University of Berlin, where the famous physicists Hermann von Helmholtz and Gustav Kirchhoff taught. Like Einstein would do some years later, Planck became interested in areas that weren't taught in courses. He studied Rudolf Clausius' work in thermodynamics from original journal papers. After graduating with an undergraduate degree, Planck wrote a thesis on the second law of thermodynamics and submitted it as a PhD dissertation to the University of Munich. The dissertation was approved, and Planck obtained his PhD in physics when he was only 21 years old.

Like most PhD physicists at the time, Planck was interested in an academic career. At the time in Germany, if one wanted to be a professor, you started as an instructor, or Privatdozent – an unpaid position with teaching duties. Privatdozents collected small fees from the students for administering exams. But he needed another job to survive. Planck was a Privatdozent in Munich from 1880 to 1885. In 1885, he was promoted to associate professor, which meant he finally got a regular salary for teaching.

With a steady income, he married his childhood girlfriend, Marie Merck. In 1889, Planck moved to the University of Berlin as a full professor, replacing Kirchhoff, who was retiring. Planck was also a gifted pianist and seriously considered a career in music before deciding on physics. He became one of the most important scientists of all time, earning the Nobel Prize in physics in 1918 for his discovery of the quantum of energy.

2.3 The source of thermal radiation

The problem that physicists hadn't been able to solve with thermal radiation was in the high-energy region, where their equations were giving them nonsense answers. The heat radiated by an object comes from the energy changes of charged particles oscillating inside the body. What physicists were observing was that at short wavelengths, the energy distributions actually grew smaller and smaller, moving toward zero at the very short wavelengths. These very short wavelengths are in the ultraviolet part of the spectrum. And physicists refer to this problem as the ultraviolet catastrophe. The model predicted that bodies should radiate more energy at short wavelengths. The observations showed that bodies emit less energy at these wavelengths. The problem was serious, because the solutions that physicists were proposing were based on a very solid theoretical framework, but the real world data wasn't matching.

Physicists were worried because their failure to explain these new observations meant that thermodynamics, the study of heat and thermal effects, was flawed.

Around 1900, when Planck was a professor of physics at the University of Munich, he decided to tackle the radiation problem. He used Maxwell's electromagnetism to develop a theory that connected the heat or thermal energy of the radiating body and the charged oscillating "particles" of electromagnetism (But up to this point, Planck thought of electromagnetism as being "wave-like" phenomena).

To do that, he had to use the statistical methods that Ludwig Boltzmann had invented for the distribution of energies in molecular collisions. When Planck began to study the problem, the prevalent model assumed that the thermal energy emitted by an object was formed by the continuous changes in energy of charged particles oscillating within the matter. In this model the distribution of energies at shorter and shorter wavelengths grew larger and larger, eventually tending toward infinity. This prediction was not only impossible; it was completely opposite to what Planck and other scientists were observing.

2.4 Energy comes in bundles

To derive his formula using Boltzmann's statistical methods, Planck first had to split the total energy being radiated from an object into a number of bundles or packets all with the same energy. He then counted the possible ways of distributing these bundles among all the oscillating particles. Planck published his results in a series of papers between 1897 and 1900. Planck's formula known today as Planck's law, agreed perfectly with the observations. Planck, however, didn't like the method he'd used to derive his equations. He wasn't happy about using statistical methods in physics. But his equations worked.

In James Clerk Maxwell's theory of electromagnetism, the charged electron moving around the nucleus gives off energy. (In the same way, Planck's oscillating charges give off the energy of radiation in a hot object.) As these charges give off energy, they should fall into the nucleus in about a microsecond. But they don't. Atoms are stable. Rutherford proposed his model of the atom in 1911. Planck had published his solutions to the ultraviolet catastrophe and the explanation of the energy distribution of objects in 1900. During his year of miracles, in 1905, Einstein generalized Planck's idea of the energy quanta into a property of light and radiation. All the elements were there to solve the riddle of the theoretical collapsing atom.

2.5 Niels Bohr

In 1913, the Danish physicist Niels Bohr came up with a new model of the atom that avoided the collapse of the electron. His model was similar to Rutherford's planetary model but had an important difference. In Bohr's model, the electrons orbit the nucleus in very specific orbits that he called "stationary orbits". In these orbits, the electrons are safe. They don't radiate any energy. In Bohr's theory, the electrons are allowed to move between orbits. When they jump to a lower orbit, they give off energy. After they arrive there, they are safe again, in another stationary orbit. If they gain energy from an outside source such as a photon, they jump to a higher orbit. Electrons can stay only in the allowed orbits. The places in between are not allowed. The energies that the electrons give off or gain when they jump around in their orbits are Planck's bundles or quanta. Electrons are allowed to give off or gain energy only in the form of these quanta.

Bohr used Planck's theory to calculate the energies of the allowed stationary orbits for the atom. When he compared his calculations with experimental data, his theory agreed exactly. Bohr's theory provided a great advance toward the understanding of the atom. But it wasn't the final word. When Bohr tried to apply his theory to other more complicated atoms, things didn't work out quite as nicely. The energies of his allowed orbits didn't match the energies measured for those atoms. Soon, physicists realized that they needed something else. The physics of Newton and Maxwell, patched up with Planck's and Bohr's discoveries, wasn't sufficient. A few physicists knew that a new physics was needed. The 26- year-old Einstein would give them the key to this new physics.

2.6 A Quantum Leap

Relativity was not Einstein's only revolutionary theory. As mentioned earlier, he also made possible "quantum theory." Quantum theory was born in March of 1905, with Einstein's first paper of his year of miracles. The title of the paper was "On a heuristic point of view concerning the production and transformation of light." The word heuristic has the meaning of "serving to guide, discover or reveal; valuable for empirical research but unproved or incapable of proof." That definition summarized Einstein's feelings about quantum theory.

After helping in the development of quantum theory, Einstein had second thoughts about its implications. He could never accept that 'God (or Nature) would play dice with the Universe,' as he put it.

Throughout his life, Einstein thought that quantum theory was not the final word and that one day it would be replaced by the true theory of the atom. However, today scientists think that quantum theory is here to stay.

2.7 Discovering the Quantum

In his March 1905 paper, Einstein started by showing why existing equations couldn't really be applied to the problem of the radiation of objects. The old equations worked fine in the low energy region, but they failed in the high energy region. Einstein explained why, in the high energy region, the equation solutions indicated there was an infinite amount of energy. Einstein then set out to study the problem in a way 'which is not based on a picture of the generation and propagation of radiation.' In other words, he wasn't using Planck's method. He decided to start from basic physics principles. When he was done, he'd shown that the radiation of hot objects behaves as if it were made up of separate quanta (bundles or packets) of energy. Einstein assumed that the energy of each quantum is related to the wavelength of the radiation emitted: The shorter the wavelength, the larger the value of the energy.

Up to this point, there was no revolution. Like Planck and everyone else had assumed, these light quanta could be interpreted as a curious property of the radiation from hot objects. But Einstein took a bold step (which eventually gained him the Nobel Prize in 1922). He declared that matter and radiation can interact only by exchanging these energy quanta. Light isn't just a wave, as Thomas Young's experiments had shown. Instead Einstein said, light is made up of quanta of energy, and these quanta (or photons) are like particles. They're not exactly like little dust particles, but light has particle properties. A photon has a fixed amount of energy, and it exerts pressure on objects. It interacts with other particles in a particle-like way, not in a wave-like way. In other words light is lumpy.

Thomas Young's experiment showed once and for all that light is a wave, and his famous interference experiment is demonstrated today in schools around the world to show students the wave nature of light. In the wave theory, waves are not considered as particles because waves and particles have very different properties. These two ideas are mutually contradictory. It's like day and night. If you have one, you can't have the other. So the question remained; is the nature of light a wave or is it a particle? But, according to Einstein light is both particle and wave. For most physicists, this statement was considered nonsense.

But in the quantum world of physics that Einstein started, it makes perfect sense. However, it would take 20 years for scientists to resolve the seemingly inherent contradiction.

2.8 Ascertaining quanta of various energies

Einstein's light-quantum idea explained the strange results that physicists were seeing in their study of the radiation from hot objects. They'd been measuring the energy radiated at different lengths and were seeing that at long and intermediate wavelengths experiments matched up with theory. But at very short wavelengths their measurements were showing very little radiation. The experimenters could not understand that result. Their equations predicted that as they looked at shorter and shorter wavelengths, the energies should be larger and larger. But the results of their equations differed greatly from the results of their experiments.

Low-energy photons have long wave-lengths, while high-energy photons have short wave-lengths. And very little short-wavelength energy was radiated out of a hot body during their experimentation, but in theory the results were different. The light quantum idea made the experiment results clear. Planck had proposed that the thermal energy emitted by an object creates charged particles that oscillate inside the object. The energy of each photon depends on the wavelength of the oscillating. At short wavelengths the radiation from a hot object is made up of high-energy photons. (Short wavelength photons have larger energies.). Very few oscillators have photons with these energies, so only a few of these high- energy photons are emitted. This is the reason why the scientists were detecting very little radiation at these short wavelengths.

Any good classical physicist starting out from Boltzmann's equations to construct a blackbody radiation formula would have completed the integration. Then, as Einstein was later to show, the adding up of the pieces of energy would have restored the ultraviolet catastrophe – indeed, Einstein pointed out that any classical approach to the problem inevitably brings about this catastrophe. It was only because Planck knew the answer that he was looking for that he was able to stop short of the full, seemingly correct, classical solution of the equation. As a result, he was left with pieces of energy that had to be explained.

He interpreted this apparent division of electromagnetic energy into individual pieces as meaning that electric oscillators inside the atom could only emit or absorb energy in lumps of certain size, called quanta. Instead of dividing the available amount of energy up a finite number of pieces among the resonators, and the energy of such a piece of radiation (E) must be related to its frequency (denoted by the Greek letter mu) according to the new formula $E = hv$, where h is a new constant, now called Planck's constant.

At long wavelengths, the photons have low energies. Many more of these low-energy photons can be emitted, but because each one has little energy, the total amount of energy from all of them is not significantly large. At the middle wavelengths, you have quite a few oscillators that emit photons of moderate size. As a result, the largest emission of energy occurs at the middle wavelengths in the spectrum, as the experiments were showing.

Planck's great insight, validated by Einstein's generalization, was to realize that the energies are related to the different wavelengths of the light emitted by the oscillators, instead of assuming that the energies are all equally distributed, like previous theories had done. Einstein's great insight was to make this a fundamental property of nature. Light and all electromagnetic radiation are made up of quanta; light is "quantized."

The quanta is determined by a constant now called Planck's constant. Radiations of shorter wavelengths carry more energy (larger steps) and are identical to each other. But those radiations are different from the ones that consist of larger wavelengths (shorter steps). Just as you can't split the steps of a staircase, you can't split photons. It was now determined that light and all electromagnetic radiation is quantized. All the photons radiated at one wavelength have the same energy and are identical.

With Einstein's new view of the nature of light, Planck's radiation law became the accepted explanation for the radiation of hot objects. This law indicated that the thermal energy emitted by an object comes from charged particles oscillating inside the object and that these oscillations have only specific energies. Einstein hadn't been satisfied with Planck's radiation law before, because it hadn't made sense. With his new insight, it made sense, and he was ready to embrace it.

2.9 Solving the photoelectric effect

Another nagging problem in physics, called the photoelectric effect, had also resisted all attempts at explanation. Heinrich Hertz had seen this effect in 1877, and physicists were puzzled by it. If you shine a beam of light on a certain material, you can detect electrons that are ejected from the metal. (That's what happens in the solar cells that power our modern devices, from calculators to the Martian rovers.) The electrons that are ejected from the electric current runs the device. If you increase the brightness of the light, you get more electricity out of the cell. However, the speeds of the electrons emitted don't change when you increase the brightness. What's worse, if you increase the wavelength of the light beyond a certain value, you won't get any electrons out of the cell, regardless of how bright the light is.

The solution again came from Einstein's "light-quantum" idea. How does the light quantum solve the first problem? When you shine a light at a material you are sending photons of a certain energy level (or frequency). Let us suppose that the light is "monochromatic" (of a single color). In this case, the wavelength of light is fixed, and all the photons that strike the surface have the same energy. When one of the photons strikes the material, all of its energy is transmitted to an electron in the material. The electron then takes in one of these photons and uses its energy to get out of the material. When you increase the brightness, you aren't increasing the energy of each photon, you are simply sending in more photons all of the same energy.

The chances of one electron taking in more than one photon are very small, so increasing the brightness releases more electrons. However, it doesn't affect their speeds.

The solution to the second problem is much simpler. It takes energy for the electron to negotiate its way out of the material. The energy that it receives from the light comes in these photons of fixed energy. Each single photon has to provide one electron with enough energy to leave the material. If you shine a light of low-energy photons, ones from a light with long wavelength, these photons may not have the minimum energy needed. Increasing the brightness of the light only increases the number of these low-energy photons, not their energy or frequency.

2.10 Waves of Matter

Physicists reacted to having these two nagging and long-standing problems with skepticism. In 1905, Einstein was an unknown. However, after the barrage of amazing papers he published that year, it was difficult not to have heard about him. Planck was one of the first physicists to acknowledge his brilliance and one of the early proponents of Einstein's special theory of relativity. But the theory of the light-quantum was another matter. Not even Planck accepted it, even though it helped explain his own discovery. As late as 1913, when Einstein was recognized as one of the top European physicists (and at that time, physics was done mostly in Europe), there was strong opposition to his quantum idea. When Einstein was proposed for membership in the Prussian Academy of Sciences in 1913, Planck and other illustrious physicists wrote in their official recommendations as follows:

'One can only say that there is hardly one among the great problems in which modern physics is so rich to which Einstein has not made a remarkable contribution. That he may sometimes have missed the target in his speculations, as, for example, in his hypothesis of light-quanta, cannot be held too much against him.' For 15 years, Einstein stood alone in his belief of the light-quantum idea. In 1918, Einstein said that he no longer doubted the reality of quanta, 'even though I am still alone in this conviction.'

Ch 3 Early Quantum Physics

3.1 The discovery of the quantum

Quantum Mechanics is among the great intellectual achievements of the 20th century, and how this came about is interesting in itself. The following only offers the briefest of sketches. The history separates naturally into two eras: 1900-1925, the development of the "Old Quantum Theory"; and 1925 to circa 1935, in which quantum mechanics and electrodynamics were discovered and their principle features expounded upon.

In 1900 Max Planck discovered that he could only account for the spectrum of thermal radiation, which was in violent contradiction with classical electrodynamics, by assuming that the material sources of radiation have a discrete (*"Quantized"*) energy spectrum. That this entitled a grave contradiction with classical physics was clear to Planck and troubled him.

In 1905, Albert Einstein produced a much deeper departure from classical concepts by applying "quantization" to the thermodynamics of the electromagnetic field itself, introduced what was later to be called the photon, and predicted correctly the basic feature of the photo-electric effect. Although this and later Einstein papers on the quantum theory had a great influence, his idea that light has *"corpuscular"* aspects was only widely accepted after the discovery of the Compton effect in 1923. The next major step came in 1913 with Niels Bohr's quantum theory of the *"hydrogen spectrum"* based on Ernest Rutherford's model of the atom. This led to a great advance in the understanding of atomic structure and spectra. The culminating theoretical advances of this first period were Wolfgang Pauli's exclusion principle and the discovery of electron spin by Goudsmit and George Uhlenbeck in 1925. The uncertainty relations do not say whether it is possible to determine one member of the pair x_i and p_i to arbitrary accuracy by surrendering all knowledge of the other.

3.2 The formal framework

Any physical theory employs concepts more primitive than those on which the theory sheds light. In classical mechanics these primitive concepts are time, and points in three-dimensional Euclidean space. In terms of these, Newton's equations are unequivocal definitions of such concepts as momentum and force, which had only been qualitative and ambiguous notions before then. Maxwell's equations and the Lorentz force law, in conjunction with Newton's equations define the concepts of electric and magnetic fields in terms of the motion of test charges. There is no corresponding clear cut path from classical to quantum mechanics. In many circumstances, though certainly not all, classical physics tells us how to construct the appropriate Schrodinger equation that describes a particular phenomenon. But the statistical interpretation of quantum mechanics is not implied by the Schrodinger equation itself. For this and other reasons the interpretation of quantum mechanics is still controversial despite its unblemished *"empirical"* success. The quantum states of all systems having the same degrees of freedom live in the same "Hilbert space." Because quantum mechanics makes only "statistical" predictions in most circumstances, there is a tendency to overstate quantum mechanical uncertainty. But in important ways, quantum mechanics provides more knowledge than classical mechanics.

3.3 The Quantum Revolution

Max Planck's formula of the *"quanta"* was announced at a meeting of the Berlin physical Society in October 1900. For the next two months he immersed himself in the problem of finding a physical basis for this law, trying out different combinations of physical assumptions to see which ones matched the mathematical equations. He later said that this was the most intensive period of work of his entire life. Many attempts failed, until at Finally, Planck was left with only one, to him an unwelcome alternative.

Planck was a physicist of the old school. In his earlier work he had been reluctant to accept the molecular hypothesis and he particularly abhorred the idea of a statistical interpretation of the property known as *"entropy,"* an interpretation introduced by Boltzmann into the science of thermodynamics. Entropy is a key concept in physics, related in a fundamental sense to the flow of time. Although the simple laws of mechanics – Newton's laws – are completely reversible as far as time is concerned, we know that the real world just isn't like that.

Think of a stone dropped on the ground. When it hits the ground, the energy of its motion is converted into heat. But if we put an identical stone on the ground and warm it by the same amount, it doesn't jump up into the air. In the case of the falling stone, an orderly form of motion (all the atoms and molecules falling in the same direction) is turned into a disordered form of motion (all the atoms and molecules jostling against one another energetically but randomly). This is in accordance with a law of nature that seems to require that disorder is always increasing and disorder is identified, in this sense, with entropy. The law is the second law of thermodynamics, and states that natural processes always move toward an increase of disorder, or that entropy always increases. If you put disordered heat into a stone it cannot, in that case, use that energy to create an orderly movement of all the molecules in the stone so that they jump upward together.

Boltzmann introduced a variation on the theme. He said that such a remarkable occurrence could happen, but it is extremely unlikely. "In the same way, as a result of the random movement of air molecules it could happen that all of the air in the room might suddenly concentrate in the corners (it has to be more than one corner because the molecules are moving in three space dimensions); but, again such a possibility is so unlikely that for all practical purposes it can be ignored.

Planck argued against this *"statistical interpretation"* of the second law of thermodynamics long and hard, both publicly and in correspondence with Boltzmann. For him, the second law was absolute; entropy must always increase, and probabilities didn't enter into it. So it is easy to understand how Planck must have felt near the end of 1900, when, having exhausted all other options he reluctantly tried to incorporate Boltzmann's statistical version of thermodynamics into his calculations of the blackbody spectrum, and found that they worked. The irony of the situation is made more piquant, however, by the fact that because of his unfamiliarity with Boltzmann's equations, Planck applied them inconsistently. He got the right answer, but for the wrong reason, and it wasn't until Einstein took up the idea that the real significance of Planck's work became clear.

It is worth stressing that it was already a major step forward in science for Planck to establish that Boltzmann's statistical interpretation of "entropy" increase is the best description of reality. Following Planck's work, it could never really be doubted that entropy increase, while very probable indeed, cannot be taken as an absolute certainty. This has interesting implications in cosmology, the study of the structure of the whole Universe, where we deal with vast stretches of time and space. The bigger the region we deal with, the more scope there is for unlikely things to happen someplace, and sometime inside it. It is even possible (though still not very likely) that the whole Universe, which is an orderly place, by and large, represents some sort of thermodynamic statistical fluctuation, a very large, very rare "hiccup" that has created a region of low entropy that is now running down. Planck's "mistake," however, revealed something even more fundamental about the nature of the Universe.

Boltzmann's statistical approach to thermodynamics involved cutting energy into chunks, mathematically, and treating the chunks as real quantities that could be handled by the probability equations. The energy divided up into portions before this part of the calculation then has to be added together (integrated) at a larger stage, to give the total energy – in this case, the energy corresponding to black body radiation. Halfway through this procedure, however, Planck realized that he already had the mathematical formula he was looking for. Before getting to the stage of integrating the pieces of energy back into a continuous whole, the blackbody equation was there in mathematics. So he took it. This was a very drastic step, and totally unjustified within the context of classical physics.

3.4 What is the meaning of *h*

It is easy to see how this resolves the ultraviolet catastrophe. For very high frequencies, the energy needed to emit one quantum of radiation is very large, and only a few of the oscillators will have this much energy (in accordance with the statistical equations) so only a few high-energy quanta are emitted. At very low frequencies (long wavelengths), very many low-energy quanta are emitted, but they each have so little energy that even added together they don't amount to much. Only in the middle range of frequencies are there plenty of oscillators that have enough energy to emit radiation in moderate-size lumps, which add together to produce the peak in the blackbody curve.

But Planck's discovery, announced in December 1900, raised more questions than it answered. Planck's own early papers on the quantum theory are not models of clarity (which perhaps reflects the confused way he was forced to introduce the idea into thermodynamics theory) and for a long time many – even most – physicists who knew about his work still regarded it simply as a "mathematical trick," a device to get rid of the ultraviolet catastrophe that had little or no physical significance. Planck himself was certainly confused. In a letter to Robert William Wood written in 1931, he looked back at his work of 1900 and said, 'I can characterize the whole procedure an act of despair... a theoretical interpretation had to be found at any price, however high it might be.' Yet he knew he had stumbled upon something significant, and according to Heisenberg, Planck's son later told how his father described his work at the time, during a long walk through the Grunewald in the suburbs of Berlin, explaining that the discovery might rank with those of Newton.

Physicists were busy in the early 1900s absorbing the new discoveries involving atomic radiation, and Planck's new "mathematical trick" to explain the blackbody curve didn't seem of overwhelming importance alongside those discoveries. Indeed, it took until 1918 for Planck to receive the Nobel Prize for his work, a very long time compared with the speed which the work of the Curies or Rutherford was recognized. (This was partly because it always takes longer to recognize dramatic new theoretical breakthroughs; a new theory isn't as tangible as a new particle, or an *X*- ray, and it has to stand the test of time and be confirmed by experiments before it achieves full recognition and acceptance).

There was also something odd about Planck's new constant, *h*. It is a very small constant, 6.6×10^{-34} joule second, but that is not so puzzling since if it were much bigger then it would have made its presence obvious long before physicists began to puzzle over blackbody radiation. The strange thing about *h* is the units in which it is measured, energy (ergs) multiplied by time (seconds). Such units are called *"actions."* And they were not an everyday feature of classical mechanics – there is no "law of conservation of action" to rank with the law of conservation of mass or energy. But an action has one particularly interesting property, which it shares, among other things, with entropy. A constant action is absolutely constant and has the same size for all observers in space time. It is a four-dimensional constant, and the significance of that only became apparent when Einstein unveiled his theory of relativity.

Because Einstein is the next actor to enter upon the quantum mechanical stage, it might be worth a small diversion to see what this means. The special theory of relativity treats the three dimensions of space and one of time as a four-dimensional whole, the space-time continuum. Observers moving through space at different speeds get a different view of things – they will disagree, for example, on the length of a stick that they measure as it passes. But he stick can be thought of as existing in four dimensions, and as it moves "through" time it traces out a four-dimensional surface, a hyper-rectangle whose height is the length of the stick and whose breath is the amount of time that has passed.

The *"area"* of that rectangle is measured in units of length *x* time, and that area comes out to be the same for all observers who measure it, even though they disagree with one another about the length and the time they are measuring. In the same way action (energy *x* time) is a four-dimensional equivalent to area and is seen to be the same for all observers, even when they disagree about the size of the energy and time components of the action. In special relativity, there is a law of conservation of action and it is every bit as important as the law of conservation of energy. Planck's constant only looked peculiar because it was discovered before the theory of relativity which describes the properties of space and time much differently than in the Newtonian description.

Of Einstein's three great contributions to science published in 1905, special relativity seems to be very different from the others, on Brownian motion and the photoelectric effect. Yet they all hang together on the framework of theoretical physics, and in spite of the publicity generated by his theory of relativity the greatest of Einstein's contributions to science was his work on quantum theory, which jumped off from Planck's work by way of the photoelectric effect.

The revolutionary aspect of Planck's work in 1900 was that it showed a limitation to classical physics. The limitation is just the fact that there are phenomena that cannot be explained solely with the classical ideas built from the work of Newton. And this was enough to herald a new era in physics. However, the original form of Planck's work was, however, much more limited that it often seems from modern accounts.

In spite of the fact that Planck invented the quantum, and freed physics from its classical chains, Planck only suggested that the electric oscillators inside atoms might be "quantized." He meant that they could only emit packets of energy of certain sizes, because something inside them prevented them from absorbing or emitting "in between" amounts of radiation. So Planck himself did not suggest that the "radiation" was quantized. He made some contributions to the science he had founded, but spent most of his working life trying to reconcile the new ideas with classical physics. It wasn't that he had changed his mind about his discovery of the quantum, but rather that he never appreciated just how far his blackbody equation was removed from classical physics. He derived the equation by combining *thermodynamics* with *electrodynamics*, and both of these were classical theories. Rather than having second thoughts, Planck's efforts to find a "halfway house" between classical physics and quantum ideas represented a major shift, for him, away from the classical ideas on which he had been taught in college. However, his grounding in classical ideas was so thorough that it took real progress in physics to being made by a new generation of physicists, less set in their ways and less committed to old ideas, enthusiastic about the new discoveries in atomic radiation and fervent in looking for new answers to both old and new questions.

3.5 The meaning of Planck's constant in terms of space and time

You can measure a "smallest distance". As fore-mentioned, physicists have a theory that a "smallest distance" exists. It's the Planck length, named after the physicist Max Planck. The length is the smallest division that, theoretically, you can divide space into. However, the Planck length — about 1.6×10^{-35} meter, or about 10^{-20} times the approximate size of a proton — is really just the smallest amount of length with any physical significance, given the current understanding of the Universe. Smaller than this, and the whole notion of distance breaks down. And so, what about time?

There may be a smallest "*time*." In the same sense that Planck length is the smallest distance, Planck time is the smallest amount of time. The Planck time is the time light takes to travel *1* Planck length, or 1.6×10^{-35} meter. If the speed of light is the fastest possible speed, you can easily make a case that the shortest time you can measure is the Planck length divided by the speed of light. The Planck length is very small, and the speed of light is very fast, which gives you a very, very short time for the *Planck time*: The Planck time is about 5.3×10^{-44} second, and the notion of time breaks down as "times" or "durations" become smaller than this. Some people say that time is broken up into quanta of time, called "chronons," and that each chronon is a Planck time in duration.

Planck Time =

$$1.6 \times 10^{-35}\ m / 3.0 \times 10^{8}\ m/s = 5.3 \times 10^{-44}\ \text{second}$$

3.6 Einstein, Light and Quanta

Einstein was twenty-one in March 1900. He took up his famous job with the Swiss patent office in the summer of 1902, and in those early years of the 20th century devoted most of his scientific attention to problems of thermodynamics and statistical mechanics. His first scientific publications were as traditional in style and in the problems they tackled as those of the previous generation, including Planck. But in the first paper he published that refers to Planck's ideas about the blackbody spectrum (it was published in 1904), Einstein began to break new ground, and to develop, and to develop a style of solving physical puzzles that was all his own. Martin Klein describes how Einstein was the first person to take the physical implications of Planck's work seriously, and treat them as more than a mathematical trick: within a year, this acceptance of the equations as having a foundation in physical reality had led a dramatic new insight, the revival of the corpuscular theory of light. The other jumping-off point for his 1904 paper, as well as Planck's work, was the investigation of the photoelectric effect by Phillip Lenard and J.J. Thomson, working independently, at the end of the 19th century. Lenard, born in 1862 in the part of Hungary that is now Czechoslovakia, received the Nobel Prize for physics in 1905 for his research on cathode rays (electrons). Among those experiments, he had shown in 1899 that cathode rays (electrons) can be produced by light shining onto a metal surface in a vacuum. Somehow, the energy in the light makes electrons jump out of the metal.

Lenard's experiments involved beams of light of a single color (monochromatic light) which means that all the waves in the light have the same frequency (color). He looked at the way the way the intensity of the light affected the way electrons were ripped out of the metal, and found a surprising result. Using a brighter light (he actually moved the same light closer to the metal surface, which has the same effect) there is more energy shining on each square centimeter of the metal surface. If an electron gets more energy, then it ought to be knocked out of the metal more rapidly, and fly off with a greater velocity. But Lenard found that as long as the wavelength of the light stayed the same all of the ejected electrons flew off with the same velocity. Moving the light closer to the metal increases the number of electrons that were ejected, but each of those electrons still came out with the same velocity as the ones produced by a weaker beam of light of the same color. On the other hand, the electrons did move faster when he used a beam of light with a higher frequency – ultraviolet, say, instead of blue or red light.

Bohr

There is a very simple way to explain this, provided you are prepared to abandon the ingrained ideas of classical physics and take Planck's equations as physically meaningful.

The importance of those provisos is clear from the fact that in the five years after Lenard's initial work on the photoelectric effect and Planck's introduction of the concept of quanta nobody took that seemingly simple step. In effect, all Einstein did was to apply the equation $E = hv$ to the electromagnetic radiation, instead of to the little oscillators inside the atom.

He said that light is not a continuous wave as scientists had thought for a hundred years, but instead comes in definite "packets," or "quanta." All the light of a particle frequency v, which means of a particular color, comes in packets that have the same energy E. Every time one of these light quanta hits an electron, it gives it the same amount of energy and therefore the same velocity. More intense light simply means that there are more light quanta, now called photons, that all have the same energy, but changing the color of the light changes its frequency, and so changes the amount of energy that is carried by each photon. This was the work for which Einstein eventually received the Nobel Prize, in 1921. Once again, a theoretical breakthrough had to wait for full recognition. The idea of photons did not gain immediate acceptance, and although the experiments of Lenard agree with the theory in a general way, it took more than a decade for the exact prediction of the relationship between the velocity of the electrons and the wavelength of the light to be tested and proved. That was achieved by the American scientist and experimenter Robert Millikan, who along the way established a very precise measurement of the value h, Planck's constant. In 1923 Millikan in turn received the Nobel Prize in Physics, for this work and his accurate measurements of the size of the charge on the electron. One paper led to the award of the Nobel Prize for Einstein; another that proved once and for all the reality of atoms; and a third that saw the birth of the theory for which he is best known: relativity. And, almost incidentally, at the same time, in 1905, he was in the *throes* of completing another little piece of work concerning the size of molecules, which he submitted as his doctoral thesis to the University of Zurich.

The doctorate itself was awarded in January 1906. Although the PhD was not then the key to a life of active research that it is today, it is still remarkable that the three great papers of 1905 were published by a man who could at that time only sign himself "Mr." Albert Einstein.

In the next few years Einstein continued to work on the integration of Planck's quantum into other areas of physics. He found that the idea explained long-standing puzzles concerning the theory of specific heat. The specific heat of a substance is the amount of heat needed to raise the temperature of a fixed amount of material by a chosen degree; it depends on the way atoms vibrate inside the material, and those vibrations turn out to be quantized.

This is a less glamorous area of science, often overlooked in accounts of Einstein's work, but the "quantum theory of matter" gained acceptance more quickly than Einstein's "quantum theory of radiation," and began the persuasion of many physicists of the old school that quantum ideas had to be taken seriously. Einstein refined his ideas on quantum radiation over the years up until 1911, establishing that the quantum structure of light is an inevitable implication of Planck's equation, and pointing out to an unreceptive scientific committee that the way to a better understanding of light would involve a fusion of the wave and particle theories that had vied with each other since the 17th century.

By 1911, his thoughts were turning to other things. He had convinced himself that quanta were real, and his own opinion was all that mattered. His new interest was the problem of gravity, and over the five years up to 1916 he developed his General Theory of Relativity, the greatest of all his works. It took until 1923 for the reality of the quantum nature of light to be established beyond all doubt, and this in turn led to a new debate about particles and waves that helped to transform quantum theory and ushered in the modern version of the theory, quantum mechanics.

The first flowering of quantum theory came in the decade during which Einstein turned away from the subject and concentrated on others matters. It came from a synthesis of his ideas with Rutherford's model of the atom, and it came largely as a result of the work of a Danish scientist, Niels Bohr, who had been working with Rutherford in Manchester. After Bohr produced his model of the atom, nobody could doubt any longer the value of quantum theory as a description of the physical world of the very small.

3.7 Bohr's Atom

By 1912 the pieces of the puzzle were ready to be fitted together. Einstein had established the broad validity of the idea of quanta, and had introduced the idea of photons even though this was not yet generally accepted. What Einstein said was that energy really does only come in packets of definite size. Rutherford had produced a new picture of the atom, with a small central nucleus and a surrounding cloud of electrons, though again this idea had yet to gain general support.

Rutherford's atom, however, simply could not be stable according to the classical laws of electrodynamics. The solution was to use quantum rules to describe the behavior of electrons within atoms. And once again the breakthrough came from a young researcher with a fresh approach to the problem – a continuing theme throughout the story of the development of quantum theory.

Niels Bohr was a Danish physicist who completed his doctorate in the summer of 1911 and went to Cambridge that September to work with J.J. Thomson at the Cavendish. He was a very junior researcher, shy and speaking imperfect English; he found it difficult to find a niche in Cambridge, but on a visit to Manchester he met Rutherford and found him very approachable and interested in Bohr and his work. So in March 1912 Bohr moved up to Manchester and began to work with Rutherford's team, concentrating on the puzzle of the structure of the atom. After six months he returned to Copenhagen, but only briefly, and he remained associated with Rutherford's group in Manchester until 1916.

3.8 Energetic Electrons

Bohr's genius was what was needed in atomic physics over the next ten to fifteen years. He didn't worry about explaining all the details in a complete theory, but was quite willing to patch together different ideas to make an imaginary "model" that worked in at least rough agreement with the observation of real atoms. Once he had a rough idea of what was going on, he could tinker with it to make the bits fit together even better and in this way work toward a more complete picture. So he took the image of an atom as a miniature "*solar system*," with electrons moving around orbits in accordance with the laws of classical mechanics and electromagnetism, and he said that the electrons could not spiral inward out of those orbits, emitting radiation as they did so, because they were only allowed to emit whole pieces of energy – whole quanta – not the continuous radiation required by classical theory. The "stable" orbits of the electrons corresponded to certain fixed amounts of energy, each a multiple of the basic quantum, but there were no in between orbits because they would require fractional amounts of energy. Pushing the solar system analogy rather more than is justified, this is like saying that the Earth's orbit around the Sun is stable, and so is that of Mars, but that there is no such thing as a stable orbit anywhere in between.

The entire idea of an orbit depends on classical physics; the idea of electron states corresponding to fixed amounts of energy or energy levels, as they came to be called, comes from quantum theory. Making a model of the atom by patching together bits of classical theory and bits of quantum theory gave no true insight into what made atoms tick, but it did indeed provide Bohr with enough of a working model to make progress. His model turns out to have been wrong in almost every respect, but it provided a transition to a genuine quantum theory of the atom, and as such it was invaluable. Unfortunately, because of its nice, simple blend of quantum and classical ideas, and the seductive picture of the atom as a miniature solar system, the model has outstayed its welcome in the pages not just of popularizations but of many school and even university texts. If you learned anything about atoms at school, you must have learned about Bohr's model, whether or not it was given that name in class. But now, at this stage of the book, you must prepare yourself to be persuaded that the Bohr's model of the atom was not the whole truth. And you should try to forget the idea of electrons as little "planets" circling around the nucleus – it's the idea Bohr had at first, but it really is misleading. An electron is simply something that sits outside the nucleus and has a certain amount of energy and other properties. It moves as we shall see, in a mysterious way.

The great early triumph of Bohr's work, in 1913, was that it successfully explained the spectrum of the light from hydrogen, the simplest atom. The science of spectroscopy goes back to the early years of the 19th century, when William Wollaston discovered dark lines in the spectrum of light from the Sun, but it was only with Bohr's work that it came into its own as a tool for probing the structure of the atom. Like Bohr mixing classical and quantum theories to make progress, however, we have to take a step back from Einstein's ideas about light quanta to appreciate how spectroscopy works. In this kind of work, it makes no sense to think of light as anything except an electromagnetic wave. White light, as Newton established is made up of all the colors of the rainbow, the spectrum. Each color corresponds to a different wavelength of light, and by using a glass prism to spread white light out into its colored components we are in effect spreading out the spectrum so that the waves of different frequencies lay beside one another a screen, or on a photographic plate. Short-wavelength blue and violet light is at one end of the optical spectrum, and long-wavelength red at the other – at both ends, though, the spectrum extends far beyond the range of colors visible to our eyes.

When the Sun's light is spread out in this way the spectrum revealed is marked by very sharp dark lines at very precise places in the spectrum corresponding to very precise frequencies. Without knowing how these lines were formed, researchers such as Joseph Fraunhofer, Robert Bunsen (whose name is immortalized in the standard laboratory burner) and Gustav Kirchhoff, working in the 19th century, established by experiments that each element produces its own set of spectral lines. When an element (such as sodium) is heated in the flame of a Bunsen burner, it produces light with a characteristic color (in this case, yellow), which is produced by a strong emission of radiation as a bright line or lines in one part of the spectrum. When white light passes through a liquid or gas containing the same element even if the element is combined with others in a chemical compound the spectrum in the light shows dark absorption lines, like those in light from the Sun, at the same frequencies characteristic of the element. This explained the dark lines in the solar spectrum.

With the aid of spectroscopy, astronomers can probe the distant stars and galaxies to find out what they are made of. And atomic physicists can now probe the inner structure of the atom using the same tool. The spectrum of hydrogen is particularly simple, which we now know is because hydrogen is the simplest element and each atom contains just one positively charged proton for its nucleus, and one negatively charged electron associated with it. The lines in the spectrum that provide the unique fingerprint of hydrogen are called the Balmer lines after Johann Balmer, a Swiss schoolteacher who worked out a formula describing the pattern in 1885, which just happens to have been there Niels Bohr was born.

3.9 The Balmer lines

Balmer's formula relates the frequencies in the spectrum at which hydrogen lines occur to one another. Starting with the frequency of the first hydrogen line, in the red part of the spectrum, Balmer's formula gives the frequency of the next hydrogen line, in the green. Starting from the green line, the same formula applied to that frequency gives the frequency of the next line, in the violet, and so on. Balmer only knew about four hydrogen lines in the visible spectrum when he worked out his formula, but other lines had already been discovered and fitted it exactly; when more hydrogen lines were identified in the ultraviolet and infrared, they too fitted this simple numerical relationship. Obviously, the Balmer formula said something significant about the structure of the hydrogen atom.

Balmer's formula was common knowledge among physicists and it was part of every physics undergraduate course by the time Bohr came on the scene. But it was just part of a mass of complicated data on spectra, and Bohr was no *spectroscopist.* When he began working on the puzzle of the structure of the hydrogen atom, he didn't immediately think of the Balmer series of lines as an obvious key to use to unlock the mystery, but when a colleague who specialized in spectroscopy pointed out to him just how simple the Balmer formula really was (regardless of the complexities of the spectra of other atoms) he was quick to see its value. A simple version of the formula says that the wavelengths of the first four hydrogen lines are given by multiplying a constant (36.456 x 10^{-5} by 9/5, 16/12, 25/21, and 36/32. In this version of the formula, the top of each fraction is given by the sequence of squares (3^2, 4^2, 5^2, 6^2 the denominators are differences of squares, 3^2, - 2^2, 4^2 - 2^2, and so on).

At that time, early in 1913, Bohr was already convinced that part of the answer to the puzzle lay in introducing Planck's constant, *h*, into the equations describing the atom. Rutherford's atom only had two kinds of fundamental numbers incorporated in its structure, the charge on the electron, *e*, and the masses of the particles involved. No matter how much you juggle with the figures, you can't get a number that has the dimensions of "length" out of a mixture of mass and charge, so that Rutherford's model had no "natural" unit of size.). But with an "action," like *h*, added to the equation it is possible to construct a number that has the dimensions of length and can be regarded, in a rough and ready sort of way, as revealing something about the size of the atom. The expression h^2/me^2 is numerically equivalent to a length, about 20 x 10^{-8} cm, which is very much in the ball park required to fit in with the properties of atoms inferred from the scattering experiments and other studies.

To Bohr, it was clear that *h* belonged in the theory of atoms. The Balmer series showed him just where it belonged. How can an atom produce a very sharp spectral line?

By either emitting or absorbing energy with a very precise frequency. Energy is related to frequency by Planck's constant ($E = hv$), and if an electron in an atom emits a quantum of energy hv then the energy of the electron must change by precisely the corresponding amount *E*. Bohr said that the electrons "in orbit" around the nucleus of an atom stayed in place because they could not radiate energy continuously, but they would, on this picture, be allowed to radiate or (absorb) a whole quantum of energy – one photon – and jump from one energy level (one orbit, on the old picture) to another. This seemingly simple idea actually marks another profound break with classical ideas.

It's as if Mars disappeared from its orbit and reappeared in the orbit of the Earth, instantaneously, while radiating off into space a pulse of energy (in this case, it would be gravitational radiation). You can see at once how poor the idea of a solar system atom is at explaining what goes on, and how much better it is to think of electrons simply as in different states, corresponding to different energy levels, inside the atom.

A jump from one state to another can occur in either direction, up or down the latter of energy. If an atom absorbs light, then the quantum *hv* is used to move the electron up an energy level; if the electron then falls back to its original state precisely the same energy *hv* will be radiated away from the atom. The mysterious constant $36.456 \, x \, 10^{-5}$ in Balmer's formula could be written naturally in terms of Planck's constant, and that meant Bohr could calculate the possible energy levels "allowed" for the single electron in the hydrogen atom, and the measured frequency of the spectral lines could now be interpreted as revealing how much energy difference there was between the different levels.

When dealing with electrons and atoms everyday energy units are rather too large for convenience, and the appropriate unit is the electron Volt (*eV*), which is the amount of energy an electron would pick up in moving across an electric potential difference of one Volt. The unit was introduced in 1912.

In more everyday terms, an electron Volt is 1.602×10^{-19} joule, and one Watt is one joule per second. An ordinary light bulb burns energy at a rate of 100 Watts, which if you want to you can express as 6.24×10^{20} *eV* per second. It certainly sounds more impressive to say my light radiates six and a quarter hundred million trillion electron volts a second, but the energy is the same as when it was a hundred-watt lamp. The energies involved in electron transitions that produce spectral lines are a few eV – it takes just 13.6 eV to knock the electron right out of a hydrogen atom. The energies of particles produced by radioactive processes are several millions of electron Volts, or MeV.

Thirteen years after Planck's desperate measure of incorporating the quantum into the theory of light, Bohr introduced the quantum into the theory of the atom. But it was to be another thirteen years before a "true" quantum theory emerged. In that time, progress was painfully slow – one step backward for every two steps forward, and sometimes two steps backward for every one that had seemed to be going in the right direction. Bohr's atom was a "hodgepodge." It mixed quantum ideas with those of classical physics, using whatever mixture seemed necessary to patch things up and keep the model going. It "allowed" many more spectral lines that can actually be seen in the light from different atoms, and arbitrary rules had to be brought in to say that some transitions between different energy states within the atom were "forbidden."

New properties of the atom – quantum numbers – were assigned *ad-hoc* to fit the observations, with no underpinning of a secure theoretical foundation to explain why these quantum numbers were required, or why some transitions were forbidden. In the middle of all this, the European world was disrupted by the outbreak of the First World War, the year after Bohr introduced his first model of the atom. Holland and Denmark, indeed, remained scientific oases at that time, and Bohr returned to Denmark in 1916 to become Professor of Theoretical Physics in Copenhagen, and then to found, in 1920, the research institute that bears his name. News from a German researcher such as Arnold Sommerfield (one of the physicists who refined Bohr's model of the atom, to such an extent that the model was sometimes referred to as the "Bohr-Sommerfield" atom) could pass neutral Denmark, and then from Bohr on to Rutherford in England. Progress continued to be made, but it was not the same.

After the war, Germany and Austrian scientists were not invited to international conferences for many years; Russia was in a revolutionary turmoil; science had lost some of its internationalism as well as a generation of young men. It fell to a completely new generation to take quantum theory from the "halfway house" of Bohr's hodgepodge atom (which had, admittedly, been refined by the diligent efforts of many researchers into a remarkably effective, if ramshackle, contrivance) on to its full glory as quantum mechanics. The names of that generation resound through modern physics – Werner Heisenberg, Paul Dirac, Wolfgang Pauli, Pascual Jordan, and others. They were members of the first quantum generation, born and bred in the years after Planck's great contribution (Pauli in 1900, Heisenberg in 1901, Dirac and Jordan in 1902), and coming into scientific research in the 1920s. They had no ingrained training in classical physics to overcome, and less need than even so brilliant a scientist as Bohr of half-measures to retain a flavor of classical ideas in their theories of the atom.

It is entirely appropriate, and perhaps no coincidence, that the time from Planck's discovery of the blackbody equation to the flowering of quantum mechanics was just twenty-six years, the time it took for a generation of new physicists to develop into research scientists. That generation, however, had two great legacies from its still active elders, apart from Planck's constant itself. The first was Bohr's atom, providing a clear indication that quantum ideas had to be incorporated into any satisfactory theory of atomic processes; the second came from the one great scientist of the time who never had seemed hamstrung by the ideas of classical physics, the exception to all the rules. In 1916, at the peak of the war and working in Germany, Einstein introduced the notion of probability into atomic theory. He did so as an expedient – another contribution to the hodgepodge that made the workings of Bohr's atom resemble the observed behavior of real atoms. But this expedient outlasted the Bohr atom to become the underpinning foundation of the true quantum theory – even though, ironically, it was later disowned by Einstein himself in his famous comment, 'God does not play dice.'

As fore-mentioned, back in the early 1900s, when Rutherford and his colleague Frederick Soddy were investigating the nature of radioactivity, they had discovered a curious and fundamental property of the atom, or rather of its nucleus. Radioactive "decay" as it became known, had to involve a fundamental change in an individual atom (we now know that it involves the breakup of a nucleus, and the ejection of parts of the nucleus), but it seemed to be unaffected by any outside influence. Heat the atoms up or cool them down, put them in a vacuum or a bucket of water, the process of radioactive decay proceeds undisturbed.

There seemed to be no way of predicting exactly when a particular atom of a radioactive substance would "decay," emitting an alpha or beta particle and gamma rays, but the experiments showed that out of a large number of radioactive atoms of the same element a certain proportion would always decay in a certain time. In particular, for every radioactive element here is a characteristic time called the "half-life," during which exactly half of the atoms in a sample decay. Radium, for example, has a half-life of 1,600 years; a radioactive form of carbon, called *carbon-14* has a half-life of a little under 6,000 years which makes it useful in archaeological dating; and radioactive potassium decays with a half-life of 1,300 million years.

Without knowing what made one atom in a vast array of atoms disintegrate while its neighbors did not, Rutherford and Soddy used this discovery as the basis of a statistical theory of radioactive decay, a theory using "actuarial" techniques like those used by insurance companies, who know that though some of the people they insure will die young and their heirs will receive far more from the insurance company than the premiums paid in, other customers will live long and pay enough in premiums to compensate. Without knowing which clients will die when, the actuarial tables enable the accounts to balance the books. In the same way, statistical tables allow physicists to balance the books of radioactive decay, provided they are dealing with large collections of atoms.

One curious feature of this behavior is that radioactivity never quite disappears from a sample of radioactive material. From the millions of atoms present, half decay in a certain time. Over the next half-life – exactly the same time span – half of the rest decay, and so on. The number of radioactive atoms left in the sample gets smaller and smaller, closer and closer to zero, but each step toward zero takes it only half the way there. In those early days, physicists such as Rutherford and Soddy imagined that eventually someone would find out exactly what made an individual atom decay, and that this discovery would explain the statistical techniques over into the Bohr model to account for details of atomic spectra. Bohr also anticipated that later discoveries would remove the need for the "actuarial tables." But they were all wrong. Because there are many steps on which the electron can lodge, and because it can jump or fall from any step (energy state) to any other, there are many lines in the spectrum of each element. Each line corresponds to a transition between steps – between energy levels with different quantum numbers.

All the transitions that end up in the ground state, for example, produce a family of spectral lines like the Balmer series; all the transitions from higher steps that end up on level two correspond to another set of lines, and so on.

In a hot gas atoms are constantly colliding with one another, so that electrons are excited to high energy levels and then fall back, radiating bright spectral lines as they do so. When light passes through a cold gas, the ground-state electrons are raised to higher energies, absorbing light as they do so and leaving dark lines in the spectrum. If the Bohr model of the atom meant anything at all, this explanation of how hot atoms radiate energy ought to tie in with Planck's law. The blackbody spectrum of cavity radiation ought to be simply the combined effect of a lot of atoms radiating energy as electrons jump from one energy level to another

In 1916 Einstein had completed his General Theory of Relativity, and he turned his attention once again to quantum theory (compared with his masterwork, this may have seemed like recreation). He was probably encouraged by the success of Bohr's model of the atom, and also at this time his own version of the corpuscular theory of light at last began to gain ground. As fore-mentioned, Robert Andrews Millikan, an American physicist from the California Institute of Technology (Cal-Tech), had been one of the strongest opponents of Einstein's interpretation of the photoelectric effect, when this interpretation first appeared back in 1905. He spent ten years testing the data in a series of superb experiments, starting out with the aim of proving Einstein wrong and ending up in 1914 with direct experimental proof that Einstein's explanation of the photoelectric effect in terms of light quanta, or photons, was correct. In the process, he derived a very accurate experimental determination of the value h and in 1923, to complete the irony, he received the Nobel Prize for this work and his measurement of the charge on the electron.

Einstein realized that the decay of an atom from an *"excited"* energy state – with an electron at a high energy level – into a state with less energy – with the electron at a lower energy level – is very similar to the radioactive decay of an atom. He used the statistical techniques developed by Boltzmann (for dealing with the behavior of collections of atoms) to deal with individual energy states working out the probability that a particular atom would be in an energy state corresponding to a particular quantum number n, and he used the probabilistic "actuarial tables" of radioactivity (also used by Rutherford earlier) to work out the "likelihood" or "probability" that an atom in state n will "decay" into another energy state with less energy (that is, with a lower quantum number). This all led, in a clear and simple way, to Planck's formula for blackbody radiation, derived entirely on the basis of quantum ideas. Soon, using Einstein's statistical ideas, Bohr was able to extend his model of the atom, taking on board the explanation that some lines in the spectrum are more pronounced than others because some transitions between energy states are more probable – more likely to happen – than others. He could not explain why this should be so, but nobody was too worried about that at the time. Like the people who studied radioactivity in those days, Einstein believed that the *"actuarial"* tables were not the last word, and that later research would determine why a particular transition occurred exactly when it did, and not at some other time.

But it was at this point that quantum theory really began to cut loose from classical ideas, and no "underlying reason" for radioactive decay or atomic-energy transitions to occur when they do has ever been found.

It really does seem that these changes occur entirely by chance on a statistical basis, and that already begins to raise fundamental philosophical questions.

In the classical world, everything has its cause. You can trace the cause of any event backward in time to find the cause of the cause, and what caused that, and so on, all the way back to the Big Bang (if you are a cosmologist) or to the moment of Creation in a religious context, if that is the model you subscribe to. But in the world of the quantum, such direct "causality" begins to disappear as soon as we look at radioactive decay and atomic transitions.

An electron doesn't move down from one energy level to another at a particular time for any particular reason. The lower energy level is more desirable for the atom, in a *statistical sense*, and so it is quite likely (the amount of likelihood can even be quantified) that the electron will match such a move sooner and occur. No outside *agency* pushes the electron, and no internal *clockwork* times the jump. It just happens, for no particular reason, now rather then. This isn't much of a break with strict causality, and although many 19th-century scientists might have been "horrified" by the idea, its doubtful if any of the readers of this book are too concerned. But it is merely the tip of an iceberg, the first clue to the real "strangeness" of the quantum world, and worth noting even though its significance was not appreciated at the time. It came in 1916, and it came from Einstein.

In June 1922, Bohr visited the University of Gottingen in Germany to give a series of lectures on quantum theory and atomic structure. Gottingen was about to become one of the three key centers in the development of the complete version of quantum mechanics, under the direction of Max Born, who became Professor of Theoretical Physics there in 1921. He had been born in 1882, son of the Professor of Anatomy at the University of Breslau, and was a student in the early 1900s, at the time Planck's ideas first appeared. At first he studied mathematics, only turning to physics (and working for a time at the Cavendish) after completing his doctorate in 1906. This, as we shall see, turned out to have been an ideal training in the years ahead. An expert on relativity, Born's work was always characterized by full mathematical rigor, in striking contrast to Bohr's patchwork theoretical edifices, built with the aid of brilliant insights and physical intuition, but often leaving others to catch up with the mathematical details. Both kinds of genius were essential to the new understanding of atoms. Bohr's lectures in June 1922 were a major event in the renewal of German physics after the war, and also in the history quantum theory.

They were attended by scientists from all over the Germany, and became known as the "Bohr Festival". And in those lectures, after carefully preparing his ground, Bohr presented the first successful theory of the periodic table of the elements, a theory that survives in essentially the same form to this day. Bohr's idea stemmed from a picture of the electrons being added to the nucleus of an atom. Whatever the atomic number of that nucleus, the first electron would go into an energy state corresponding to the ground state of hydrogen. The next electron would go into a similar energy state, giving an outward appearance rather like the helium atom, which has two electrons. But, said Bohr, there was no room for any more electrons at that level in the atom, and the next one to be added would have to go into a different kind of energy level. So an atom with three protons in its nucleus and three electrons outside the nucleus should have two of those electrons more tightly tied to the nucleus and one left over; it ought to behave rather like a one-electron atom (hydrogen) as far as chemistry is concerned.

The $Z = 3$ element is lithium, and it does indeed show some chemical similarities to hydrogen. The next element in the periodic table with similar properties to lithium is sodium, with $Z = 11$, eight places beyond lithium. So Bohr argued that there must be eight places available in the set of energy levels outside the inner two electrons, and that when these were filled the next electron, the eleventh in all, had to go into another energy state still less tightly tied to the nucleus, again mimicking the appearance of an atom with only one electron. These energy states are called "shells" and Bohr's explanation of the "periodic table" involved successively filling up the shells with electrons as Z increased. You can think of the shells as onion skins wrapped around one another; what matters for chemistry is the number of electrons in the uttermost shell of the atom.

What goes on deeper inside plays only a secondary role in determining how the atom will interact with other atoms. Working outward through the electron shells, and incorporating all the evidence from spectroscopy. Bohr explained the relationship between the elements in the periodic table in terms of atomic structure. He had no idea why a shell containing eight electrons should be full ("closed"), but he left none of his audience in any doubt that he had discovered the essential truth. As Heisenberg said later, Bohr "had not proven anything mathematically... he just knew that this was more or less the connection."

Chemistry is concerned with the way atoms react and combine to make molecules. Why does carbon react with hydrogen in such a way that four atoms of hydrogen attach to one of carbon to make one molecule of methane? Why does hydrogen come in the form of molecules, each made of two atoms, while helium atoms do not form molecules? And so on. The answers came with stunning simplicity from the shell model.

Each hydrogen atom has one electron, whereas helium has two. The "innermost" shell would be full if it had two electrons in it, and (for some unknown reason) filled shells are more stable – atoms "prefer" to have filled shells. When two hydrogen atoms get together to from a molecule

they share their two electrons in such a way that each feels the benefit of a closed shell. Helium, having a full shell already, is not interested in any such proposition and disdains to react chemically with anything. Carbon has six protons in its nucleus and six electrons outside. Two of these are in the inner closed shell, leaving four associated with the next shell, which is half empty. Four hydrogen atoms can each claim a part share in one of the four outer carbon electrons and contribute their own electron to the deal. Each hydrogen atom ends up with a semi-closed *"pseudo-closed"* shell of two inner electrons, while each carbon atom has a semi-closed second shell of eight electrons.

Atoms combine, said Bohr, in such a way that they get as close as they can to making a closed outer shell. Sometimes, as with the hydrogen molecule, it is best to think of a pair of electrons being shared by two *nuclei*; in other cases, an appropriate picture is to imagine an atom that has an odd electron in its outer shell (*sodium*, perhaps) giving the electron away to an atom that has an outer shell containing seven electrons and one "vacancy" (in this case, it might be *chlorine*). Each atom is happy – the sodium, by losing an electron, leaves a deeper, but filled shell; the chlorine, by gaining an electron fills its outermost shell. The net result, however, is that the sodium atom has become a positively charged "ion" by losing one unit of negative charge, while the chlorine atom has become a negative ion. Since opposite charges attract, the two stick together to form an electrically "neutral molecule" of *sodium chloride*, which is common *salt*.

All chemical reactions can be explained in this way, as a sharing or swapping of electrons between atoms in a bid to achieve the "stability" of filled electron shells. Energy transitions involving outer electrons produce the characteristic spectral *fingerprint* of an element, but energy transitions involving deeper shells (and therefore much more energy, in the *X*-ray part of the spectrum) should be the same for all elements, as indeed they prove to be. Like all the best theories, Bohr's model was confirmed by a successful prediction. With the elements arranged in a periodic table, even in 1922 there were a few gaps, corresponding to undiscovered elements with atomic numbers 43, 61, 72, 75, 85 and 87. Bohr's model predicted the detailed properties of these "missing" elements and suggested that element 72, in particular, should have properties similar to *zirconium*, a forecast that contradicted predictions made on the basis of alternative models of the atom. The prediction was confirmed within a year with the discovery of hafnium, element 72, which turned out to have spectral properties exactly in line with those predicted by Bohr.

This was the high point of the *old quantum theory.* Within three years, it had been swept away, although as far as chemistry is concerned you need little more than the ideas of electrons as tiny particles orbiting around atomic nuclei in shells that would "prefer" to be full (or empty, but preferably not in between). And if you are interested in the physics of gases, you need little more than the image of atoms as hard, indestructible billiard balls. Nineteenth-century physics will do for everyday purposes; and the physics of 1923 will do for most of chemistry; and the physics of the 1930s takes us about as far as anyone has yet gone in the search for "ultimate truths." There has been no great break-through that time the rest of science has been catching up with the insights of a handful of geniuses. The success of the *Aspect experiment* in Paris in the early 1980s marked the end of that catching-up period, with the first direct experimental proof that even the most strange aspects of quantum mechanics are a literal description of the way things are in the real world. The time has come to discover just how strange the world of the quantum really is. The "little more" that is needed to explain more complex molecules was developed in the late 1920s and early 1930s using the fruits of the full development of quantum mechanics.

The person who did most of the work was Linus Pauling, more familiar today as a peace campaigner and proponent of vitamin C, who received the first of his two Nobel Prizes of the work, being cited in 1954 "for his research into the nature of the *chemical bond* and its application to the elucidation of the structure of complex substances." Those "complex substances" elucidated with the aid of quantum theory by Pauling, a physical chemist, opened the way to a study of *molecules of life.* The key significance of quantum chemistry of molecular biology is acknowledged by Horace Judson in his epic book "The Eighth Day of Creation.

3.10 Photons and electrons

In spite of the success of Planck and Bohr in pointing the way toward a physics of the very small that differed from classical mechanics, quantum theory as we know it today only really began with the acceptance of Einstein's idea of the light quantum, and the realization that light had to be described both in terms of particles and waves. And even through Einstein first introduced the light quantum in his 1905 paper on the photoelectric effect, it was not until 1923 that the idea became accepted and respectable. Einstein himself moved cautiously, well aware of the revolutionary implications of his work, and in 1911 he told the participants at the first Solvay Congress, "I insist on the provisional character of this concept, which does not seem reconcilable with the experimentally verified consequences of the wave theory."

Although Millikan proved by 1915 that Einstein's equations for the photoelectric effect was correct, it still seemed unreasonable to accept the reality of particles of light, and looking back on his work from the 1940s Millikan commented on his tests of this equation. 'I was compelled in 1915 to assert its unambiguous verification in spite of its unreasonableness . . . it seemed to violate everything we knew about the interference of light.' At the time, he expressed himself more forcefully. Reporting the experimental verification of the accuracy of

Einstein's equation of the photoelectric effect, he went on to say, 'The semi-corpuscular theory by which Einstein arrived at his equation seems at present wholly untenable.' This was written in 1915. In 1918 Rutherford commented that there seemed to be "no physical explanation" for the connection between energy and frequency that Einstein had explained thirteen years before with his hypothesis of light quanta. It wasn't that Rutherford didn't know of Einstein's suggestion, but that he wasn't convinced by it. Since all experiments were designed to test the wave theory of light to be made up of waves, how could light be made of particles?

3.10 Particles of Light

In 1909 about the time that he ceased being a patent clerk and took up his first academic post, as an associate professor in Zurich, Einstein made a small but significant step forward, referring for the first time to point-like quanta with energy hv. Particles like electrons are represented by "*point-like*" objects in classical mechanics, and this is a far cry from any description in terms of waves, except that the frequency of the radiation v, tells us the energy of the particle. 'It is my opinion,' said Einstein in 1909, 'that the next phase in the development of theoretical physics will bring us a theory of light that can be interpreted as a kind of fusion of the wave and the emission theory.' That comment strikes to the heart of modern quantum theory. In the 1920s, Bohr expressed this new basis of physics as the "*principle of complementarity,*" which holds that the wave and particle theories of light are not mutually exclusive to one another but complementary. But concepts are necessary to provide a complete description, and this shows up strikingly in the need to measure the energy of the light "particle" in terms of its frequency or wavelength. Soon after he made these remarks, however, Einstein left off serious thinking about quantum theory while he developed his General Theory of Relativity. When he returned to the quantum fray in 1916, it was with another logical development on the light-quantum theme. His statistical ideas helped, as we have seen, to tidy up the picture of the Bohr atom, and improved Planck's description of blackbody radiation. These calculations of the way matter absorbs or emits radiation also explained how momentum is transferred from radiation to matter provided that each quantum of radiation hv carries with it a momentum hv/c. The work harks back to another of the great 1905 papers, on Brownian motion. Just as pollen grains are *buffeted* by the atoms of a gas or liquid, so that their motion proves the reality of the atoms, so the atoms are themselves buffeted by the "particles" of blackbody radiation.

This "Brownian motion" of atoms and molecules could not be observed directly, but the buffeting causes statistical effects that could be measured in terms of properties such as the pressure of a gas. It was the statistical effects that Einstein explained in terms of blackbody *radiation particles* which carry momentum.

However the same expression for the momentum of a "particle" of light comes straight out of special relativity, in a very simple way. In relativity theory, the energy (E), momentum (p) and rest mass (m) of a particle are related by the simple equation.

$$E^2 = m^2 * c^4 + p^2 * c^2$$

Since the particle of light has *no* rest mass, this equation very quickly reduces to

$E^2 = p^2 * c^2$ *or* simply $p = E/c$.

It may seem surprising that it took Einstein so long to spot this, but then he had other things like general relativity on his mind. Once he did make the connection, however, the agreement between the statistical arguments and relativity theory certainly made the case a lot stronger. (From another point of view, since the statistics show that $p = E/c$, you can argue that the relativistic equations then establish that the particle of light has zero rest mass.) It was this work that convinced Einstein himself that light quanta were real. The name "photon" for the particle of light was not introduced until 1926 (by Gilbert Lewis, based in Berkeley California), and it only became part of the language of science after the Solvay Congress was held under the title "Electrons and Photons" in 1927. But although in 1917 Einstein stood alone in his belief in the reality of what we now call photons, this seems an appropriate time to introduce the name. It was another six years before incontrovertible, direct experimental proof of the reality of photons was obtained, by the American physicist Arthur Compton. Compton had been working with *X*-rays since 1913. He worked in several American universities and at the Cavendish in England.

Through a series of experiments in the early 1920s he was led inexorably to the conclusion that the interaction between *X*-rays and electrons could only be explained if the *X*-rays were treated in some ways as particles – photons.

The key experiments concern the way in which the *X*-radiation is scattered by an electron – or, in particle language, the way a photon and an electron interact when they collide. When an *X*-ray photon hits an electron, the electron gains energy and momentum and moves off at an angle. The photon itself loses energy and momentum and moves of at a different angle, which can be calculated from the simple laws of particle physics. The collision is like the impact of a moving billiard ball on a stationary ball, and the transfer of momentum occurs in just the same way. In the case of the photon, however, the loss of energy means a change in frequency of the radiation, by the amount hv given up to the electron. You need both descriptions, *particle* and *wave*, to get a complete explanation of the experiment.

When Compton made the experiments, he found the interaction behaving exactly in accordance with this description – the scattering angles, the wavelength changes and the recoil of the electron all fitted perfectly with the idea that *X*-radiation comes in the form of particles with energy *hv*.

The process is now called the "Compton effect." And in 1927 Compton received a Nobel Prize for this work. After 1923, the reality of photon as particles carrying both energy and momentum was established (although Bohr struggled hard for a time to find an alternative explanation of the Compton effect; he did not immediately see the need to include both particle and wave descriptions in a good theory of light, and saw the particle theory as a rival to the wave theory incorporated in his model of the atom). But all the evidence for the wave nature of light remained. As Einstein said in 1924, "there are therefore now two theories of light, both indispensable . . . without any logical connection." The connection between those two theories formed the basis of the development of quantum mechanics in the next few hectic years. Progress was made on many different fronts simultaneously, and new ideas and new discoveries did not come neatly in the order they were required to build up the new physics. To tell a coherent story, I have to make the account more orderly than science itself was at the time,, and one way of doing this is to lay the groundwork of relevant concepts before describing quantum mechanics itself, even though the theory of quantum mechanics began to be developed before some of those concepts were understood. Even the full implications of the *particle/wave* duality were not appreciated when quantum mechanics began to take shape – but in any logical description of quantum theory the next step after the discovery of the dual nature of light must be the discovery of the dual nature of matter.

3.11 Particles/Wave duality

The discovery stemmed from a suggestion made by a French nobleman, Louis de Broglie. It sounds so simple, yet it struck to the heart of the matter. 'If light waves also behave like particles,' we can imagine de Broglie musing 'why shouldn't electrons also behave like waves?' If he had stopped there, of course, he would not have been remembered as one of the founders of quantum theory, nor would he have received a Nobel Prize in 1929. As an idle speculation the idea doesn't amount to much, and similar speculations had been aired about *X*-rays long before Compton's work, at least as early as 1912, when the great physicist (and yet another Nobel Laureate) W.H. Bragg said of the state of *X*-ray physics at the time. 'The problem becomes, it seems to me, not to decide between two theories of *X*-rays, but to find . . . one theory which possesses the capacity of both.'

De Broglie's great achievement was to take the idea of particle/wave duality and to carry it through mathematically, describing how matter waves ought to behave and suggesting ways in which they might be observed.

He has one great advantage as a relatively junior member of the theoretical physics community, an elder brother, Maurice, who was a respected experimental physicist and who steered him toward the discovery. Louis de Broglie said later that Maurice stressed to him in conversations the 'importance and undeniable reality of the dual aspects of particle and wave.' This was an idea whose time had come, and Louis de Broglie was lucky to be around at the time when a conceptually simple piece of intuition could transform theoretical physics. And he certainly made the most of his intuitive leap.

De Broglie had been born in 1892. When he entered the University of Paris in 1910 he became fired with interest in science, especially quantum mechanics, a world opened to him partly by his brother (seventeen years his senior) who had obtained his doctorate in 1908 and who as one of the scientific secretaries to the first Solvay Congress passed on news to Louis. But after a couple of years his study of physics was interrupted in 1913 by what should have been a short period of compulsory military service but lasted, because of the first world war, until 1919. Picking up the threads after the war, de Broglie returned to the study of quantum theory, and began working along the lines that were to lead to his discovery of the underlying unity of particle and wave theories. The break-through came in 1923, when he published three papers on the nature of light quanta in the French journal "Comptes Rendus," and wrote an English summary of the work which appeared in the "Philosophical Magazine" in February 1924.

These short contributions made no great impact, but de Broglie immediately set about ordering his ideas and presenting them in a more complete form for his doctoral thesis. His examination at the Sorbonne was held in November 1924, and the thesis was published early in 1925 in the *Annales de Physique.* It was in that form that the basis of his work became clear and sparked one of the major advances in physics during the 1920s. In his thesis, de Broglie started out from the two equations that Einstein had derived for light quanta,

$$E=hv; \; p=hv/c$$

In both those equations, properties that belong to particles (energy and momentum) appear on the left, and properties that belong to waves (frequency) appear on the right. He argued that the failure of experiments to settle once and for all whether light is wave or particle must therefore be because the kinds of behavior are inextricably tangled – even to measure the particle property of momentum you have to know the wave property called frequency. Yet this duality did not apply only to photons. Electrons were thought at the time to be well-behaved particles, except for the curious way they occupied distinct energy levels inside atoms. But de

Broglie realized that the fact that electrons only existed in "orbits" defined by whole numbers (integers) also looked in some ways like a wave property. "The only phenomena involving inters in physics were those of interference and of normal modes of vibration." He wrote in his thesis. "This fact suggested to me the idea that electrons too could not be regarded simply as corpuscles, but that periodicity must be assigned to them."

Normal modes of vibration are simply the vibrations that make the notes of a violin string or a sound wave in an organ pipe. A tightly stretched string, for example, might vibrate in such a way that each end is fixed while the middle wiggles to and fro. Touch the center of the string, and each half will vibrate in the save way, with the center at rest – and this higher "mode" of vibration also corresponds to a high note, a harmonic, of the untouched full sting. In the first mode, the wavelength is twice as long as in the second, and higher modes of vibration, corresponding to successively higher notes, can fit the vibrating string provided always that the length of the string is a whole number of wavelengths (1, 2, 3, 4, and so on). Only some waves, with certain frequencies, fit the string.

This indeed, similar to the way electrons, "fit" into atoms in states corresponding to quantum-energy levels 1, 2, 3, 4, and so on. Instead of a stretched straight string, imagine one bent into a circle, an "orbit" around an atom. A standing vibration wave can run happily around the string, provided that the length of the circumference is a whole number of wavelengths. For any wave that did not precisely "fit" the string in this way, the wave would be unstable and dissipate as it interfered with itself. The head of the snake must always catch hold of its tail, or the string, like the analogy, falls apart. Could this explain the quantization of energy states in the atom, with each one corresponding to a resonating electron wave of a particular frequency? Like so many of the analogies based on the Bohr atom – indeed, like all physical pictures of the atom – the image is far from the truth, but helped toward a better understanding of the world of the quantum.

De Broglie thought of the waves as being associated with particles, and suggested that a particle such as a photon is in fact guided on its way by the associated wave to which it is tied. The result was a thorough mathematical description of the behavior of light, which incorporated the evidence from both wave and particle experiments.

The examiners who studied be Broglie's thesis appreciated the math, but did not believe that the proposal for a similar wave associated with a particle like the electron had any physical meaning – they regarded it as just a quirk of the mathematics. De Broglie did not agree. When asked by one of the examiners if an experiment could be devised to detect the matter waves, he said that it should be possible to make the required observations by diffraction a beam of electrons from a crystal.

The experiment is just like the diffraction of light through not just two but an array of slits, with the gaps between the regularly spaced atoms in the crystal providing an array of "slits" narrow enough to diffract the high-frequency (small wavelength, compared with light or even *X*-rays) electron waves.

De Broglie knew the right wavelength to look for, since by combining Einstein's two equations for light particles, he obtained the very simple relation $p=hv/c$, which we've already met. Since wavelength is related to frequency by $=c/v$, this means p*x=h or in plain words momentum multiplied by wavelength gives Planck's constant. The smaller the wavelength, the bigger the momentum of the corresponding particle, making electrons, with their small mass and correspondingly small momentum, the most "wavelike" of the particles then known. Just as in the case of light, or waves on the surface of the sea, diffraction effects only show up if the wave passes through a hole much smaller than its wavelength, and for electron waves that means a very small hole indeed, about the size of the gap between atoms in a crystal.

What de Broglie did not know was that effects that can best be explained in terms of diffraction of electrons had been observed when beams of electrons were used to probe crystals as long ago as 1914. Two American physicists, Clinton Davisson and his colleague Charles Kunsman, had, indeed, been studying this peculiar behavior of electrons, scattering from crystals, during 1922 and 1923 while de Broglie was formulating his ideas. Ignorant of all this, de Broglie tried to persuade experimenters to carry out a test of the "electron-wave hypothesis." Meanwhile, de Broglie's thesis supervisor, Paul Langevin, had sent a copy of the work to Einstein, who, hardly surprisingly, saw it as much more than a mathematical trick or analogy, and realized that matter waves must be real. He in turn passed the news on to Max Born in Gottingen, where the head of the experimental physics department, James Franck, commented that Davisson's experiments "had already established the existence of the expected effect".

Even in 1925, in spite of the existing experimental evidence, the idea of *"matter waves"* remained no more than a vague notion. It was only when Erwin Schrodinger came up with a new theory of atomic structure incorporating de Broglie's idea but going far beyond it that the experimenters felt an urgent need to check the electron-wave hypothesis by performing the diffraction experiments. When the work was done, in 1927, it proved de Broglie to have been entirely correct – electrons are diffracted by *crystal lattices* just as if they are a form of wave. The discovery was made independently in 1927 by two groups, Davisson and a new collaborator, Lester Germer, in the U.S., and George Thomson (son of J.J.) and research student Alexander Reid, working in England and using a different technique.

In 1906, J.J. Thomson had received the Nobel Prize for proving that electrons are particles; in 1937 he saw his son awarded the Nobel Prize for proving that electrons are waves. Both father and son were correct, and both awards were fully merited. Electrons are particles and electrons are waves. From 1928 onward, the experimental evidence for de Broglie's wave/particle duality became overwhelming. Other particles, including the proton and the neutron, were subsequently found to possess wave properties, including diffraction, and in a series of beautiful experiments in the late 1970s and 1980s Tony Klein and colleagues at the University of Melbourne repeated some of the classic experiments that had established the wave theory of light in the 19th century, but using a beam of neutrons (particles of matter) instead of a beam of light.

Ch 4 Quantum Physics

Pauli

4.1 Revolution of the new physics

One of the puzzles of atomic spectroscopy that the simple Bohr model of the atom failed to explain involves the splitting of the spectral lines that "ought" to be individually closely spaced "multiplets." Because each *spectral line* is associated with a transition from one energy state to another, the number of lines in the spectrum reveals how many energy states there are in the atom or how many "steps" there are on the quantum staircase, and how deep each step is. From their studies of spectra, the physicists of the early 1920s came up with several possible explanations for the multiplet structure.

What proved to be the best explanation came from Wolfgang Pauli, and it involved assigning four separate quantum numbers to the electron. This was in 1924, when physicists still thought of the electron as a particle and tried to explain its quantum properties in terms familiar from the everyday world. Three of these numbers were already included in the Bohr model, and were thought of as describing the angular momentum of an electron (the speed with which it moved around its orbit), the shape of the orbit, and its orientation. The fourth number had to be associated with some other property of the electron, a property that came in only two varieties, to account for the observed splitting of spectral lines.

It didn't take long for people to latch on to the idea that Pauli's fourth quantum number described the electron's "spin" which could be thought of as pointing either up or down, giving a nice double-valued quantum number. The first person to propose this was the young physicist Ralph Kronig. He proposed that the electron had an intrinsic spin of one half in the natural units (h/2*П), and that this spin could line up either parallel to the magnetic field of the atom or antiparallel.

To his surprise, Pauli himself strongly opposed the idea, largely because it could not be reconciled with the idea of the electron as a particle within the framework of relativistic theory. Just as an electron in orbit around the nucleus "ought" not to be stable according to classical electromagnetic theory, so a spinning electron "ought" not to be stable according to relativity theory. Perhaps Pauli should have been a little more open minded, because the result was that Kronig gave up the idea and never published it. Less than a year later, however, the same idea occurred to George Uhlenbeck and Samuel Goudsmit, of the Institute for Theoretical Physics, in Leyden. They published the suggestion in the German journal *Die Naturwissenschaften* late in 1925, and in *Nature* early in 1926.

The theory of the "spinning electron" was soon refined to explain fully the previously unexplained splitting of spectral lines, and by March 1926 Pauli himself was convinced. But what is this phenomena called spin? If you try to explain it in ordinary language, the concept, like so many quantum concepts, the understanding of it slips away. In one "explanation," for example, you might be told that electron spin is not like the spin of a child's top because the electron has to spin around twice to get back to where it started. Then again, how can an electron wave "spin" at all?

Nobody was happier than Pauli when Bohr was able to establish, in 1932, that electron spin cannot be measured by any classical experiment such as deflection of beams of electrons by magnetic fields. It is a property that only appears in quantum interactions, such as the ones that produce the splitting of the spectral lines, and it has no "classical" meaning whatsoever. How much easier it might have been for Pauli and his colleagues struggling to understand the atom in the 1920s, if they had talked about the electron's "gyre" instead of its "spin" in the first place.

Today, we are stuck with the term "spin", and no campaign for the abolition of classical terminology in quantum physics is likely to succeed. Nobody truly understands what "really" goes on in atoms, but Pauli's four quantum numbers do explain some very crucial features or properties of the atom.

4.2 Pauli and the Exclusion principle

Wolfgang Pauli was one of the most remarkable of the remarkable body of scientists who co-founded quantum theory. Born in Vienna in 1900, he enrolled at the University of Munich in 1918, but brought with him a reputation as a precocious mathematician and a finished paper on general relativity theory that immediately aroused Einstein's interest, and was published in January 1919.

Swallowing up physics from classes at the university and the Institute of Theoretical Physics, as well as his own reading, his command of relativity was so great that in 1920 he was assigned the task of writing a major review article on the subject for a definitive encyclopedia of mathematics. This masterly article by the twenty-year-old student spread his fame throughout the scientific community, where the work was praised highly by the likes of Max Born, whom he joined in Gottingen, as assistant, in 1921.

From Gottingen he soon moved on, first to Hamburg and later to Bohr's Institute in Denmark. But Born did not suffer from the loss – his new assistant, Werner Heisenberg, was just as gifted and played a key role in the development of quantum theory. Even before Pauli's fourth quantum number was labeled "spin," he had been able, in 1925, to use the fact of the four numbers to resolve one of the great puzzles of the Bohr atom. In the case of hydrogen, the single electron naturally sits in the lowest energy state available, at the bottom of the quantum staircase. If it is excited by a collision, perhaps it may jump up to a higher step on the quantum energy staircase, then fall back to the ground state, emitting a quantum of radiation as it does so. But when more electrons are added to the system, for more massive atoms, they do not all fall into the ground state, but distribute themselves up at higher energy levels. Bohr talked of the electrons as being in "shells" around the nucleus with "new" electrons going into the lowest shell with the least energy until it was full, and then into the next higher shell, and so on. In this way he built up the "periodic table" of the elements which explained many chemical mysteries. But he did not explain how or why a shell becomes full or why the first shell could contain only two electrons, and so on.

Each of Bohr's shells corresponded to a set of quantum numbers, and Pauli realized in 1925 that with the addition of his fourth quantum number for the electron, the number of electrons in each full shell exactly corresponded to the number of different sets of quantum numbers belonging to that shell. He formulated what is now known as the "Pauli Exclusion Principle," that no two electrons can have the same set of quantum numbers, and thereby provided a reason for the way the shells fill up more and more massive atoms.

The exclusion principle, and the discovery of electron spin, really arrived ahead of their time, and were only fully fitted into the new physics in the late 1920s, after the modern physics had itself been invented. Because of the almost headlong progress in physics in 1925 and 1926, the importance of exclusion sometimes get overlooked, but is in, in fact, a concept as fundamental and far reaching as the fundamental concepts of reality, and it has broad applications across physics. The Pauli Exclusion

Principle, it turns out, applies to all particles that have a "half-integral" amount of spin such as $(1/2)\hbar$, $(3/2)\hbar$, $(5/2)\hbar$ and so on. Particles that have no spin at all (like photons) or integer spin such as $\hbar$, $2\hbar$, $3\hbar$, and so on, behave in a completely different way, following a different set of rules. The rules that are obeyed by the "half-spin" particles are called Fermi-Dirac statistics is named after Enrico Fermi and Paul Dirac, who worked them out in 1925 and 1926. Such particles are called "fermions."

The rules obeyed by "full-spin" particles are called Bose-Einstein statistics, after the two men who worked them out, and the particles are called "bosons". In 1924, Einstein became interested in the work of the Indian physicist Saryendra Bose.

Einstein was so impressed by the work that he translated it into German himself and passed it on with a strong recommendation, ensuring its publication in August 1924.

He came up with a very simple derivation of the law involving mass-less particles that obey particles that obey a special kind of statistics, and sent a copy of his work, in English, to Einstein with a request that he should pass it on for publication in the Zeitschrift fur Physik.

Using a new way of counting particles that he invented, Bose was able to derive Max Planck's formula for the radiation of bodies. Einstein extended Bose's work and applied it to atoms and molecules. This method became part of the modern development of "quantum statistical mechanics." With their new method, Einstein and Bose predicted the existence of a new state of matter, the *Bose-Einstein condensate.* This new state of matter was discovered recently through experimentation.

Satyendra Bose was born in Calcutta in 1894, and in 1924 he was Reader in Physics at the then new Dacca University. Following the work of Planck, Einstein, Bohr, and Sommerfeld from afar, and aware of the still imperfect basis of Planck's law, he set about deriving the blackbody law in a new way, starting out from the assumption that light comes in the form of photons, as they are now called.

By removing all elements of classical theory and deriving Planck's law from a combination of light quanta – regarded as relativistic particles with zero mass – and statistical methods, Bose finally cut quantum theory free from its classical antecedents. Radiation could now be treated as a quantum gas, and the statistics involved counting particles, not counting wave frequencies.

Bose-Einstein statistics were being developed at the same time, as all the excitement about de Broglie waves, Compton effect, and the electron spin between 1924-1925. They mark Einstein's last great contribution to quantum theory (indeed, his last great piece of scientific work), and they too represent a complete break with classical ideas.

As fore-mentioned, Einstein developed the statistics further, and applied them to what was then the hypothetical case of a collection of atoms, gas or liquid, obeying the same rules. The statistics turned out to be inappropriate for real gases at room temperature, but they are exactly right to account for the bizarre properties of super-fluid helium, a liquid cooled close to the absolute zero of temperature, -273 C. With Fermi-Dirac statistics coming on the scene by 1926, it took some time for physicists to sort out which rules applied where, and to appreciate the significance of the half-integer spin. All the "material" that we are used to – electrons, protons, and neutrons – are *fermions*, and without the exclusion principle the variety of the chemical elements and all the features that make up our physical world would not exist. The *bosons* are more ghostly particles, such as photons, and the *blackbody law* is a direct result of all the photons trying to get into the same energy state. Helium atoms can mimic the properties of bosons, under the right conditions, and become superfluid because each atom of 4-He contains two protons and two neurons, with their half-integer spins arranged to add up to zero. Fermions are also conserved in interactions between particles. It is impossible to increase the overall number of electrons in the Universe, whereas bosons, as anyone who has ever switched a light on knows, can be manufactured in vast quantities.

Although it sounds reasonably neat and tidy from the perspective of 2011, by 1925 quantum theory was in "disarray." There was no straight forward progress, but rather many individuals each taking a separate path through the quantum theory jungle. Top researchers knew this only too well, and expressed their concern publicly; but the great leap forward was yet to come, with one exception, from the new generation who entered research after the First World War and were, perhaps as a result, open to new ideas. In 1924.

Heisenberg, after an unsuccessful attempt to calculate the structure of the helium atom, commented to Pauli early in 1923, 'What a misery!' – a phrase Pauli repeated in a letter to Sommerfeld in July of that year, saying 'The theory . . . with atoms having more than one electron, it is such a great misery.'

Heisenberg

In May 1925 Pauli wrote to Kronig saying that 'physics at the moment is again very muddled,' and by 1925 Bohr himself was similarly gloomy about the many problems concerning his model of the atom.

As late as June 1926, Wilhelm Wien, whose blackbody law had been one of the springboards for Planck's leap in the dark, wrote to Schrodinger about the "morass of integral and half-integral quantum "discontinuities" and the arbitrary use of the classical theory." All of the big names in quantum theory were aware of the problems in 1925. Henri Poincare; Lorentz, Planck, J.J.

Thompson, Bohr, Einstein, and Born were still going strong, while Pauli, Heisenberg, Dirac, and others were beginning to make their mark.

The two great authorities in quantum theory were Einstein and Bohr, but by 1925 they had begun to differ strikingly in their scientific views. First, Bohr was one of the strongest adversaries of the light quantum; then, as Einstein began to be concerned about the role of "probability" in quantum theory Bohr became its great supporter. The statistical methods, ironically introduced by Einstein, became the cornerstone of quantum theory, but as early as 1920 Einstein wrote to Born, "that business about causality causes me a lot of trouble, too . . . I must admit that . . . I lack the courage of my convictions." Thereafter, the dialogue between Einstein and Bohr on this theme continued for thirty-five years, until Einstein's death.

4.3 Matrices and Waves

Werner Heisenberg was born in Wurzburg on 5 December 1901. In 1920 he entered the University of Munich, where he studied physics under Arnold Sommerfeld, one of the leading physicists of the time who had been closely involved with the development of the Bohr model of the atom. Heisenberg was plunged straight into research on quantum theory, and set the of finding quantum numbers that could explain some of the splitting of spectral lines into pairs, or doublets. He found the answer in a couple of weeks - the whole pattern could be explained in terms of half-integer quantum numbers. The young, unprejudiced student had found the simplest solution to the problem, but his colleagues and his supervisor Sommerfeld were horrified. To Sommerfeld, steeped in the Bohr model, integral quantum numbers were established doctrine, and the young student's speculations were quickly quashed. The fear among the experts was that by introducing half-integers into the equations they would open the door to quarter-integers, then eights and sixteenths, destroying the fundamental basis of quantum theory.

Within a few months, the older and more senior physicist Alfred Lande came up with the same idea and published it; it later turned out that "half-integer" quantum numbers are crucially important in the full quantum theory and play a key role in describing the property of electrons called "spin." Objects that have integer or zero spin, like photons, obey the *Bose-Einstein* statistics, while those that have half-integer spin (1/2, or 3/2, and so on) obey the *Fermi-Dirac* statistics. The half-integer spin of the electron is directly related to the structure of the atom and the periodic table of the elements. It is still true that quantum numbers change only by whole integers, but a jump from ½ to 3/2, or 5/2 to 9/2, is just as legitimate as a jump from 1 to 2, or 7 to 12.

So Heisenberg missed a chance for credit for a new idea in quantum theory. Heisenberg certainly made up for missing out on one minor scientific "first" with his work over the next few years. At the time, Wolfgang Pauli, a close friend to Heisenberg's equally precocious and another former student of Sommerfeld's, was just moving on from a spell as Born's assistant in Gottingen, and Heisenberg took over the post in 1924. It was a job that gave him the opportunity to work for several months with Bohr in Copenhagen, and by 1925 the precocious mathematical physicist was better equipped than anyone to find the logical quantum theory that every physicist expected to be found eventually, but no one expected to find so soon.

Heisenberg's breakthrough was founded on an idea he picked up from the Gottingen group but nobody now is quite sure who suggested it first, that a physical theory should only be concerned with the things that can actually be observed by experiments. This sounds trite, but it is actually a very deep insight. An experiment that "observes" electrons in atoms, for example, doesn't show us a picture of little hard balls orbiting around the nucleus. There is no way to observe the orbit, and only the evidence from spectral lines tells us what happens to electrons when they move from one energy state (or orbit) to another. All of the observable features of electrons and atoms deal with two states, and the concept of an orbit is something tacked on to the observations by "analogy" with the way things move in our everyday world. Heisenberg stripped away the clutter of the everyday analogies, and worked intensively on the mathematics that described not one "state" of an atom or electron, but the "associations" between pairs of states.

4.4 Breakthrough in Quantum Theory

The story is often told of how Heisenberg was struck down by a severe bout of hay fever in May 1925, and went off to recuperate on the rocky island of Helgoland, where he painstakingly tackled the task of interpreting, what was known about quantum behavior in these terms. With no distractions on the island, and his hay fever gone. Heisenberg was able to work intensively on the problem.

Returning to Gottingen, Heisenberg spent three weeks preparing his work in a form suitable for publication and sent a copy of the paper first to his old friend Pauli, asking if he thought it made sense. Pauli was enthusiastic, but Heisenberg was exhausted by his efforts and not yet sure that the work was ready for publication. He left the paper with Born to dispose of as he felt appropriate, and departed, in July 1925, to give a series of lectures in Leyden and Cambridge. Ironically, he did not choose to speak about his new work to the audiences there, who had to wait for news to reach them by other channels.

Born was happy enough to send Heisenberg's paper off to the "Zeitschrift fur Physik," and almost immediately realized what it was that Heisenberg had stumbled upon. The mathematics involving two states of an atom couldn't be dealt with by ordinary numbers, but involved arrays of numbers, which Heisenberg had thought of as tables. The best analogy is with a chess board. There are 64 squares on the board, and in this case you could identify each square by one number, in the range 1 to 64.

However, chess players prefer to use a notation that labels the "columns" of squares across the board by the letters *a, b, c, d, e, f, g,* and *h*, with the "rows" numbered up the board 1, 2, 3, 4, 5, 6, 7, 8. Now, each square on the board can be identified by a unique pair of identifying lables: a1 is the home square of a rook; g2 is the home square of a knight's pawn, and so on. Heisenberg's tables, like a chess board, involved two-dimensional arrays of numbers, because he was doing calculations involving two states and their interactions.

Those calculations involved, among other things, multiplying two such sets of numbers, or arrays, together, and Heisenberg had laboriously worked out the right mathematical tricks to do the job. But he had come up with a very curious result, so puzzling that it was one of the reasons for his diffidence about publishing his calculations. When two of these arrays are multiplied together, the "answer," you get depends on the order in which you do the multiplication.

This is strange indeed. It is as if 2 * 3 is not the same as 3 * 2, or in algebraic terms (a * b) not-equal-to (b * a). Born worried at this peculiarity day and night, convinced that something fundamental lay behind it. Suddenly, he saw the light.

The mathematical arrays and tables of numbers, so laboriously constructed by Heisenberg, were already known in mathematics. A whole calculus of such numbers existed; they were called *"matrices"* and Born had studied them in the early years of the 20th century, when he was a student in Breslau.

It isn't really surprising that he should have remembered this obscure branch of mathematics more than twenty years later, for there is one fundamental property of matrices that always makes a deep impression of students when they first learn of it – the answer you get when you multiply matrices depends on the order in which you do the multiplying, or in mathematical language, matrices do not commute. Each square on a chessboard can be identified by a paired number and letter, such as *b4* or *f7.* Quantum mechanical states are also defined by pairs of numbers. The "state" of each square on the chessboard is determined by the chess piece occupying that square. In this notation, a pawn is denoted by 1, rook by 2, and so on; positive numbers are white pieces, negative black. We can describe a change in the state of the whole board by an expression such as "pawn" to queen four, or by the algebraic notation *e2-e4.*

Quantum transitions are described in similar notation linking paired initial and final states; in either case do we have any indication of how the transition from one state to another is carried out, a point brought out most strongly by the knight's move and by castling.

In the chess analogy, we might image the smallest possible change in the board, e2-e3, as corresponding to the input of a quantum of energy, hv, while the "transition" *e3-e2* would then correspond to the release of the same quantum. The analogy is inexact but highlights the way different forms of notation describe the same event. Heisenberg, Dirac, and Schrodinger similarly found different forms of mathematical notation to describe the same quantum events.

4.5 Quantum Mathematics

In the summer of 1925, working with Pascual Jordan, Born developed the beginnings of what is now known as *"matrix mechanics,"* and when Heisenberg returned to Copenhagen in September he joined them through correspondence in producing a comprehensive scientific paper on *"quantum mechanics."* In this page, far more clearly and explicitly than in Heisenberg's original paper, the three authors stressed the fundamental importance of the non-commutativity of quantum variables. Already, in his joint paper with Jordan, Born had found the relation $pq - qp = \hbar/i$, where *p* and *q* are matrices representing quantum variables, the equivalent in the quantum world of momentum and position. Planck's constant appears in the new equation, along with *I*, the square root of minus one; in what became known as the "three-man paper," the Gottingen team stressed that this is the "fundamental quantum-mechanical relation." But what did it mean in physical terms? Planck's constant was by now familiar enough, and physicists knew equations involving *I* which was a clue to what was to come, if they had but realized it, since such equations generally involve oscillations, or waves.

But matrices were totally unfamiliar to most mathematicians and physicists in 1925, and the *non-commutativity* seemed as strange to them as Planck's introduction to *h* had seemed at first sight to their predecessors in 1900.

The equations of Newtonian mechanics were replaced by similar equations involving matrices and, says Heisenberg, 'It was a strange experience to find that many of the old results of Newtonian mechanics, like conservation of energy, etc., could be derived also in the same scheme.' In other words, matrix mechanics included Newtonian mechanics within itself, just as Einstein's relativistic equations include Newton's equations as a special case.

Dirac

Paul Dirac was a few months younger than Heisenberg, having been born on 8 August 1902. He is generally regarded as the only English theorist who can rank with Newton, and he developed the most complete form of what is now called quantum mechanics. Yet he did not turn to theoretical physics until after he graduated from Bristol University in 1921, with a degree in engineering. Unable to find a position in engineering, he was offered a studentship to study mathematics in Cambridge, but was unable to take it up due to lack of money. Staying in Bristol and living with his parents, he took the three-year course in mathematics in only two years, thanks to his engineering degree, and completed a *BA* in applied mathematics in 1923. Now, at last he could go to Cambridge to take up research, supported by a grant from the Department of Scientific and Industrial Research – and only on arriving in Cambridge did he learn for the first time about quantum theory.

So it was as an unknown and inexperienced research student that Dirac heard Heisenberg talk in Cambridge in July 1925. Although Heisenberg did not talk publicly about his new work then, he did mention it to Ralph Fowler, Dirac's supervisor, and as a result he sent Fowler a copy of the proof of the paper in the middle of August, before it appeared in the Zeitschrift. Fowler passed the paper on to Dirac, who had it in front of him before anybody outside Gottingen (except Heisenberg's friend Pauli) had had a change to study the new theory. In this first paper, although he pointed out the non-commutativity of the variables in quantum mechanics, the matrices, Heisenberg did not develop the idea, but tried to fudge around it. When he got to grips with the equations, Dirac soon appreciated the fundamental importance of the simple fact that (a * b) is not-equal-to (b * a).

Unlike Heisenberg, Dirac already knew of mathematical quantities that behaved in this fashion, and within a few weeks he was able to rework Heisenberg's equations in terms of a branch of mathematics developed by William Hamilton a century earlier. By one of the most delicious scientific ironies, the Hamiltonian equations that proved so useful in the new quantum theory, which dispensed altogether with electron orbits, had been developed in the 19th century largely as an aid to the calculation of the orbits of bodies in a system, like the solar system, where there are several interacting planets.

So Dirac discovered, independently of the Gottingen group, that the equations of quantum mechanics have the same mathematical structure as the equations of classical mechanics, and that classical mechanics is included within quantum mechanics as a special case, corresponding to large quantum numbers or to setting Planck's constant equal to zero. Following his own direction,

Dirac developed yet another way of expressing the dynamics mathematically, using a special form of algebra, which he called *"quantum algebra,"* involving the addition and multiplication of quantum variables, or "*q* numbers." These *q* numbers are strange beasts, not least because in this mathematical world developed by Dirac it is impossible to say which of two numbers is larger, *a* or *b*; the concept of one number being bigger or smaller than another has no place in this algebra.

But, again, the rules of that mathematical system exactly fitted the observations of the behavior of atomic processes. Indeed, it is correct to say that *quantum algebra* includes *matrix mechanics* within itself, but does much more besides.

In the first half of 1926, Dirac carried this work through in a series of four definitive papers, with the whole package forming the thesis for which he was duly awarded his doctorate. While all this was going on, Pauli had used matrix methods to predict, correctly, the Balmer series for the hydrogen atom, and by the end of 1925 it had become clear that the splitting of some spectral lines into doublets could indeed be best explained by assigning the new property called "spin" to the electron. The pieces fitted together very well indeed, and the different mathematical tools used by the different exponents of matrix mechanics were clearly just different aspects of the same reality.

Again, the game of chess can help to make this clear. There are several different ways of describing a chess game. One way is to print a representative "chessboard" with the positions of all the pieces marked, but that would take up a lot of space if we wanted to record a whole game. Another way is to name the pieces being moved: "King's pawn to King's pawn four." And in the most concise algebraic notation the same move becomes simply "d2-d4." Three different descriptions provide the same information about a real event, the transition of a pawn from one "state" to another (and, just as in the quantum world, we know nothing about how the pawn got from one state to other, a point that is even more clear if you consider the knight's move).

The different formulations of quantum mechanics are like this. Dirac's *"quantum algebra"* is the most elegant and "beautiful" in the mathematical sense; the *"matrix methods"* developed by Born and his collaborators in the wake of Heisenberg are more clumsy but none the less effective. Some of Dirac's most dramatic early results came when he tried to include special relativity in his quantum mechanics. He was delighted with the idea of light as a particle (photon). Dirac was also pleased to find that by including time as a (q) number along with all the rest in his equations he was inevitably led to the "prediction" that an atom must suffer a recoil when it emits light, just as it should do if the light is in the form of a particle carrying its own momentum, and he went on to develop a quantum-mechanical interpretation of the Compton effect. Dirac's calculations went in two parts, first the numerical manipulations involving the (q) numbers, and secondly the interpretation of the equations in terms of what might be physically observed. This process exactly fits the way nature seems to "make the calculation" and then present us with an observed event, and electron transition, say, but unfortunately instead of following this idea through completely in the years after 1926 physicists were seduced away from quantum algebra by the discovery of yet another mathematical technique that could solve the long-standing problems of quantum algebra, had started out from the picture of an electron as a particle making a transition from one quantum state to another.

But what about de Broglie's suggestion that electrons, and other particles, must also be thought of as waves? In Dirac's version of quantum mechanics, a key expression in the Hamiltonian equations is replaced by the quantum mechanical expression $(ab-ba)/i\hbar$, which is just another form of the expression Born, Heisenberg, and Jordan called the "fundamental quantum-mechanical relation," in their three-man paper, written before Dirac's first paper on quantum mechanics appeared, but published after Dirac's paper. With characteristic, and genuine, modesty, Dirac has described how easy it was to make progress once it was known that the correct quantum equations were simply classical equations put into Hamiltonian form. For any of the little puzzles that beset quantum theory, all that was necessary was to set up the equivalent classical equations, turn them into Hamiltonians, and solve the puzzle. "It was a game, and a very interesting one could play.

4.6 Schrodinger's Theory

While matrix mechanics and quantum algebra were making their relatively unsung debuts on the scientific scene, there was plenty of other activity in the field of quantum theory. By late 1925, de Broglie's theory of electron waves had already appeared on the scene, but the definitive experiments that proved the wave nature of the electron had not yet been carried out.

Quite independently of the work of Heisenberg and his colleagues, this led to another discovery, a quantum mathematics based on the wave idea.

The idea came from de Broglie, via Einstein. De Broglie's work might have remained obscure for years, regarded as no more than an interesting mathematical quirk with no "physical reality," if it hadn't come to Einstein's attention. It was Einstein who told Born about the idea and thereby setting off the train of experimental work that proved the reality of *"electron waves,"* and it was in one of Einstein's papers, published in February 1925, where Erwin Schrodinger read Einstein's comment on de Broglie's work, 'I believe that it involves more than merely an analogy.' In those days physicists hung on Einstein's every word and a nod from the great man was enough to set Schrodinger off on an investigation of the implications of taking de Broglie's idea at face value.

Schrodinger is the odd one out among the physicists who developed the new quantum theory. He was born in 1887 and was thirty-nine years old when he completed his greatest contribution to science – a remarkably advanced age for original scientific work of such importance. He had received his doctorate back in 1910, and since 1921 he had been Professor of Physics in Zurich, a pillar of scientific respectability, and not an obvious source of revolutionary new ideas.

But, as we shall see, the nature of his contribution to quantum theory was very much what we might expect from a member of the older generation in the mid-1920s. Where the Gottingen group, and Dirac even more so, made quantum theory more abstract and cut it free from everyday physical ideas, Schrodinger tried to restore easily understood physical concepts, describing quantum physics in terms of waves, which are familiar features of the physical world, and fighting to the end of his life against the new ideas of indeterminacy and the instantaneous jumping of electrons from one state to another. He gave physics an invaluable practical tool for solving problems, but in conceptual terms his wave mechanics was a step backward, a return to 19th-century ideas.

De Broglie had pointed the way with his idea that electron waves in "orbit" around an atomic nucleus had to fit a whole number of wavelengths into each orbit, so that in-between orbits were "forbidden." Schrodinger used the mathematics of waves to calculate the energy levels allowed in such a situation, and was disappointed at first to get answers that did not agree with the known patterns of atomic spectra. In fact, there was nothing wrong with his technique, and the only reason for his initial failure was that he had not taken into account the spin of the electron – hardly surprising, since at that time in 1925 the concept of electron spin had not yet emerged.

So he put the work aside for several months, and thereby missed being the first person to publish a complete, logical, and consistent mathematical treatment of quanta. He came back to the idea when he was asked to give a colloquium explaining de Broglie's work, and it was then that he found that if he left out the relativistic effects from his calculations he could get a

good agreement with observations of atoms in situations where relativistic effects were not important. As Dirac was later to show, *"electron spin"* is essentially a relativistic property (nothing like the property associated with rotating objects in the everyday world). So Schrodinger's great contribution to quantum theory was published in a series of papers in 1926, hot on the heels of the papers from Heisenberg, Born and Jordan, and Dirac. The equation in Schrodinger's variation on the quantum theme are members of the same family of equations that describe real waves in the everyday world, for example, waves on the surface of the ocean, or the sound waves that carry noised through the atmosphere.

The equation in Schrodinger's variation on the quantum theme are members of the same family of equations that describe real waves in the everyday world, for example, waves on the surface of the ocean, or the sound waves that carry noised through the atmosphere. The world of physics greeted them with enthusiasm, precisely because they seemed so comfortable and familiar.

No two approaches to the problem could have been more different. Heisenberg deliberately discarded any picture of the atom and dealt only in terms of quantities that could be measured by experiment; at the heart of his theory, though, was the idea that electrons are particles. Schrodinger, on the other hand, started out from a clear physical picture of the atom as a "real" entity. At the heart of his theory was the idea that electrons are waves. Both approaches produced sets of equations that exactly described the behavior of things that could be measured in the quantum world. At first sight, this was astonishing. Yet before long Schrodinger himself, the American Carl Eckart, and then Dirac proved mathematically that the different sets of equations were in fact exactly equivalent to one another; different views of the same mathematical world. Schrodinger's equations include both the *non-commutativity* relation, and the crucial factor $\hbar/i$, in essentially the same way that they turn up in *"matrix mechanics"* and *"quantum algebra."* The discovery that the different approaches to the problem were indeed mathematically equivalent to one another strengthened the confidence of physics in them all. It seems that, whatever sort of mathematical formalism you like to use, when you tackle the fundamental problems of quantum theory you are led inexorably to the same "answers." Mathematically speaking, Dirac's variation on the theme is the most complete, since his quantum algebra includes both matrix mechanics and wave mechanics as special cases. Naturally enough, however, the physicists of the 1920s chose to use the most familiar version of the equations, Schrodinger's waves, which they could understand in everyday terms, and whose equations were familiar from the problems of everyday physics - optics, hydrodynamics, and the like.

But the very success of Schrodinger's version of the story may have backed any fundamental understanding of the quantum world for decades

4.7 The original founder of Quantum Mechanics

Historically, Dirac did not discover (or invent) wave mechanics. The quantum equations, developed by Hamilton that proved so useful in quantum mechanics, has their origin in a 19th-century attempt to unify the wave and particle theories of light. Sir William Hamilton was born in Dublin in 1805, and became regarded by many as the foremost mathematician of his age. His greatest achievement, although not regarded as such at the time, was a unification of the laws of optics and dynamics in one mathematical framework, one set of equations that could be used to described both the motion of a wave and the motion of a particle.

The work was published in the late 1820s and early 1830s, and both aspects were taken up by others. The mechanics and the optics were each useful to researchers in the second half of the 19th-century, but scarcely anybody took notice of the coupled mechanics/optics system that was Hamilton's real concern. The clear implication of Hamilton's work is that just as light "rays" have to be replaced by the concept of waves in optics, so particle tracks have to be replaced by wave motions in mechanics. But such an idea would have been so alien to 19th-century physics that nobody, not even Hamilton, articulated it. It wasn't that the idea was raised and rejected as absurd; it was literally too bizarre even to occur to anyone. This would have been an impossible conclusion for any 19th-century physicist to have reached, and it was inevitable that the idea would only become established after the inadequacy of classical mechanics as a description of atomic processes had been proved.

Sir William Hamilton is the forgotten founder of quantum mechanics. He also invented the form of mathematics now called *matrix algebra.* Had he been around at the time, he would have been quick to see the connection between matrix mechanics and wave mechanics. Dirac could have done so, but it isn't really surprising that he missed the connection at first. He was, after all, a young student deep in his first major piece of research, and there is a limit to how much one person can do in a few weeks.

Perhaps more importantly, though, he was dealing with abstract ideas, and following up Heisenberg's attempt to cut quantum physics free from the familiar everyday picture of electrons orbiting atomic nuclei, and had no expectation of finding a nice, intuitive physical picture of the atom.

What people did not immediately appreciate was that wave mechanics itself did not, in spite of Schrodinger's expectation, provide such a cozy picture. Schrodinger thought that he had eliminated quantum jumps from one state to another by putting waves into quantum theory.

He envisaged the "transitions" of an electron from one energy state to another as something like the change in the vibration of a violin string from one note to another (one harmonic to another), and he thought of the wave in his wave equation as the matter wave invoked by de Broglie.

But as other researchers sought to find the underlying significance of the equations, these hopes of restoring classical physics to the center stage evaporated. Bohr, for example, was baffled by the wave concept.

How could a wave, or a set of interacting waves, make a Geiger counter click just as if it recorded a single particles? What was actually "waving" in the atom? And, crucially, how could the nature of blackbody radiation be explained in terms of Schrodinger's waves? So in 1926 Bohr invited Schrodinger to spend some time in Copenhagen, where they tackled these problems and came up with solutions that were not very tasteful to Schrodinger.

First, the waves themselves turned out, on close inspection, to be as abstract as Dirac's *q* numbers. The mathematics showed that they couldn't be real waves in space, like ripples on a pond, but represented a complex form of vibrations in an imaginary mathematical space called "*configuration space*." Worse than that, each particle, like an electron, for instance, needs its own three dimensions. One electron on its own can be described by a wave equation in three-dimensional configuration space. To describe two electrons requires a six-dimensional configuration space while three electrons require nine dimensions, and so on. As for the blackbody radiation, even when everything was converted into wave-mechanical language the need for discrete quanta, and quantum jumps remained.

Schrodinger was disgusted, and made the remark which has often been quoted, with slight variations in the translation: 'Had I known that we were not going to get rid of this damned quantum jumping, I never would have involved myself in this business.'

As Heisenberg put in his book "*Physics and Philosophy*, . . ." The paradoxes of the dualism between wave picture and particle picture were not solved; they were hidden somehow in the mathematical scheme.

Without doubt, the appealing picture of "physically" real waves circling around atomic nuclei that had led Schrodinger to discover the wave equation that now bears his name is wrong. Wave mechanics is no more a guide to the reality of the atomic world than matrix mechanics, but unlike matrix mechanics, wave mechanics gives an "*illusion*" of something familiar and comfortable.

It is that illusion that has persisted to the present day and that has disguised the fact that the atomic-world is totally different from the macro-world we live in. Several generations of students, who have now grown up to become professors themselves, might have achieved a much deeper understanding of quantum theory if they had been forced to come to grips with the abstract nature of Dirac's approach, rather than being able to imagine that what they knew about the behavior of waves in the everyday world gave a picture of the way atoms behave. And that is why it seems that although there have been enormous strides in the application of quantum mechanics to many interesting problems, we are scarcely today, more than eighty years on, any better placed than the physicists of the late 1920s concerning our fundamental understanding of quantum physics. The very success of the Schrodinger equation as a practical tool has stopped people from thinking deeply about how and why the tool works.

4.8 Developed Quantum physics

The basics of quantum theory or practical quantum physics since the 1920s depend upon ideas developed by Bohr and Born in the late 1920s. Bohr gave us a philosophical basis with which to reconcile the dual particle/wave nature of the quantum world, and Born gave us the basic rules to follow in preparing our quantum recipes. Bohr said that both the theoretical pictures, particle physics and wave physics, are equally valid, complementary descriptions of the same reality.

Neither description is complete in itself, but there are circumstances where it is more appropriate to use the particle concept and circumstances where it is better to use the wave concept. A fundamental entity such as an electron is neither a particle nor a wave, but under some circumstances it behaves as if it were a wave, and other circumstances it behaves as if were a particle. But under no circumstances can you invent an experiment that will show the electron behaving in both fashions at once.

This idea of wave and particle being two complementary facets of the electron's complex personality is called complementarity.

Born found a new way of interpreting Schrodinger's waves. The important thing in Schrodinger's equation that corresponds to the physical ripples on the pond in the everyday world is a "*wave function*," which is usually denoted by the Greek letter *Psi*. Working in Gottingen alongside experimental physicists who were performing new electron experiments confirming the particle nature of the electron almost every day, Born simply could not accept that this *Psi*function corresponded to a "real" electron wave, although like almost all physicists at the time (and since) he found the wave equations the most convenient for solving many problems. So he tried to find a way of associating a wave function with the existence of particles. The idea he picked up on was one that had been aired before in the debate about nature of light, but which he now took over and refined.

The particles were real, said Born, but in some sense they were guided by the wave, and the strength of the wave (more precisely, the value of Psi^2 at any point in space was a measure of the probability of finding the particle at that particular point. We can never know for sure where a particle like an electron is, but the wave function enables us to work out the probability that, when we carry out an experiment designed to locate the electron, we will find it in a certain place. The strangest thing about this idea is that it means that any electron might be anywhere at all, it's just that it is extremely likely to be in some locations and very unlikely to be in others. But like the statistical rules that say that it is possible for all the air in the room to gather in the corners, so Born's interpretation of *Psi* removed some of the certainty from the already uncertain quantum world.

The ideas of both Bohr and Born tied in very well with Heisenberg's discovery, late in 1926, that uncertainty is indeed inherent in the equations of quantum mechanics. The mathematics that sys that *pq* not-equal-to *qp* also says that we can never be certain just what *p* and *q* are. If we call *p* the momentum of, say, an electron, and use *q* as a label of its position, we can imagine measuring either *p* or *q* very accurately. The amount of "error" in our measurement might be called *Λ-p* or *Λ-q*, since mathematicians use the Greek letter *Λ*, to symbolize small pieces of variable quantities.

What Heisenberg showed was that if you tried in this case, to measure both the position and momentum of an electron you could never quite succeed because *Λ-p* x *Λ-q* must *"always"* be bigger than *h-bar*, Planck's constant divided by *2Pi*. The more accurately we know the position of an object the less certain we are of its momentum. And if we know its momentum very accurately, then we can't be quite sure where it is. The uncertainty relation has far-reaching implications. The important point to appreciate, however, is that it does not represent any deficiency in the experiments used to measure the properties of the electron. It is a cardinal rule of quantum mechanics that in principle it is impossible to measure precisely certain pairs of properties, including position/momentum, simultaneously. There is no absolute truth at the quantum level.

4.9 Heisenberg's uncertainty relation

The Heisenberg uncertainty relation measures the amount by which the complementary descriptions of the electron, or other fundamental entitles, overlap. Position is very much a particle property – particles can be located precisely, waves, on the other hand, have no precise location, but they do have momentum. The more you know about the wave aspect of reality, the less you know about the particle, and vice versa.

Experiments designed to detect particles always detect particles; experiments designed to detect waves always detect waves. No experiment shows the electron behaving like a wave and a particle at the same time.

Bohr stressed the importance of experiments in our understanding of the quantum world. We can only probe the quantum world by doing experiments, and each experiment, in effect, asks a question of the quantum world. The questions we ask are highly colored by our everyday experience, so that we seek properties like "momentum" and "wavelength," and we get "answers" that we interpret in terms of those properties. The experiments are rooted in classical physics, even though we know that classical physics does not work as a description of atomic processes. In addition, we have to interfere with the atomic processes in order to observe them at all, and, said Bohr, that means that it is meaningless to ask what the atoms are when we are not looking at them. All we can do, as Born explained, is to calculate the probability that a particular experiment will come up with a particular result.

This collection of ideas – uncertainty, complementarity, probability, and the disturbance of the system being observed by the observer – are together referred to as the "Copenhagen interpretation" of quantum mechanics, although nobody in Copenhagen (or anywhere else) ever set down in so many words a definitive statement labeled the *"Copenhagen interpretation,"* and one of the key ingredients, the statistical interpretation of the wave function, actually came from Max Born in Gottingen.

The Copenhagen interpretation is many things to many people, if not quite all things to all people. Bohr first presented the concept in public at a conference in Tomo Italy, in September 1927. That marked the completion of the consistent theory of quantum mechanics in a form where it could be used by any competent physicist to solve problems involving atoms and molecules, with no great need for thought about the fundamentals but a simple willingness to follow the recipe book and turn out the answers.

In the following decades, many fundamental contributions were made by the likes of Dirac and Pauli, and the pioneers of the new quantum theory were duly honored by the Nobel Committee, although the allocation of the awards followed the committee's own curious logic. Heisenberg received his in 1932, and was mortified that the prize had not gone also to his colleagues Born and Jordan; Born himself remained bitter about this for his years, often commenting that Heisenberg didn't even know what a matrix was until he (Born) had told him, in writing to Einstein in 1953. It was he who reaped all the rewards of our work together, such as the Nobel Prize." Schrodinger and Dirac shared the physics prize in 1933, but Pauli had to wait until 1945 to receive his, for the discovery of the exclusion principle, and Born was at last honored in 1954 with a Nobel Prize for his work on the probabilistic interpretation of quantum mechanics.

There is no "real" model of what the atom and elementary particles are really like, and nothing that tells us what goes on when we are not looking at them. But the equations of wave mechanics (the most popular and widely used variation on the theme) can be used to make predictions on a statistical basis. If we make an observation of a quantum system and get the answer *A* to our measurement, then the quantum equations tell us what the probability if of getting answer *B* (or *C*, or *D*, or whatever) if we make the same observation a certain time later.

Quantum theory does not say what atoms are like, or what they are doing when we are not looking at them. Unfortunately, most of the people who use the wave equations today do not appreciate this, and only pay lip service to the role of probabilities. Students learn what Ted Bastin has called 'a crystallized form of the play of ideas current in the late twenties . . . what the average physicist who never actually asks himself what he believes on foundational questions, is able to work with in solving his detailed problems.' They learn to think of the waves as real, and few of them get through a course in quantum theory without coming away with a picture of the atom in their imagination.

People work with the probabilistic interpretation without really understanding it, and it is a testimony to the power of the equations developed by Schrodinger and Dirac in particular, and the interpretation provided by Born, that even without understanding why the theory works people are able to perform quantum physics so effectively with quanta.

The first "quantum designer" was Dirac. Just as he had been the first person outside Gottingen to understand the new matrix mechanics and develop it further, so he was the person who took Schrodinger's wave mechanics and put it on a more secure footing while developing it further. In adapting the equations to fit the requirements of relativity theory, adding time as the fourth dimension, Dirac found in 1928 that he had to introduce the term that is now regarded as representing the electron's spin, unexpectedly providing the explanation of the *"doublet splitting"* of spectral lines that had baffled theorists throughout the decade. The same improvement of the equations threw up another unexpected result, one that opened the way for the modern development of *"particle physics."*

4.10 Antimatter

According to Einstein's equations, the energy of a particle that has mass *m* and momentum *p* is given by

$$E^2 = m^2*c^4 + p^{-2}*c^2$$

which reduces to the well-known $E=mc^2$ when the momentum is zero. But this isn't quite the whole story. Because the more familiar equation comes from taking the square root of the full equation, in mathematics we have to say that *E* can be either positive or negative. Just as 2*x*2=4, so does -2x-2=4, and strictly speaking.

$$E=+/- mc^2.$$

When such *"negative root"* crop up in the equations, as often as not they can be dismissed as meaningless, and it is "obvious" that the only answer that we are interested in is the positive root. Dirac, being a genius, did not take this obvious step, but puzzled over the implications. When energy levels are calculated in the relativistic version of quantum mechanics, there are two sets, one all positive, corresponding to $m*c^2$, and the other all negative, corresponding to $-mc^2$. Electrons ought, according to the theory, fall into the lowest unoccupied energy state, and even the highest negative energy state is lower than the lowest positive energy state. So what do the negative energy levels mean, and why didn't all the electrons in the Universe fall into them and disappear?

Dirac's answer hinged upon, the fact that electrons are fermions, and that only one electron can go into each possible state (two per energy level, one with each spin). It must be, he reasoned, that electrons didn't fall into the negative energy states because all those states are already full. What we call "empty space" is actually a sea of negative energy electrons. And he didn't stop there. Give an electron energy, and it will jump up the ladder of energy states. So, if we give an electron in the negative sea enough energy it ought to jump up into the real world and become visible as an ordinary electron.

To get from the state $-mc^2$ to the state $+mc^2$ clearly requires an input of energy of $2mc^2$, which, for the mass of an electron, is about 1 MeV and can be provided quite easily in atomic processes or when particles collide with one another. The negative energy electron promoted into the real world would be normal in every respect, but it would leave behind a hole in the negative energy sea, the absence of a negatively charged electron. Such a hole, said Dirac, ought to behave like a positively charged particle (much as a double negative makes an affirmative, the absence of a negatively charged particle in a negative sea ought to show up as a positive charge).

When he first thought of the idea, he reasoned that because of the symmetry of the situation this positively charged particle ought to have the same mass as the electron. But in a moment of weakness when he published the idea he suggested that the positive particle might be the proton, which was the only other particle known in the late 1920s. As he describes in *"Directions in Physics,"* this was quite wrong, and he should have had the courage to predict that experimenters would find a previously unknown particle with the same mass as the electron but a positive charge.

Nobody was quite sure how to take Dirac's work at first. The idea that the positive counterpart to the electron was the proton was dismissed, but nobody took the idea very seriously, until Carl Anderson, an American physicist, discovered the trace of a positively charged particle during his pioneering observations of cosmic rays in 1932. Cosmic rays are energetic particles that arrive at the Earth from space. They had been discovered by the Austrian Victor Hess before the First World War, and he shared the Nobel Prize with Anderson in 1936. Anderson's experiments involved tracking charged particles as the moved through a cloud chamber, a device in which the particles leave a trail like the condensation trail of an aircraft, and he found that some particles produced tracks that were bent by a magnetic field the same amount as the track of an electron, but in the opposite direction. They could only be particles with the same mass as the electron but positive charge, and they were dubbed "positrons". Anderson received a Nobel Prize for the discovery in 1936, three years after Dirac had received his own Prize, and the discovery transformed physicists' view of the particle world. They had long suspected the existence of a neutral atomic particle, the neutron, which James Chadwick found in 1932 (for which he received the 1935 Nobel), and they were fairly happy with the idea of an atomic nucleus made up of positive protons and neutral neutrons, surrounded by negative electrons. But positrons had no place in this scheme of things, and the idea that particles could be created out of energy changed he concept of a fundamental particle entirely.

Any particle can, in principle, be produced by the Dirac process from energy, provided that it is always accompanied by the production of its antiparticle counterpart, the "hole" in the negative energy sea. Although physicists prefer more erudite versions of the particle-creation story today, the rules are much the same, and one of the key rules is that whenever a particle meets its antiparticle counterpart it "falls into the hole," liberating energy $2mc^2$ and disappearing, not so much in a puff of smoke as in a burst of gamma rays. Before 1932, many physicists had observed particle tracks in cloud chambers, and many of the tracks they had observed must have been due to positrons, but until Anderson's work it had always been assumed that such tracks were due to electrons moving into an atomic nucleus, rather than positrons moving outward Physicists were biased against the idea of new particles.

Today the situation is reversed and, says Dirac, "People are only too willing to postulate a new particle on the slightest evidence, either theoretical or experimental." (*Directions in Physics*, page *18*.) The result is that the *"particle zoo"* comprises not just the two fundamental particles known in the 1920s, but more than 200, all of which can be produced by providing sufficient energy in particle accelerators, and most of which are highly unstable, "decaying" very rapidly into a shower of other particles and radiation. Among that zoo, the antiproton and antineutron, discovered in the mid-1950s, are almost lost, but none the less significant confirmation of the correctness of Dirac's original ideas.

Whole books have been written about the *particle zoo*, and many physicists have built their careers as particle taxonomists. But it seems to me that there cannot be anything very fundamental about such a profusion of particles, and the situation is rather like that in spectroscopy before quantum theory, when *spectroscopists* could measure and catalogue the relationships between lines in different spectra, but had no idea of the underlying causes of the relationships they observed. Something more truly fundamental must provide the ground rules for the creation of the *plethora* of known particles, a view that Einstein expressed to his biographer Abraham Pais in the 1950s. 'It was apparent that he felt that the time was not ripe to worry about such things and that these particles would eventually appear as solutions to the equations of a unified field theory.' Sixty years on, it looks very much as if Einstein was right, and his sketchy outlines of one possible unified theory which incorporates the particle zoo physicists work with. Here, it is sufficient to point out that the great explosion of particle physics since the 1940s has its roots in Dirac's development of "quantum theory."

4.11 Inside the nucleus

In terms of its radius, the nucleus is 100,000 times smaller than the atom; since volume is proportional to the cube of radius, it is more meaningful to say that the atom is a *thousand million million (10^{15})* times bigger than the nucleus. Simple things like mass and charge of the nucleus can be measured, and these measurements lead to the concept of isotopes, nuclei that have the same number of protons, and therefore form atoms with the same number of electrons (and the same chemical properties) but different numbers of neutrons, and therefore different mass.

Since all the protons packed into the nucleus have positive charge, and therefore repel each other, there must be some stronger form of "glue" holding them together, a force that only works across the very short ranges corresponding to the size of the nucleus, and this called the strong nuclear

force (there is also a weak nuclear force, which is weaker than the electric force but plays an important part in some nuclear reactions) And it looks as if the neutrons play a part in the stability of the nucleus as well because simply by counting the numbers of protons and neutrons in stable nuclei physicists come up with a picture rather like the shell picture of electrons around the nucleus. The largest number of protons found in any naturally occurring nucleus is 92, in uranium. Although physicists have succeeded in manufacturing nuclei whit up to 106 protons; these are unstable (except for some isotopes of plutonium, atomic number 94) and break up into other nuclei. Altogether, there are some 260 known stable nuclei; the state of our knowledge about those nuclei, even today, is rather less adequate than the Bohr model is as a description of the atom, but there are clear signs of some sort of structure within the nucleus.

Nuclei that have 2, 8, 20, 28, 50, 82, and 126 nucleons (neutrons or protons) are particularly stable, and the corresponding elements are much more abundant in nature than elements corresponding to atoms with slightly different numbers of nucleons, so these are sometimes called "magic numbers." But protons dominate the structure of the nucleus, and for each element there is only a limited range of possible isotopes corresponding to different numbers of neutrons – the possible number of neutrons is generally a little bigger than the number of protons, and gets bigger for heavier elements. Nuclei that possess *"magic numbers"* of both protons and neutrons are particularly stable, and theorists predict on the basis that super-heavy elements with about 114 protons and 184 neutrons in their nuclei ought to be stable – but these massive nuclei have never been found in nature or made in particle accelerators by sticking more nucleons onto the most massive nuclei that occur in nature.

The most stable nucleus of all is *iron-56*, and lighter nuclei would like to gain nucleons and become iron, while heavier nuclei would like to lose nucleons and move toward the most stable form. Inside stars, the lightest nuclei, *hydrogen* and *helium*, are converted into heavier nuclei in a series of nuclear reactions that fuse the light nuclei together, making elements such as carbon and oxygen along the road to iron, and releasing energy as a result.

When some stars explode as supernovae, a great deal of gravitational energy is put into the nuclear processes, and this pushes the fusion beyond iron to make heavier elements, including things like *uranium* and *plutonium*. When heavy elements move back toward the most stable

configuration, by ejecting nucleons in the form of alpha particles, electrons, positrons, or individual neutrons, they too release energy, which is essentially the stored-up energy of a long-gone supernova explosion. An alpha particle is essentially the nucleus of a helium atom and contains two protons and two neutrons. By ejecting such a particle, a nucleus reduced its mass by four units, and its atomic number by tow. And it does so in accordance with the rules of quantum mechanics and the uncertainty relations discovered by Heisenberg.

The nucleons are held together inside the nucleus by the strong nuclear force, but if an alpha particle was just outside the nucleus it would be strongly repelled by the electric force. The combined effect of the two forces is to make what physicists call a "potential well". Although the potential well still provides a barrier, it is not insurmountable, and there is a definite, if small, probability that an alpha particle might actually be outside, not inside, the nucleus.

In terms of uncertainty, one of the Heisenberg relations involves energy and time, and says that any particle's energy can only be defined within a range $\Lambda\text{-}E$ over a period of time $\Lambda\text{-}t$, such that $\Lambda\text{-}E \; x \; \Lambda\text{-}t$ is bigger than $\hbar$. For a short time, a particle can "borrow" energy from the uncertainty relation, gaining enough energy to jump over the *"potential barrier"* before giving it back. When it returns to its "proper" energy state, it is just outside the barrier instead of just inside, and rushes away.

Or you can look at it in terms of the uncertainty of position. A particle that "belongs" just inside the barrier may appear just outside, because its position is only fuzzily determined in quantum mechanics. The larger the energy of the particle, the easier it is for it to escape, but it does not have to have enough energy to climb out of the potential well in the way classical theory requires. The process is as if the particle tunneled out through the barrier, and it is purely a "quantum effect." This is the basis of radioactive decay; but to explain nuclear fission we have to turn to a different model of the nucleus.

Forget about the individual nucleons in their shells for the time being, and consider the nucleus as a droplet of liquid. Just as a drop of water wobbles in a changing pattern of shapes, so some of the collective properties of the nucleus can be explained as due to the changing shape of the nucleus. A large nucleus can be thought of as "wobbling" in and out, changing shape from a sphere to something like a dumbbell and back again. If energy is put into such a nucleus, the oscillation may become so extreme that it breaks the nucleus in two, splitting off into two smaller nuclei and a spattering of tiny droplets, alpha and beta particles and neutrons. For some nuclei, this splitting can be triggered by the collision of a fast-moving neutron with the nucleus, and a chain reaction occurs when each nucleus fissions in this way produced enough neutrons to ensure that at least two more nuclei in its neighborhood also fission.

For *uranium-235*, which contains 92 and 143 neutrons, two unequal nuclei with atomic numbers in the range 34 to 58, and adding up to 92, are always produced, with a scattering of free neutrons. Each fission releases about 200 *MeV* of energy, and each one sets off several more, provided that the lump of uranium is big enough that the neutrons do not escape from it altogether. Left to run away exponentially, this is the process of the atomic bomb; moderated by using a material that absorbs neutrons to keep the process just ticking over, we have a controlled fission reactor that can be used to heat water into steam and generate electricity. Once again, the energy we extract is away.

In the fusion process, however, we can mimic the energy production of a star like the Sun here on Earth. So far, we have only been able to copy the first step up the fusion ladder, from hydrogen to helium, and we have not been able to control the reaction, only to let it run away in the hydrogen, fusion bomb. The trick with fusion is the opposite of the trick with fission. Instead of encouraging a large nucleus to break up, you have to force small nuclei together, against the natural *"electrostatic repulsion"* of their positive charges, until they are so close that the strong nuclear force, which only has a very short range, can overwhelm the electric force and pull them together. As soon as you get a few nuclei to fuse in this way, the heat generated in the process causes an outward rush of energy that tends to blow apart any other nuclei on the point of fusing, and stops the whole process in its tracks.

The hope of unlimited energy for the future from *"nuclear fusion"* depends on finding a way to hold enough nuclei together in one place for long enough to get a useful amount of energy out. It is also rather important to find a process that releases more energy than we have used pushing the nuclei together in the first place. It's easy enough in a bomb – essentially, you just surround the nuclei you want to fuse with uranium, than trigger the uranium into a fission explosion.

The inward pressure from the surrounding explosion will then bring enough hydrogen nuclei into contact to set off the second pulses of light from laser beams which physically squeeze the nuclei together. Lasers, of course, are manufactured in accordance with another procedure from the quantum theory.

A *"potential well"* is at the heart of an atomic nucleus. A particle at A has to stay inside the well unless it can gain enough energy to jump "over the top" to *B*, when it will rush away "downhill." Quantum uncertainty allows a particle, occasionally, to "tunnel" through from *A* to *B* (or *B* to *A*) without having enough energy of its own to climb the hill.

The same process operates in reverse when nuclei "fuse" together. When two light nuclei are pushed together by the pressure inside a star, they can only fuse if they overcome the potential barrier from the outside. The amount of energy each nucleus possesses in that situation depends on the temperature at the heart of the star, and in the 1920s astrophysicists were puzzled to find that the calculated temperature inside the Sun is a little less than it ought to be – the nuclei at the heart of the Sun do not have enough energy to overcome the potential barrier and fuse together, according to classical mechanics.

The answer is that some of them tunnel through the barrier at a slightly lower energy, in line with the rules of quantum mechanics. Among other things, quantum theory explains why the Sun shines, when classical theory says it cannot. One way to gain energy from fusion is by combining an *isotope* of hydrogen which has one proton and one neutron (*deuterium*), with one which has one proton and two neutrons (*tritium*). The result is a helium nucleus (two protons, two neutrons), a free neutron, and 17.6 MeV of energy. Stars work by more complicated processes involving nuclear reactions between hydrogen and nuclei such as carbon that are present in small quantities inside the star. The net effect of such reactions is to fuse four protons into a helium nucleus, with two electrons and 26.7 *MeV* of energy being released, and the carbon put back into circulation to catalyze another cycle of reactions. But it is processes involving tritium and deuterium that are being investigated in fusion laboratories here on Earth.

4.12 Lasers and masers

Remember that when an atom gains a quantum of energy an electron jumps up to a different orbit, on this picture, and that when such an excited atom is left alone then, sooner or later, the electron will fall back to the ground state, releasing a very precisely defined quantum of radiation with a definite wavelength. The process is called *"spontaneous emission"* and is the counterpart of *"absorption."*

When Einstein was investigating such processes in 1916 and laying the statistical ground rules for quantum theory, which he later found so abhorrent, he realized that there is another possibility. An excited atom can be triggered into releasing its extra energy and going back into the ground state, if it is nudged, as it were, by a passing photon. This process is called *"stimulated emission,"* and it only happens if the passing photon has exactly the same wavelength as the one that the atom is primed to radiate. Rather like the cascade of neutrons that is involved in a chain reaction of nuclear fission, we can imagine an array of excited atoms with just one photon of the right wavelength coming along and stimulating one atom to radiate; the original photon plus the new one then stimulated two more atoms to radiate, the four photons together trigger four more, and so on. The result is a cascade of radiation, all with precisely the same frequency. Furthermore, because of the way the emission is triggered, all of the waves move precisely in step with one another – all the waves go "up" together, and all the troughs go "down" together, producing a very pure beam of what is called coherent radiation. Because none of the peaks and troughs in such radiation is canceling each other out all of the energy released by the atoms is present in the beam and can be delivered onto a small area of material that the beam is shone upon.

When a collection of atoms or molecules is excited by heat they fill up a band of energy levels and, left to their own devices radiate different wavelengths of energy in an incoherent and jumbled fashion carrying much less effective energy that the atoms and molecules release. But there are tricks that can be used to fill up a narrow band of energy levels preferentially and then trigger the return of the excited atoms in the band to their ground state.

The trigger for the cascade is a weak input of radiation of the right frequency; the output is a much stronger, amplified beam with the same frequency. The techniques were first developed in the late 1940s, independently by teams in the U.S. and the U.S.S.R., using radiation in the radio band of the spectrum from about 1 cm to 30 cm, called the "microwave band"; the pioneers received in this Nobel Prize for their work in 1954. Because the radiation in this band is called microwave radiation, and because the process involves amplification of microwaves by the stimulated emission of radiation in line with Einstein's ideas of 1917, pioneers coined the name microwave amplification by stimulated emission of radiation, and the acronym *MASER*, for the process.

It was ten years before anyone succeeded in finding a way to make the trick work for optical frequencies of radiation. Today, there are several different kinds of laser, the simplest being the optically pumped solid lasers. In this design, a rod of material (such as ruby) is prepared with polished, flat ends and surrounded by a bright light source, a gas discharge tube that can flash rapidly on and off, producing pulses of light with sufficient energy to excite the atoms in the rod. The whole apparatus is kept cool to ensure the minimum amount of interference from thermal excitation of the atoms in the rod, and the bright flashes from the lamp are used to stimulate (or pump) the atoms into an excited state. When the laser is triggered, a pulse of pure ruby light, carrying thousands of watts of energy, emerges from the flat end of the rod.

Variations on the theme include liquid lasers, fluorescent-dye lasers, gas lasers, and so on. All share the same essential features – incoherent energy is put in, and coherent light comes out in a pure pulse carrying a lot of energy. Some, like the gas lasers, give a continuous, pure beam of light that is the ultimate "straight edge" for surveying purposes, and that has found widespread use at rock concerts and in advertising. Others produce short-lived but powerful pulses of energy that can be used to drill holes in hard objects (and might one day have military applications). Laser cutting tools are used in situations as diverse as the clothing industry and microsurgery. And laser beams can be used to carry information far more effectively than radio waves, since the amount of information that can be passed each second increases as the frequency of the radiation used increases. The bar codes on many supermarket products (and on the cover of this book) are read by a laser scanner; the video disks and compact audio disks that came on the market in the early 1980s are scanned by laser. Genuine three-dimensional photographs, or *holograms*, can be made with the aid of lasers; and so on. The list is virtually endless, even before we include the applications of masers in amplifying faint signals (for example, from communications satellites), radar and the like it is all thanks to Albert Einstein and Niels Bohr who laid down the principles of stimulated emission more than eighty-five years ago.

4.13 Chance and Uncertainty

Heisenberg's uncertainty principle is seen today as a central feature – perhaps the central feature – of quantum theory. It wasn't picked up immediately by his colleagues, but took nearly ten years to achieve this exalted position. Since the 1930s, however, its position may have been a little too exalted.

The concept grew out of Schrodinger's visit to Copenhagen in September 1926, the occasion of his famous remark to Bohr about "damned quantum jumping." Heisenberg realized that one of the main reasons why Bohr and Schrodinger sometimes seemed to be at loggerheads was a conflict of concepts. Ideas like "position" and "velocity" (or "spin," which came later) simply don't have the same meaning in the world of microphysics as they do in the everyday world. So what meaning do they have, and how can the two worlds be related? Heisenberg went back to the fundamental equation of quantum mechanics,

$$pq - qp = \hbar/i$$

and showed from this that the product of the uncertainties in position *(Λ-q)* and momentum *(Λ-p)* must always be bigger than $\hbar$ *(not h)*. The same uncertainty rule applies to any pair of what are called conjugate variables, variables that multiply out to have the units of action like $\hbar$; the units of action are energy *x* time, and the other most important pair of such variables is indeed energy *(E)* and time *(t)*. The classical concepts of the everyday world still existed in the micro-world, said Heisenberg but they could only be applied in the restricted sense revealed by the uncertainty relations. The more accurately we know the *position* of a particle, the less accurately we know its *momentum* (speed & direction) and vice versa.

4.14 The meaning of uncertainty

These starling conclusions were published in the *"Zeitschrift fur Physik"* in 1927, but while theorists such as Dirac and Bohr, familiar with the new equations of quantum mechanics, appreciated their significance at once, many experimenters saw Heisenberg's claim as a challenge to their skills. They imagined that he was saying that their experiments weren't good enough to measure both position and momentum at the same time, and tried to conceive experiments to prove him wrong. But this was a futile aim, since that wasn't what he meant at all.

This misconception still arises today, partly because of the way the idea of uncertainty is often taught. Heisenberg himself used the idea of observing an electron to make his point. We can only see things by looking at them, which involves bouncing photons of light off them and into our eyes. A photon doesn't disturb an object like a house very much, so we don't expect the house to be affected by looking at it. For an electron, though, things are rather different. To start with, because an electron is so small we have to use electromagnetic energy with a short wavelength in order to see or interact with it (with the aid of experimental apparatus) at all.

Such gamma radiation is very energetic, and any photon of gamma radiation that bounces or scatters off an electron and can be detected by our experimental apparatus will drastically change the position and momentum of the electron – if the electron is in an atom, the very act of observing it with a *gamma ray microscope* may knock it out of the atom altogether.

All this is true enough, and it does give a general idea of the impossibility of measuring precisely both the position and momentum of an electron. But what the uncertainty principle tells us is that, according to the fundamental equation of quantum mechanics, there is no such thing as an electron that possesses both a precise momentum and a precise position.

This has far-reaching implications. As Heisenberg said at the end of his paper in the *Zeitschrift*, "We cannot know as a matter of principle, the present in all its details." This is where quantum theory cuts free from the determinacy of classical physics. To Newton, it would be possible to predict the entire course of the future if, we knew the position and momentum of every particle in the Universe; to the modern physicist, the idea of such perfect prediction is meaningless because we cannot know the position and momentum of even one particle precisely.

The same conclusion comes out of all the different versions of the equations, the wave mechanics, the Heisenberg-Born-Jordan matrices, and Dirac's *q* numbers, although Dirac's approach, which carefully avoids any physical comparisons with the everyday world, seems the most appropriate. Indeed, Dirac very nearly came to the uncertainty relation

before Heisenberg. In a paper for the Proceedings of the Royal Society in December 1926 he pointed out that in quantum theory it is impossible to answer any question that refers to numerical values of both *q* and *p*, although, "one would expect, however, to be able to answer questions in which only the *q* or only the p are given numerical values."

It was only in the 1930s that the philosophers took up the implications of these ideas for the concept of *causality* – the idea that every event is cause by some other specific event – and the puzzle of predicting the future. Meanwhile, although the uncertainty relations had been derived from the fundamental equations of quantum mechanics, some influential experts began to teach quantum theory by starting out from the uncertainty relations. Wolfgang Pauli was probably the key influence in this trend. He wrote a major encyclopedia article on quantum theory that began with the uncertainty relations, and he encouraged a colleague, Herman Weyl, to begin his textbook "*Theory of Groups and Quantum Mechanics*" in much the same way.

This is a peculiar accident of history. After all, the basic equations of quantum theory lead on to the uncertainty relations, but if you start with uncertainty there is no way to work out the basic quantum equations. What's worse, the only way to introduce uncertainty without the equations is to use examples like the *gamma-ray* microscope for observing electrons, and this immediately makes people think that uncertainty is all about experimental limitations, not a "fundamental truth" about the nature of the Universe. So, in the common tradition of learning quantum theory, one has to learn one thing, then backtrack to learn something else, then move forward to discover just what it was that you learned about in the first place.

Science is not always "logical" nor are scientists and science professors. The result has been generations of confused students and misconceptions about the uncertainty principle – misconceptions that you, the reader, don't share, because you have discovered things in the right order.

4.15 The Copenhagen interpretation

An important aspect of the uncertainty principle, which doesn't always get the attention it deserves, is that it does not work in the same sense forward and backward in time. Very few things in physics "care" which way time flows, and it is one of the fundamental puzzles of the Universe we live in that there should indeed be a definite "arrow of time," a distinction between the past and the future. The uncertainty relations tell us that we cannot know position and momentum at the same time, and therefore we cannot predict the future – the future is inherently unpredictable and uncertain. But it is quite within the rules of quantum mechanics to set up and experiment from which it is possible to calculate backward and work out exactly what the position and momentum of an electron, say, was at some time in the past. The future inherently uncertain – we do not know exactly where we are going; but the past is clearly defined – we do know exactly where we have come from.

To paraphrase Heisenberg, 'We can know, as a matter of principle, the past in all its details.' This precisely fits in with our everyday experience of the nature of time, moving from a known quantum world at its most fundamental. This may be linked to the arrow of time we perceive in the Universe at large; its more bizarre possible implications will be discussed in the next chapter.

While the philosophers slowly began to grapple with such intriguing implications of the uncertainty relation, to Bohr they came like a shaft of light illuminating the concepts he had been groping toward for some time. The idea of "*complementarity*," that both wave and particle pictures are necessary to understand the quantum world (although in fact an electron, say, is neither wave nor particle), found a mathematical formulation in the uncertainty relation that said that position and momentum could not both be known precisely, but formed complementary and in a sense mutually exclusive aspects of reality.

From July 1925 until September 1927 Bohr published hardly anything on quantum theory, and then he presented a lecture in Como, Italy, which introduced the idea of "complementarity" and what is known as the "Copenhagen interpretation" to a wide audience. He pointed out that whereas in classical physics we imagine a system of interacting particles to function, like clockwork, regardless of whether or not they are observed, in quantum physics the observer interacts with the system to such an extent that the system cannot be thought of as having independent existence. By choosing to measure position precisely, we force a particle to develop more uncertainty in its momentum, and vice versa; by choosing an experiment to measure wave properties, we eliminate particle features, and no experiment reveals both particle and wave aspects at the same time; and so on.

In classical physics, we can describe the positions of particles precisely in space-time, and forecast their behavior equally precisely; in quantum physics we cannot, and in this sense even relativity is a "classical" theory.

It took a long time for these ideas to be developed and for their significance to sink in. Today, the key features of the Copenhagen interpretation can be more easily explained, and understood, in terms of what happens when a scientist makes an experimental observation. First, we have to accept that the very act of observing a thing changes it, and that we, the observers, are in a very real sense part of the experiment – there is no clockwork that ticks away regardless of whether we look at it or not.

Secondly, all we know about are the results of experiments. We can look at an atom and see an electron in energy state *A* then look again and see an electron in energy state *B*. We guess that the electron jumped from *A* to *B*, perhaps because we looked at it.

What we can learn from experiments, or from the quantum theory, is the probability that if we look at a system once and get answer A then the next time we look we will get answer B. We can say nothing at all about what happens when we are not looking, and how the system gets from *A* to *B*, if indeed it does. The "damned quantum jumping" that had disturbed Schrodinger is purely our interpretation of why we get two different answers to the same experiment, and it is a "false" interpretation.

Sometimes things are found to be in state *A*, sometimes in state *B*, and the question of what lies in between, or how they get from one state to another, is completely meaningless. This is the really fundamental feature of the quantum world. It is interesting that there are limits to our knowledge of what an electron is doing when we are looking at it, but it is absolutely mind-blowing to discover that we have no idea at all what it is doing when we are not looking at it. Since then, neutrinos have indeed been "discovered" in three different varieties (plus their three different anti-varieties) and other kinds are postulated. Can Eddington's doubts really be taken at face value? Is it possible that the nucleus, the positron and the neutrino did not exist until experimenters discovered the right sort of chisel with which to reveal their form? Such speculations strike at the roots of sanity, let alone our concept of reality. But they are quite sensible questions to ask in the quantum world. If we follow the quantum formula book correctly, we can perform an experiment that produces a set of pointer readings that we interpret as indicating the existence of a certain kind of particle.

Almost every time we follow the same formula, we get the same set of readings. But the interpretation in terms of particles is all in the mind, and may be no more than a consistent delusion. The equations tell us nothing about what the particles do when we do not look at them, and before Rutherford nobody ever looked at a nucleus, before Dirac nobody even imagined the existence of a positron. If we cannot say what a particles does when we are not looking at it, neither can we say if it exists when we are not looking at it, and it is reasonable to claim that nuclei and positrons did not exist prior to the 20th century, because nobody before 1900 ever saw one.

In the quantum world, what you see is what you get, and nothing is real; the best you can hope for is a set of delusions that agree with one another. Unfortunately, even those hopes are dashed by some of the simplest experiments. Remember the double-split experiments that "proved" the wave nature of light? How can they be explained in terms of photons?

One of the best, and best-known, teachers in quantum mechanics over the past fifty years has been Richard Feynman, of the California Institute of Technology. His three volume *"Feynman Lectures on Physics,"* published in the early 1960s, provides a standard against which other undergraduate texts must be compared, and he has also been involved in popular lectures on the subject, such as his series on *BBC* television in 1965, published as the *"Character of Physical Law."*

Born in 1918, Feynman was at the peak of his prowess as a theoretical physicist in the 1940s, when he was involved in setting up the equations of the quantum version of electromagnetism, called quantum electro-dynamics; he received the Nobel Prize for this work in 1965. Feynman's special place in the history of quantum theory is as a representative of the first generation of physicists to grow up with all of the basics of quantum mechanics established, and all the ground rules laid. Whereas Heisenberg and Dirac had to work in a changing environment, where new ideas did not always appear in their correct sequence and the logical relation of one concept to another (as in the case of spin) was not necessarily immediately obvious to Feynman's generation, for the first time, all the pieces of the puzzle were present and the logic of their ordering could be seen, if not quite at a glance then certainly after a little thought and intellectual effort.

4.17 Matter Waves

The fact that light, which is conventionally thought of as a wave, has particle properties led the French physicist Louis de Broglie to speculate that other objects we commonly think of as particles may have wave properties. Thus, a beam of electrons, which is most naturally imagined as a stream of very small bullet-like particles, would in some circumstances behave as if it were a wave. This radical idea was first directly confirmed in the 1920s by

Davidson and Germer : they passed an electron beam through a crystal of graphite and observed an interference pattern that was similar in principle to that produced when light passes through a set of slits. As we saw, this property is central to the evidence for light being a wave, so this experiment is direct confirmation that this model can also be applied to electrons. Later on, similar evidence was found for the wave properties of heavier particles, such as neutrons, and it is now believed that wave-particle duality is a universal property of all types of particle.

Even everyday objects such as grains of sand, gloves or cars have wave properties, although in these cases the waves are completely unobservable in practice, partly because the relevant wavelength is much too small to be noticeable, but also because classical objects are composed of atoms, each of which has its own associated wave and all these waves are continually chopping and changing. We saw above that in the case of light the vibration frequency of the wave is directly proportional to the energy of the quantum. In the case of matter waves, the frequency turns out to be hard to define and impossible to measure directly. Instead there is a connection between the wavelength of the wave and the momentum of the object, such that the higher is the particle momentum the shorter is the wavelength of the matter wave.

In classical waves, there is always something that is "waving." Thus in water waves, the water surface moves up and down, in sound waves the air pressure oscillates and in electromagnetic waves the electric and magnetic fields vary. What is the equivalent quantity in the case of matter waves? The conventional answer to this question is that there is no physical quantity that corresponds to this. We can calculate the wave using the ideas and equations of quantum physics and we can use our results to predict the values of quantities that can be measured experimentally, but we cannot directly observe the wave itself, so we need not define it physically and should not attempt to do so.

To emphasize this, we use the term 'wave function' rather than wave, which emphasizes the point that it is a mathematical function rather than a physical object. Another important technical difference between wave functions and the classical waves we discussed earlier is that, whereas the classical wave oscillates at the frequency of the wave, in the matter-wave case the wave function remains constant in time.

However, although not physical in itself, the wave function plays an essential role in the application of quantum physics to the understanding of real physical situations. Firstly, if the electron is confined within a given region, the wave function forms standing waves similar to those discussed earlier; as a result, the wavelength and therefore the particle's momentum takes on one of a set of discrete quantized values.

4.18 The experiment with two holes

If we carry out experiments to detect the presence of the electron near a particular point, we are more likely to find it in regions where the wave function is large than in ones where it is small.

This idea was placed on a more quantitative basis by Max Born, whose rule states that the probability of finding the particle near a particular point is proportional to the square of the magnitude of the wave function at that point. Atoms contain electrons that are confined to a small region of space by the electric force attracting them to the nucleus. From what we saw earlier, we could expect the associated wave functions to form a standing-wave pattern and we shall see shortly how this leads to an understanding of important properties of atoms.

4.19 Varying potential Energy

Particles move in a region where the potential energy is constant, so, if we remember that the total energy is conserved, the kinetic energy and hence the particle's momentum and speed must be the same wherever it goes. In contrast, a ball rolling up a hill, for example, gains potential energy, loses kinetic energy and slows down as it climbs. Now we know that the de Broglie relation connects the particle's speed to the wavelength of the wave, so if the speed stays constant, this quantity will also be the same everywhere, which is what we have implicitly assumed.

In general, the analysis of a situation where the potential energy varies requires a study of the mathematical equation that controls the form of the wave in the general case. As mentioned earlier, this equation is known as the 'Schrodinger equation.' In the examples discussed above, where the potential is uniform, the solutions of the Schrodinger equation have the form of travelling or standing waves and our fairly simple approach is justified. A full understanding of more general situations is mathematically quite challenging and not appropriate to this book.

4.20 Quantum Tunneling

Considering the form of the matter-wave, on the basis of our earlier discussion, we expect particles approaching the step to be represented by travelling waves moving from left to right, whereas after they bounce back, the wave will be travelling from right to left. In general we do not know what the particle is doing at any particular time, so the wave function to the left of the step will be a combination of these, and this is confirmed when the Schrodinger equation is solved mathematically.

However, there is now a travelling wave of comparatively small, but finite, amplitude to the right of the barrier. Interpreting this physically, we conclude that there is a small probability that a particle approaching the barrier from the left will not bounce back but will emerge from the other side. This phenomenon is known as 'quantum mechanical tunneling' because the particle appears to tunnel through a barrier that is impenetrable classically.

There is a wide range of physical phenomena that demonstrate quantum tunneling in practice. For example, in many radioactive decays, where "alpha particles" are emitted from the nuclei of some atoms, the probability of this happening for a particular atom can be very low – so low in fact that a particular nucleus will wait many millions of years on average before decaying. This is now understood on the basis that the alpha particle is trapped inside the nucleus by the equivalent of a potential barrier, similar in principle to that discussed above. A very low amplitude wave exits outside the barrier, which means that there is a very small (but non-zero) probability of the particle tunneling out.

4.21 A Quantum Oscillator

In this, we consider a particle moving in a parabolic potential. The size, or "amplitude," of the oscillation is determined by the particle's energy: at the foot of the well all this energy is kinetic, while the particle comes to rest at the limits of its motion, where all the energy is potential. The wave functions are obtained by solving the Schrodinger equation. The wave function corresponding to the lowest energy state is represented by a single hump that reaches a maximum in the center; the next highest state has two humps, one positive and the other negative with the wave function crossing the axis and so on.

4.22 The hydrogen atom

The simplest atom is that of the element hydrogen, which consists of a single negatively charged electron bound to a positively charged nucleus by the electrostatic (or 'Coulomb') force which is strong when the electron is close to the nucleus and steadily reduces in strength when the electron is further away. As a result, the potential energy is large and negative near the nucleus and gets closer to zero as we move away from it.

An important simplifying feature of the hydrogen atom is that the Coulomb potential is 'spherically symmetric' – i.e. it depends only on the distance between the electron and the nucleus – whatever the direction of this separation. A consequence is that many of the wave functions associated with the allowed energy levels have the same symmetry.

The Coulomb potential confines the electron to the vicinity of the nucleus and the oscillator potential confine the particle. When an atom moves from one energy level to another, the energy is absorbed or emitted as a photon of radiation, whose frequency is related to the energy change by the Planck relation. The pattern of frequencies calculated in this way from the above pattern of energy levels is the same as that observed experimentally when electrical discharges are passed through hydrogen gas. A full solution of the Schrodinger equation predicts a value for the constant R in terms of the electron charge and mass and Planck's constant, and this value agrees precisely with that deduced from the experimental measurements. We now therefore have complete quantitative agreement between the predictions of quantum physics and the experimental measurements of the energy levels of the hydrogen atom.

We have used the principle of wave-particle duality to obtain the quantized energy levels, but how are we to interpret the wave function that is associated with each level? The answer to this question lies in the Born rule stated earlier: the square of the wave function at any point represents the probability of finding the electron near that point. A model of the atom consistent with this is that, in this content, the electron should be thought of not as a point particle but as a continuous distribution spread over the volume of the atom. We can envisage the atom as a positively charged nucleus surrounded by a cloud of negative charge whose concentration at any point is proportional to the square of the wave function at the point. This model works well in many situations, but should not be taken too literally: If we actually look for the electron in the atom, we will always find it as a point particle. On the other hand, it is equally wrong to think of the electron as being a point particle when we are not observing its position. In quantum physics, we use models, but do not interpret them too literally.

4.23 Other Atoms

Atoms other than hydrogen contain more than one electron, which causes further complications. Before we can address these, we have to consider another quantum principle, known as the 'Pauli exclusion principle' after its inventor Wolfgang Pauli. This states that any particular quantum state can contain more than one particle of a given type, such as an electron. This principle, although easily stated, can be proven only by using very advanced quantum analysis and we shall certainly not try to do this here. Before we can correctly apply the exclusion principle, however, we have to know about a further property possessed by quantum particles and known as 'spin'.

We know that the Earth spins on its axis as it moves in orbit around the Sun, so if the atom were a classical object, we might well expect the electron to be spinning in a similar manner.

This analogy holds to some extent, but once again there are important differences between the classical and quantum situations. There are two quantum rules that govern the properties of spin: first, for any given type of particle (electron, proton, neutron) the rate of spin is always the same; and, second, the direction of spin is either clockwise or anticlockwise about some axis.

This means that an electron in an atom can have one of only two spin states. Thus any quantum state described by a standing wave can contain two electrons provided they spin in opposite directions. As an example of an application of the "*Pauli exclusion principle*," consider what happens if we place a number of electrons in a container. To form the state with lowest total energy, all the electrons must occupy the lowest possible energy levels. Thus, if we think of adding them to the container one at a time, the first goes into the ground state, as does the second with spin opposite to that of the first. This level is now full, so the third electron must go into the next highest energy level along with the fourth and the spins of these two electrons must also be opposite. We can continue adding up to two electrons to each energy state until all have been accommodated.

We now apply this process to atoms, first considering helium, which has two electrons. If we initially ignore the fact that the electrons exert a repulsive electrostatic force on each other, we can calculate the quantum states in the same way as we did for hydrogen, but allowing for the fact that the nuclear charge is doubled. This doubling means that all the energy levels are considering helium, which has two electrons. If we initially ignore the fact that the electrons exert a repulsive electrostatic force on each other, we can calculate the quantum states in the same way as we did for hydrogen, but allowing for the fact that the nuclear charge is doubled. This doubling means that all the energy levels are considerably reduced (i.e. made more negative), but otherwise the set of standing waves is quite similar to those in hydrogen and it turns out that this pattern is not greatly altered when the interactions between the electrons are included. The lowest energy state therefore has both electrons with opposite spin in the lowest.

In the case of lithium with three electrons, two of these will be in the lowest state, while the third must be in the next higher energy state. The latter state can actually contain a total of six electrons: two of these occupy a state of spherical symmetry while the others fill three separate non-spherical states.

A set of states with the same value of n is known as a 'shell' and if electrons occupy all these states, it is called a "closed shell." Thus lithium has one electron outside a closed shell, as does sodium with eleven electrons – i.e. two in the $n=1$ closed shell, eight in the $n=2$ closed shell and one electron in the $n=3$ shell. It is known that many of the properties of sodium are similar to those of lithium and similar correspondences between the properties of other elements underlie what is known as the "periodic table" of the elements.

4.24 Summary of Quantum properties

Examples of classical waves are water waves, sound waves and light waves. They are all typified by a frequency, which determines how many times per second any point on the wave vibrates, and a *wavelength*, which measures the repeat distance along the wave at any time. Waves have the form of *travelling waves* or *standing waves*. Travelling waves move at a speed determined by the frequency and the wavelength. Because standing waves result from a wave being confined to a region in space the wavelength and hence the frequency of a standing wave is restricted to have one of a set of allowed values. This is exemplified in the notes produced by musical instruments. Although there is evidence that light is a wave, in some circumstances it behaves as if it were a stream of particles, known as light quanta or 'photons.' Similarly, quanta particles such as electrons behave in some contexts as if they were waves.

When an electron is confined by a potential, such as a 'box', the matter waves are standing waves with particular wavelengths, which in turn cause the electron energy to be quantized – i.e. to have one of a set of particular values. When a quantum system moves from one energy level to another, the change in energy is provides by an incoming photon or given to an outgoing photon. The wave properties of quantum particles enable them to tunnel through potential barriers that could not surmount classically. The calculated and measured energy levels of the hydrogen atom agree precisely, which is strong evidence for the correctness of quantum physics.

The Pauli *principle state* that no two electrons can occupy the same quantum state. Because an electron can be in one of two spin states, this means that each standing wave can contain up to two electrons. A change in the electron speed implies a change in its momentum, and hence its wavelength and therefore it quantum state, which is possible only if allowed by the exclusion principle. The *Pauli exclusion principle* allows both electrons to occupy the ground state, provided they have opposite spin.

First, it increases because of the electrostatic repulsion between the two positively charged protons; secondly, it decreases because each electron is now subject to attraction by both protons; thirdly it increases because of the repulsion between the two negatively charged electrons. In addition, the kinetic energy of the electrons decreases because the electrons are able to move around and between the two nuclei, so the size of the effective "box" confining them is increased. The net effect of all these changes depends on how far the atoms are apart: when they are widely separated, there is little change in the total energy, and when they are very close, the electrostatic repulsion between the nuclei dominates. However, at intermediate distances, there is an overall reduction in the total energy and this reduction is at its greatest when the protons are about 7.4×10^{-10} m apart. At this point, the difference between the energy of the molecule and that of the widely separated hydrogen atoms equals about one third of the ground-state energy of the hydrogen atom. Where does this surplus energy end up? The answer is that some of it goes into the kinetic energy of the moving molecule, while the rest is given off in the form of photons.

Both are effectively forms of heat, so the overall effect is a rise in temperature, which is just what we expect from a fuel. The above example illustrates the principle of how energy can be released by bringing atoms together to form molecules, but the particular case of hydrogen is not in practice a useful source of energy, because any hydrogen gas we have on Earth is already composed of molecules.

4.25 The Higgs particle - God particle

Take quantum electrodynamics, the theory explaining the interaction of matter and light. This is how it was described in an open lecture by Richard Feynman the brilliant twentieth-century American physicist whom many would consider one of the few true geniuses to flourish in the scientific community. Once two particles become entangled, it doesn't matter where those particles are; they retain an immediate and powerful connection that can be harnessed to perform seemingly impossible tasks. "Quanta" – are the tiny packets of energy and matter that are the building blocks of reality. A quantum is usually a very small speck of something, a uniform building block normally found in vast numbers, whether it's a photon of light, an atom of matter, or a subatomic particle like an electron.

Dealing in *quanta* implies that we are working with something that comes in measured packages, fixed amounts rather than delivered as a continuously variable quantity. In effect, the difference between something that is quantized and something continuous is similar to the difference between digital information, based on quanta of 0s and 1s, and analog information the can take any value. In the physical world, a quantum is usually a very small unit, just as a quantum leap is a very small change - quite different from its implications in everyday speech.

The phenomenon at heart of this book is a linkage between the incomprehensibly small particles that make up the world around us. At this quantum level, it is possible to link particles together so completely that the linked objects (photons, electrons, and atoms, for instance) become, to all intents and purposes, part of the same thing. Even if these entangled particles are then separated to opposite sides of the Universe, they retain this strange connection. Make a change to one particle, and that ability of entanglement to provide an intimate link between particles at a distance seems just as odd to physicists as it does to rest of us. Albert Einstein, who was directly responsible for the origins of quantum theory that made entanglement inevitable, was never comfortable with the way entanglement acts at a distance without anything connecting the entangled particles. He referred to the ability of quantum theory to ignore spatial separation as literally spooky or ghostly distant actions, in a letter written to a fellow scientist. Entanglement, as a word, seems to have entered the language of physics at the hand of scientist Erwin Schrodinger.

The English term has subtly negative connotations. It gives a sense of being out of control and messed up. But the German word is more structured and neutral - it is crossing over in an orderly manner. A piece of string that is knotted and messed up is entangled, where a carefully woven tapestry has none.

For Einstein, the prediction that entanglement should exist was a clear indicator of the lack of sense in quantum theory. The idea of entanglement was an anathema to Einstein, a challenge to his view on what "reality" truly consisted of. And this was all because entanglement seemed to defy the concept of locality.

Locality, it's the kind of principle that is so obvious we usually assume it without even being aware of it. If we want to act upon something that isn't directly connected to us - to give it a push, to pass a piece of information to it, or whatever - we need to get something from us to the object we wish to act upon. Often this "something" involves direct contact - One reaches over and pick up my coffee cup to get it moving toward my mouth. But if we want to act on something at a distance without crossing the gap that separates us from that something, we need to send an intermediary from one place to another.

Imagine that you are throwing stones at a can that's perched on a fence. If you want to knock the can off, you can't just look at it and make it jump into the air by some sort of mystical influence; you have to throw a stone at it. Your hand pushes the stone, the stone travels through the air and hits the can; as long as your aim is good (and the can isn't wedged in place), the can falls off and you smile smugly. Similarly, if one wants to speak to someone across the other side of a room, my vocal chords vibrate, pushing against the nearest air molecules.

These send a train of sound waves through the air, rippling molecules across the gap, until finally those vibrations get to the other person's ear, start her eardrum vibrating, and result in my voice being heard. In the first case, the ball was the intermediary, in the second the sound wave, but in both cases something physically traveled from *A* to *B*. This need for travel - travel that takes time - is what locality is all about. It says that you can't act on a remote object without that intervention.

Scientific evidence shows that we are programmed from birth to find the ability to influence objects at a distance to be unnatural. Research on babies has shown that they don't accept action at a distance, believing that there needs to be contact between two objects to allow one to act on the other. The brain is entirely convinced that action at a distance is not real. However, though action at a distance looks unreal, this doesn't rule out the possibility of its truly happening. We are used to having to overcome appearances, to take a step away from what looks natural, given extra knowledge.

We know gravity works from a great range, yet there is no obvious linkage between the two bodies that are attracted to each other. Gravitation seems to offer a prime challenge to the concept of locality. This idea of gravitational attraction emerged with the Newtonian view of the world, but even as far back as the ancient Greeks, before an idea of gravity existed, there was awareness of other apparent action at a distance. Amber rubbed with a cloth attracts lightweight objects, such as fragments of paper, toward it. Lodestones, natural magnets, attract metal and spin around, when set on a cork to float on water, until they are pointing in a particular direction.

In each case the action has no obvious linkage to make it work. The attracted object moves toward the magnet - the floating lodestone spins and the static-charged amber summons its retinue of paper scraps as if by magic.

The Greeks had competing schools of thought on what might be happening. One group, the atomists, believed that everything consists of eight different types of atoms along with the void - and, as nothing could act across a void, there had to be a continuous chain of atoms that linked *"cause"* to *"effect."* Other Greek philosophers put action at a distance down to each other as one person attracts another. This was little more than a variant on the third possibility open to the Greek mind - *supernatural* intervention. In effect, this theory said there was something out there that provided an occult nudge to make things happen. This idea was widely respected in ancient times as the mechanism of the long-lasting and scientifically unsupported concept of astrology, in which supernatural influence by planets was thought to shape our lives.

Even though, nearly two thousand years later, Newton was able to exhibit pure genius in his description of what happened as a result of one apparent action at a distance - gravity - he was no better than the Greeks in

explaining how one mass influenced another without anything connecting them. In his masterpiece, published in 1688, Newton was saying that gravity exists, but he wasn't going to provide a "non-empirical" guess at how it works. Some would continue to believe that gravity had some "occult" mechanism, on a par with astrology, but mostly the workings of gravity were swept under the carpet until Einstein came along.

One fundamental that came out of Einstein's work was that nothing could travel faster than light. It had been known since 1676 that light traveled at a finite speed when Danish astronomer Ole Roeme made the first effective determination of a velocity now set at around 186,000 miles per second. Einstein showed that an action could not escape this constraint. Nothing, not even gravity, could travel faster than the speed of light. It was the ultimate cosmic limit.

4.26 Light Quanta's entanglement property

Planck was no enthusiast for this promotion of light quanta (light particle) as a concept toward reality and went so far as to criticize Einstein in a deeply condescending fashion. When Planck recommended the younger man for the Prussian Academy of Sciences in 1913, he asked that they wouldn't hold it against Einstein that he sometimes "missed the target in his speculations, as for example, in the theory of light quanta . . ." The same year that Planck made his remark, Einstein's idea would be absorbed and amplified by the man who later became Einstein's chief sparring partner over quantum theory, and particularly over "quantum entanglement," Niels Bohr. The New Zealand-born physicist Ernest Rutherford had come to fame by discovering the atomic nucleus.

Einstein could not accept this randomness he felt there should be a strict, causal process underlying what was observed. As far as he was concerned, the electron jumped out of the metal it was in at a time and in a direction that could have been predicted, had all the facts been available. Quantum theory disagreed, saying it wasn't ever possible to know when the electron would pop out, or in what direction. Similarly, quantum theory assumed that a particle didn't have a precise position until a measurement on it was made - it was the act of measurement that transformed its position from a probability to an "actual" value. And the same went for the other properties of the particle.

The ability to send messages anywhere, instantly, has a powerful appeal, and it's a possibility that occurs to almost everyone who hears about quantum entanglement. After all, instant communication is exactly what the entangled particles are doing. A change in one is reflected in the other, wherever they are, at whatever distance. Even on the relativity small scale of the Earth, the speed of light, the fastest rate at which we can send a message (around 300,00 kilometers or 186,000 miles second in a vacuum, and slower elsewhere), causes delays. Light speed intrudes irritating pauses into a satellite phone conversation, and provides practical headaches for engineers building large-scale communication networks. The farther information has to travel, the greater the lag becomes.

Sound has the useful ability to work its way around obstacles but dissipates very quickly and is highly dependent on wind direction. It's dangerous to expect sound to travel reliably for more than about half a mile, which would imply having a huge number of stations to carry a sound message across a country.

How, Einstein asks, can we really know that the two lightning strikes were simultaneous? What does simultaneous mean for two events that are spatially separated? We can't be in both places at the same time, nor can we be sure that clocks at the two locations are exactly synchronized (whatever that means). The only way we can check for simultaneity, he suggests, is to have an observer midway between the two strikes with a pair of mirrors so that he can see in both directions at once.

When we describe light or an electron as a wave or a particle, what we really mean is that we are using the model of waves (like the actual, real ripples we see on the sea) or the model of particles (like a stream of very, very tiny bits of dust). But what shouldn't be implied is that light or an electron is a wave or that it is a particle. Attractive though it is to have something that we are familiar with to hang on to, we always have to remember that we are dealing with a model, not the real thing.

What is light or an electron, then? Light is light. An electron is an electron. They happen to exhibit strong similarities to waves and particles some of the time, which can be very useful when we are trying to predict what they will do, but that isn't what they are. If you can take the step of seeing this, the whole concern about quantum theory is better put in proportion. The only reason the behavior of a photon appearing to pass through two slits at once, or entanglement working at any distance, as odd, is that we are letting the models take over, giving them more weight than they deserve.

This doesn't mean that anything goes. Our models are very useful, and we can continue to get better and better predictions of what the world can do, of what it is like (as opposed to what it is), but we shouldn't expect absolute understanding ever to come from working with a model.

This generates a trap for would-be speculators. The fact that we're dealing with models of reality makes it easy to come up with ideas that don't have any direct basis in experiment. For example, entanglement could be responsible for all kinds of phenomena. Some have argued, for instance, that consciousness is a quantum phenomenon involving entanglement, but the biological evidence for the mechanisms of the brain suggest that there is no need to resort to such complex and delicate phenomena to explain what appears to be a very robust and invariant capability.

Max Planck's original attempts a hundred years ago to explain the precise amount of energy radiated by hot bodies began a process of capturing in mathematical from a mysterious, elusive world containing 'spooky interactions at a distance', real enough however would lead to inventions such as the laser and transistor. Quantum theory is now being fruitfully combined with theories of information and computation. These developments may lead to an explanation of processes still not understood within conventional science such as "*telepathy*," an area where Britain is at the forefront of research.

A real understanding of quantum mechanics eludes us: we tend to have too local a view of nature. Quantum theory, he points out, is our most accurate description of how atoms work, and underlies all that we know about chemistry. Chemistry, in turn, provides the operational mechanisms for biology, including those that drive our metabolisms and our reproduction. 'So, might it be not only that quantum effects are responsible for the behavior of inanimate matter, but that the magic of entanglement is also crucial in the existence of life?' Stated by Vedral. And he concludes by saying, 'Our very existence could be made possible by entanglement.'

The reason entanglement doesn't spread uncontrollably, according to Penrose, is that the act of observation destroys it. Other physicists, like Professor Tony Sudbery of York University, suggest it's not so much that observation destroys the entanglement but rather, because we ourselves are part of the entangled world, we can't see beyond our own selves of the entanglement. A boson, like a photon, is one of two types of quantum particles (the other being a fermion such as an electron). Bosons can share quantum states, whereas fermions can't. The Higgs boson is a hypothetical particle that no one has yet seen. There have been plenty of attempts, and occasionally the whole idea has been declared defunct, but as yet proof has not gone either way.

The idea was dreamed up by Peter Higgs, of Edinburgh University, in the early 1960s to explain where mass comes from, and why different particles have such varying mass. Each natural force has a corresponding field, a sort of texture of the environment in which it works, which is communicated by a boson. The electromagnetic field's carrier is the photon. Higgs imagined that there was also a field, the "*Higgs field*," that was responsible for mass. According to this theory, a particle's mass comes from its interactions with a Higgs equivalent of a photon, the *Higgs boson*. These Higgs bosons are still controversial. They work very well to explain the current theories and observations, but no one has ever seen one of these so-called "*God-particles*" (they were given this name because of their fundamental role of giving the other particles mass). Back in 2001, there was some speculation that they didn't exist because experiments that should have been revealing them weren't delivering, but there is still widespread acceptance that they may be real - the experimenters say we just haven't got powerful enough particle colliders to capture the elusive Higgs. And governments regularly cancel collider projects because they are so huge.

Despite seventy years or more of attacks from doubters, entanglement is not going to go away. Every experiment takes us a step closer to realizing just how strange the world is at the quantum level. The natural inclination of even as original a thinker as Albert Einstein was to deny the viability of quantum theory, but this has not proved an acceptable strategy. Quantum theory works. It delivers a tangle of secrets. Entanglement provides a secret link, an unfathomable bond between two particles. It perhaps isn't surprising that once practical entanglement became a reality it came to the attention of those whose business is keeping secrets from prying eyes. An initial interest was inspired by the spooky connection between entangled particles itself. If entanglement could be used to carry a message, it would be totally impenetrable to any interception.

4.28 Force Carriers: Particles that make things happen

Here's one way to describe what physics is all about. It's about what is (things) and what happens (actions). The particles we see – leptons, baryons, and mesons – together with those we don't see but nevertheless study – quarks – constitute what is. There is another class of particles, called force carriers that determine what *happens*. I should say right now that what *doesn't* happen is just as interesting as what does happen. There are a great many processes that don't occur (and, we think, can't occur) –the creation of electric charge from nothing, for instance, or the appearance or disappearance of energy, and perhaps the radioactive decay of a proton.

4.29 Gravitational Interaction

There are the force carriers of four different kinds of interaction, in order of increasing strength. The weakest force is gravity. Because so many billions upon billions of gravitons participate in even the puniest gravitational interaction that we can measure, we see only the collective effect of gravitons in a large "area number," never the effect of any one alone. Thus, we have no current hope of detecting the graviton. How can it be that the weakest force in nature as it holds us on Earth, keeps Earth in its orbit around the Sun.

There are two reasons. One is that gravity is attractive only, whereas the much stronger electric forces are both attractive and repulsive, in which case causes a sort of balanced force.

Electrically, our Earth is so finely balanced in positive and negative charge that even if you get yourself charged up by shuffling across a rug on a dry day, you will feel no appreciable electric force pulling you to the ground. If all of the negative charge could be magically stripped away from the Earth, leaving only the positive charge and if you carried about the some modest negative charge that shuffling across a rug produces), you would be instantly crushed to death by the enormous electric pull. If, conversely, all of the Earth's positive charge could be magically stripped away, leaving only the negative charge, you would be propelled electrically into space far faster than any rocket. This same fine balancing – an almost perfect cancellation of attractive and repulsive forces – prevails throughout the Universe, leaving gravity as the ruling force.

The other reason that gravity is so evident to us despite its weakness is that there is a lot of mass pulling us down. We're held to the ground by the gravity of six thousand billion billion tons of matter. Actually each piece of matter is pulled gravitationally to every other. When we look at the gravity of ordinary-sized objects, gravity's weakness becomes plainer. Another consequence of gravity's weakness is that it plays no known role in the subatomic world.

Acting between the proton and the electron in a hydrogen atom, the electric force out pulls the gravitational force by a factor that can truly be called humongous: more than 10^{39}. (How big is 10^{39}? Those many atoms, stacked end to end, would stretch to the edge of the Universe and back a thousand times.) Yet despite it being smaller than the size of a proton, it does become intertwined with quantum theory in ways that we can now only dimly imagine?

4.30 Electromagnetic Interaction

The photon, has had an interesting history, from its "invention" by Albert Einstein in 1905 through its somewhat shadowy existence in the 1920s as a "corpuscle," not quite a real particle; on through its central role in the 1930s and 1940s, when physicists linked it with the electron and positron to build the powerful theory called quantum electrodynamics; up to the present day, when we see it as a fundamental particle of truly zero mass and truly zero size serving as the force carrier for electromagnetism. Indeed, you literally "see" protons nearly every waking moment, day and night. And there are lots of photons that you don't see –those that carry radio and television and wireless-phone signals, heat from warm walls, and *X*-rays through your body.

The Universe is filled with low-energy photons, the so-called cosmic background radiation, left over from the Big Bang. All in all, there are about a billion photons in the Universe for every material particle.

4.31 Feynman Diagrams

The American physicist Richard Feynman – celebrated for his wit and his writing as much as for his brilliant contributions to physics - invented a method of diagramming events in the subatomic world that is a great aid to visualizing what is going on there. In particular, these "Feynman diagrams" reveal what we think is "really" happening when a force carrier gets exchanged between two other particles, accounting for an interaction between them. In the hands of a theoretical physicist, Feynman diagrams amount to more than an aid to visualization. They provide a way to catalog possible reactions among particles and even to calculate the likelihood of various reactions. Here, however, I want to use them just as a visual aid, so that you can see more clearly how the exchange forces produced by the force carriers work. A Feynman diagram is a miniature "space-time map." To sneak up on this idea, let's start with a plain ordinary map, a "space map," such as might be found in a road atlas. Typically, such a map is two-dimensional, with north pointing up and east pointing to the right. A line on the map traces a path through space – or, as a mathematician would say, a projection onto the ground of a path through space.

To include time as well as space, we would need to go to four dimensions – that is, into the realm of space-time, which gets beyond our powers of visualization. (Physicists well versed in relativity are no better than the average person in visualizing four dimensions.) So a space-time map tells us about both the where and the when of a "thing". Each segment of the world line tells all there is to know. But when one segment gives way to another segment, at a point where an interaction occurs, that's a different story. In relativity theory, an *"event"* is something that happens at a particular space-time point – that is, at a point in space and at an instant of time. So in the particle world, events seem to occur at exact space-time points, not spread over space and not spread over time. Indeed, experiments indicate that everything that happens in the subatomic world happens ultimately because of little explosive events at space-time points – events, moreover, in which nothing survives. What comes into the point is different from what leaves it.

There are two important features of this diagram that show up in all Feynman diagrams; one is obvious, and one is not so obvious. The obvious one is that at the interaction points *A* and *B*, three particle lines meet. A point such as *A* or *B* is called a vertex, specifically a three-prong vertex.

It is the space-time point where the interaction occurs. If you glance at other similar diagrams, you'll see that all of them have three-prong vertices. Moreover, they are vertices of a particular type, where two fermion lines and one boson line meet. The fermions in these diagrams are either leptons or quarks, and the bosons are force carriers – photons, *W* bosons, or gluons. Here is the stunning generality that physicists now believe to be true.

Every interaction in the world results ultimately from the emission and absorption or bosons (the force carriers) by leptons and quarks at space-time points.

Three-prong vertices lie at the heart of every interaction. This is a interaction that is a truly catastrophic event in which every particle is either annihilated or created. At point *A*, the incoming electron is destroyed, a photon is created, and a new electron is created. The electron flying upward to the left in the figure cannot be said to be the same as the electron that entered from the lower left. They are identical, because they are both electrons, but saying that the one leaving is the same as the one that arrived has no meaning. Figure 15.5 shows one way in which an electron and its antiparticle, a positron, can meet and annihilate to create two photons, a process represented by

$e^{-} + e^{-}$ à 2(γ).

Once again there are two interaction vertices, *A* and *B*, where two fermion lines and one boson line meet. At *A*, an incoming electron emits a photon and creates a new electron, which flies to *B*, where it encounters an incoming positron and emits another photon. To think of this process proceeding forward in time, you could again imagine a horizontal ruler moved slowly upward, and ignore the arrows. You may reasonably ask: Why are the arrows there if I'm being asked to ignore them - because they are labels. Their purpose is to tell whether the line is that of a particle or an antiparticle.

So the line on the right with a downward-pointing arrow represents a positron moving forward in time – that is, upward in the diagram. (And here is where the Wheeler-Feynman vision kicks in) the forward-in-time positron is equivalent to a backward-in-time electron. So it is also possible to interpret this diagram more in the way it appears to the eye. An electron comes along from the left, moving forward in time, emits photons at A and B, and then reverses its course through time. Wheeler and Feynman showed that the descriptions in terms of a forward-in-time positron and a backward-in-time electron are both "correct" because they are mathematically equivalent and indistinguishable. Yet – you might protest – you and I don't have the option of moving forward or backward in time. We move inexorably forward, like the horizontal ruler that slides upward over the diagram. What we see in figure 9.1 is a positron moving to the left on its way to a collision with an electron, even though, instructed in the ways of the quantum world, we are prepared to believe that what we are seeing can also be described as an electron backpedaling through time and moving to the right as time unwinds.

4.32 Quantum Lumps

Max Planck did not set out to be a revolutionary. When he presented his theory of radiation to the Prussian Academy in Berlin in December 1900 and introduced his now-famous constant *h*, he thought he was offering a refinement to classical theory, fixing a little flaw in a solid edifice. (When the quantum revolution he started gained momentum in the years that followed, Planck wanted no part of it. He couldn't embrace what he had initiated.)

Planck was fixing a problem that showed up when the theories of electromagnetism and thermodynamics were blended. Electromagnetism deals with electricity, magnetism and light while thermodynamics deals with "radiation." Thermodynamics also deals with temperature and the flow and distribution of energy in complex systems. These two theories, the pillars of 19th-century physics, were not up to the task of explaining "cavity radiation" – the radiation inside a closed container at a fixed temperature.

When Planck introduced the quantum idea in 1900, he was forty-two, well beyond the age at which many theoretical physicists do their most notable work. Carl Anderson put this age at twenty-six. Among the physicists who developed the full theory of quantum mechanics a quarter-century later, in the period 1924-1928, some hadn't even been born when Planck launched the revolution.

'If your head doesn't swim when you think about the quantum,' Niels Bohr reportedly said, 'you haven't understood it.' And Richard Feynman, the brash and brilliant American physicist who understood quantum mechanics as deeply as anymore, wrote: 'My physics students don't understand it either. That is because I don't understand it.'

One thing that has stood the test of time since that day in Berlin in December 1900 is Planck's constant. It remains the fundamental constant of quantum theory, with ramifications that go far beyond its original role in relating radiated energy to radiated frequency. It is the constant that sets the scale of the subatomic world and that distinguishes the subatomic world from the "classical" world of everyday experience. This chapter is about quantum lumps, of which I have discussed only one so far: the lump (or quantum) of radiated energy that becomes the lump (or particle) or radiant energy known as the photon.

There are two kinds of lumpiness in nature: the *granularity* of things and the granularity (discreteness) of certain *properties* of those things. Let's start with the granularity of things. Everyone knows that you can't subdivide a piece of matter indefinitely. If you divide it finely enough, you get to atoms (the word "atom" was originally chosen to mean "indivisible"), and if you pull atoms apart, you get to electrons and atomic nuclei and eventually to *quarks* and *gluons*. So far as we know, that's as far as you can go. We know of no size and no structure for electrons and quarks. "Well," you might ask, "isn't that only because we haven't yet learned to probe more deeply? Why shouldn't there be worlds within worlds within worlds?" Scientists have a couple of reasons for believing that the onion of reality has a core with only so many layers to uncover, and that we have reached, or are very close to reaching, that core.

One reason for this conclusion is that it takes only a few quantities to completely describe a fundamental particle. An electron, for instance, is described by its mass, charge, flavor, and spin, and by the strength of its interaction with force-carrying bosons of the weak interaction – and that's about it. Physicists are confident that if there are any properties of the electron still to be discovered, these properties are few in number. So it takes only a short list to specify everything there is to specify about an electron.

Another reason for believing that we are close to a genuine core of matter – a reason closely related to simplicity of description – is the identity of particles. Even with the strictest standards of manufacture, no two ball bearings can ever be truly identical all red up-quarks are truly identical, and so on. The fact that electrons obey the Pauli exclusion principle (which says that no two of them ever exist together in exactly the same state of motion) can be understood if the electrons are identical but cannot be understood if electrons differ from one another in any way. If there were infinitely many layers of matter to uncover, we would expect electrons to be as complex as ball bearings, no two of them alike, and each one requiring a vast array of information for its exact description. This isn't the case. The simplicity of the fundamental particles and their identity give strong reason to believe that we may be close to reaching the ultimate "reality" of matter.

4.33 Charge and Spin

One of the "quantized" properties of matter that has already been introduced is electric charge. Some observed particles have no charge. All others carry a charge that is an integral multiple (positive or negative) of the charge *e* of the proton. Spin is another such property. It is either zero or an integral multiple of the spin of the electron, which, in angular-momentum units, is $(^{1}/_{2})\hbar$. For the particles – including the composite particles – the multipliers *e* and $(^{1}/_{2})\hbar$ are typically 0, 1, or 2, but everyday objects may have charges vastly greater than *e* and angular momenta vastly greater than $(^{1}/_{2})\hbar$.

Lumpiness, or quantization, means that there is a finite difference between allowed values of the property. It does *not* mean that there are only a finite number of possible values. Just as the even numbers 2, 4, 6, and so on have a finite separation but go on without limit, charges and spins have a finite separation but an infinite number of possible values. We have no understanding of the particular magnitude of the charge quantum *e*. It is relatively small, even by the standards of the particle world. It measures the strength of the interaction between charged particles and photons. This interaction (the electromagnetic interaction) is about a hundred times weaker than the quark-gluon interaction (understandably called the *strong interaction*). On that basis we call the electron charge small, although both of these interactions are enormously strong relative to the weak interaction. But the point here is that the size of the lump of charge is simply a measured quantity. We don't know why it has the value it has.

Similarly, the magnitude of the quantity $\hbar$ (which sets the size of the lumps of spin) is a measured quantity without a theoretical underpinning. The theory of quantum mechanics, as developed in the 1920s, accounts for the existence of spin and angular-momentum quantization, but not for the magnitude of the quantum unit.

1. Fermions (such as leptons and quarks) have half-odd-integral spin in units of $\hbar$ ($^1/_2$, $^3/_2$, $^5/_2$, and so on), while bosons (such as photons and gluons) have integral spin in its unit (0, 1, 2, and so on).
2. Orbital angular momentum is always an integral multiple of $\hbar$ (0, 1, 2, and so on).
3. Angular momentum, either spin or orbital, can point in only certain directions, and the projections of angular momentum along any chosen axis differ from one to the next exactly $\hbar$ (that is, one unit).

4.34 Mass

In a way, the most obviously lumpy quantity is mass, for every particle has its own specific mass. Indeed, every composite entity – an atomic nucleus, for example, or a protein molecule – has its own definite (and therefore quantized) mass. But because of the contribution of energy to mass, the mass of a composite particle is not equal to the sum of the masses of its constituent particles. A single neutron, for example, has a mass that is considerably greater than the sum of the masses of the three quarks plus any number of massless gluons) of which is composed. Energy is part of the mix, and that energy contributes to the mass. Or consider a deuteron (the nucleus of heavy hydrogen), composed of a proton and a neutron. Its mass is a little less than the sum of the masses of a proton and a neutron. Energy again – but this time is a binding energy, a negative contributor to mass.

In order to break up a deuteron into its constituent particles, energy must be added – just enough of it to offset the binding energy.

Charge and angular momentum are simpler. The charge of a neutron (zero) is the sum of the charges of the quarks that compose it. The deuteron's spin is the sum of the spins of the proton and of the neutron that make it up (added as vectors, taking account of direction as well as magnitude). And so on, for all combinations of particle. The fact that component masses do not add is a reminder that at the deepest level, a composite entity is not a simple combination of parts – it is a new entity entirely. Why is the mass of the muon more than two hundred times the mass of the electron? Why is the mass of the top quark some fifty thousand times the mass of the up quark. No one knows. Quantized mass is just there, awaiting explanation.

Albert Einstein, generally regarded as the greatest physicist of the 20th century – and, ironically, one of the architects of quantum theory – never liked quantum probability. He often remarked that he did not believe God played dice, and in 1953 he wrote, "In my opinion it is deeply unsatisfying to base physics on such a theoretical outlook, since relinquishing the possibility of an objective description . . . cannot but cause one's picture of the physical world to dissolve into fog." Another famous Einstein quote is carved in stone in the Princeton University building where Einstein once had his office. It says, "God is subtle but not malicious." Einstein accepted God's subtlety: the laws of nature are not immediately evident, he was saying; it takes great effort to figure them out. But surely no grand designer of the Universe would be so malicious as to build unpredictability into the basic laws. If a reason for quantum unpredictability is one day found, the fog may lift.

Here's a quiz: Is an atom of sodium 23, whose nucleus contains 11 protons and 12 neutrons, a boson or a fermion? You might think it is a fermion because its nucleus contains 23 fermions – an odd number. Or, if you count quarks – three for each proton and three for each neutron – the nucleus contains 69 fermions, still an odd number. But wait. Circling the nucleus are 11 electrons. So quarks plus electrons add up to 80, an even number. (Or protons, neutrons, and electrons add up to 34, also an even number.) The atom of sodium 23 is a boson.

Bosons and fermions possess many properties that don't distinguish one from the other. For instance, they can be fundamental or composite. They can be charged (positive or negative) or neutral. They can interact strongly or weakly. They can have a wide range of masses, including (at least for the photon) zero mass. But when it comes to spin, they are different. Bosons have integral spin (0, 1, 2, and so on); fermions have half-odd-integral spin ($^1/_2$, $^3/_2$, $^5/_2$, and so on).

Their greatest difference, however, is how they behave when two or more of them are together. Fermions are "antisocial." They obey the *exclusion principle*, which states that no two identical fermions (for instance, no two electrons) can occupy the same state of motion at the same time. Bosons are "social." Not only can two identical bosons occupy the same state of motion at the same time; they *prefer* to do so (again I assign intention to what is only mathematics).

Why do the two classes of particle divide according to their social or antisocial "instincts"? The answer lies in a subtle yet relatively simple feature of quantum mechanics that I will discuss at the end of the chapter. It is a feature specific to quantum mechanics – a feature with no counterpart whatsoever, not even approximately, in classical physics, yet with the most far-reaching consequences in the large-scale world we inhabit.

4.35 Fermions

The twenty-five-year-old Wolfgang Pauli postulated the exclusion principle in 1925. Pauli, an Austrian, had been educated in Germany and later settled in Switzerland, also spending a good deal of time in the United States. In 1921 he had earned his Ph.D. degree at Munich and also published a definitive survey of relativity theory, astonishing Albert Einstein by his grasp of the subject. In 1926, a year after introducing the exclusion principle, he led the way in applying Werner Heisenberg's brand-new quantum theory to the atom. He proposed the neutrino in 1930, by which time he was all of thirty and a professor at Zurich.

In his later years, he was famous for his ability to intimidate physicists who were lecturing on their research results; he would sit in the first row, shaking his head and scowling. Sending Pauli their latest thoughts on particle theory, his first reaction to their letters was always the same: *'Alles Quatsch'* (total nonsense). After a second round of correspondence, he allowed that there was perhaps a little sense in their ideas. By the third round, he was congratulating them on their insight. Pauli formulated the exclusion principle near the end of a dozen year period of confusion and frustration for physicists. After the twenty-seven-year-old Niels Bohr provided a quantum theory of the hydrogen atom in 1913, physicists knew that Planck's constant, *h*, must play an essential role within the atom, and they assumed that the ideas Bohr had laid out – electrons occupy stationary states and make quantum jumps between these states, while emitting and absorbing photons – were probably correct for all atoms. But a real quantum theory was lacking until Werner Heisenberg in 1925 (at age thirty-three) and Erwin Schrodinger in 1926 (at the advanced age of thirty-eight) tied the loose ends together and created the quantum theory that has stood till this day. Pauli's exclusion principle was part of what launched this revolution.

Before discussing the implications of the exclusion principle, we need a definition of "state of motion" and "quantum number." What does it means to say that an electron is in a certain state of motion or that it has certain quantum numbers. An automobile traveling due west at constant speed on an absolutely straight road can be said to be in a certain state of motion. This "state" is defined not by where the car is but by how it is moving – at what speed and in what direction. Another car moving at the same speed in the same direction on the same highway is in the same state of motion as the first car – even if they are far apart. To give another automotive example, and Indianapolis race car tearing around the track, varying its speed and acceleration in exactly the same way on each circuit, is in a certain state of motion. Two cars are in the same state of motion if they have identical patterns of speed and acceleration, whatever their separation might be.

A satellite circling the Earth has a state of motion defined not by where it is at a certain time but by its energy and angular momentum. Another satellite following a long, skinny elliptical path might have the same energy as the first satellite but less angular momentum. It is in a different state of motion. So the state of motion of an object is a "global" property related to the totality of its motion, not to some specific part of the motion.

For an electron in an atom, physicist can't follow the details of the motion; nature doesn't allow it. The only available information is global information. It's as if the Indianapolis race car is moving so fast that you can't make out anything other than a general blur. You know it's confined to the race track and you know it's moving with a certain average speed, but you don't know where it is at any time. The electron in a state of motion within an atom is likewise a blur, with some probability of being in one place and another probability of being in a different place.

But not everything about the electron is fuzzy. It may have a definite energy, a definite angular momentum, and a definite orientation of the axis of its orbital motion. This leads to the possibility of assigning numbers to define the state – a number for the particular energy of the state, a number for its angular momentum, and a number for the orientation of the angular momentum. Since these three physical quantities are *quantized* – meaning they can take only certain discrete values – the numbers that characterize the state are also quantized.

Accordingly, they called *quantum numbers*. For instance, the principal quantum number, n, is chosen to be 1 for the lowest-energy state, 2 for the next state, and so on. It tells where the state lies in the ladder of allowed energies.

The angular-momentum quantum number, l, measuring the angular momentum in units of $\hbar$, can be zero or any positive integer. Finally, the orientation quantum number, m, can take on negative and positive values ranging from $-l$ to $+l$.

4.36 Bosons

In 1924, Satyendra Nath Bose ("BO-suh"), a thirty-year-old physics professor at the university of Dacca, sent of a letter to Albert Einstein in Berlin. Enclosed was a paper of his entitled "Planck's Law and the Hypothesis of Light Quanta," which had been rejected by a leading British journal, *Philosophical Magazine*. Undaunted by the rejection, Bose had decided to approach the world's most celebrated physicist, emboldened perhaps by the fact that he himself ad translated Einstein's relativity text from German to English for the Indian market, or perhaps simply because he was quite confident that his paper had something significant to say. The "Planck's Law" in the title of Bose's paper is the mathematical law introduced by Max Planck in 1900 that gives the distribution of energy among different frequencies of radiation within an enclosure at constant temperature – the so-called blackbody radiation, or cavity radiation, that was discussed in an earlier chapter.

Recall that in Planck's hands the formula $E = hf$ referred not to the energy of a photon of frequency f but to the minimum energy that a material object could give to or take from radiation. Planck assumed, as did most physicists after him for nearly a quarter of a century, that what was quantized was energy transfer to and from radiation, not the radiation itself - this despite the fact that Einstein had proposed the photon in 1905 (actually so named only much later) and that Arthur Compton had seen evidence for photon-electron scattering in 1923. When Bose wrote his paper in 1924, he still referred to "light quanta" (photons) as hypothetical entities. His paper would play a major role in transforming the photon from hypothesis to accepted reality.

Gilbert Lewis coined the term "photon" in 1926. Bose had no inkling of the exclusion principle, which was advanced in the following year and was applied by Pauli to electrons, not photons. In the paper he sent to Einstein, Bose showed that he could derive Planck's law by postulating that the radiation consists of a "gas" of "light quanta" that do not interact with one another and which can occupy any energy state irrespective of whether another "light quantum" is already in that state. Einstein at once realized that Bose's derivation was a giant step forward from Planck's original derivation and that it provided strong, if indirect, evidence for the reality of the light quantum. Einstein personally translated Bose's paper from English into German and forwarded it to a leading German journal, *Zeitschrift fur Physik*, with his recommendation that it be published. It was promptly.

Einstein, intrigued by Bose's work, temporarily turned his attention away from his effort to unify electromagnetism and gravitation and considered how atoms would behave if they followed the same rules as photons. His paper was published hard on the heels of Bose's paper in the same year. So came into being what we now call *Bose-Einstein statistics*. A few years later, Paul Dirac suggested that particles obeying these statistical rules be named bosons. One thing that Einstein worked out was what would happen to a gas of atoms at extremely low temperature. (He assumed that his atoms satisfied Bose-Einstein statistics, which, as it turns out, about half of all atoms do. Let me go back to the apartment-house model. For boson residents, there is no limit to how many may occupy the first floor or on any other floor. You might think this would make all the bosons in a given assemblage gather in the lowest-energy state. They do have some tendency to do so, but it is a tendency that is fully realized only at extremely low temperatures. Einstein realized that the bosons would not only occupy the energy-state – but would distribute themselves identically over that energy-state. Every boson in the collection would be in identically the same state of motion as every other one, so that they would be totally overlapping and interpenetrating.

Each one would occupy the whole energy-state. Every atom spreads itself out with a probability distribution the same as that of every other atom (now we call it a "*Bose-Einstein condensate.*" It took seventy years for experimenters to catch up with theory and produce a Bose-Einstein condensate in the lab. The principal reason for the delay was the great difficulty in pushing temperature to the extraordinary low value required. Neither Bose nor Einstein lived to see the verification of this remarkable behavior of bosons.

Dirac, a famously modest man, also named *fermions*, whose properties both he and Fermi discovered. One difference between fermions and bosons has to do with their numbers. Evidence suggests that the total number of fermions in the Universe is constant (provided anti-fermions are assigned negative particle number), whereas the number of bosons can change. The fermion rule comes into play in every individual particle reaction.

For instance, in the decay of a negative muon into an electron, a neutrino, and an antineutrino, there is one fermion present before the decay and one afterward (assigning negative particle number to the antineutrino). Similarly, in the decay of a neutron, there is one fermion before the decay and a net of one afterward. When an electron and positron annihilate to create a pair of photons, the count of fermions is zero before and after. Similarly, when a proton collides with another proton in an accelerator, various bosons may be created. For example,, three bosons emerge where there were none before. Note that the number of fermions, two, is preserved.) To give just one more example, in the decay of a negative pion into a muon and an antineutrino, the boson count goes from one to zero, and the fermion count from zero to zero (again assigning negative particle number to the antineutrino). No one has an answer to the deep questions: Why do fermions preserve their numbers? Why do bosons come and go in arbitrary numbers?

4.37 Why Fermions and Bosons?

How does Nature contrive that every particle is social or antisocial – either wanting to cluster in the same state of motion or refusing to do so? Classical theory cannot provide an answer to this question, not even an approximate answer. That's why it's such an interesting question. To give an answer, we have to turn to a feature of quantum theory that is mathematical, but hopefully comprehensible. The answer depends also on the fact that there exist in nature particles that are identical.

The exclusion principle does *not* say that no two fermions can occupy the same state of motion. It says that no two fermions of the same kind (two electrons, or two protons, or two red up quarks) can occupy the same state of motion. Similarly, it is only two bosons of the same kind (two *photons* or two positive *pions* or two negative *kaons*) that prefer to occupy the same state of motion. If every particle in the Universe differed from every other one, in no matter how small a particular, it wouldn't matter much whether a particle was a fermion or a boson, for then there would be no bar to fellow bosons to do so. Particles would then be like baseballs – all slightly different and with no incentives or disincentives to aggregating. So the fact that in the subatomic realm we find entities that are truly identical has, literally, cosmic consequences. If, for instance, electrons were not all precisely identical, they would not fill sequential shells in atoms, there would be no periodic table, and there would be no you and no me.

One thing that sets quantum theory apart from theories that went before is the fact that it deals with *unobservable* quantities. One of the un-observables is called the *wave function* or *wave amplitude*. The probability that a particle is in a certain place or is moving in a certain way is proportional to the square of the wave function. So the wave function multiplied by itself relates to what is observable, but the wave function alone does not. This means that whether the wave function is positive or negative has no observable consequences, since the squares of both positive and negative numbers are positive.

The full story is a bit more complicated than this. The unobservable wave function can be a complex number, meaning a combination of a real and an imaginary number. That moves it even further from *observability* than it would be if merely negative. Something called the "*absolute square*" of a complex number is a positive quantity, and that's what is observable.

4.38 Quantum Mechanics and Gravity

I have already mentioned quantum foam, the roiling of space-time at distance of 10^{-35} *m* and times of 10^{-45} *s*. In the world that scientists have studied so far, down to dimensions far smaller than the size of an atomic nucleus, gravity and quantum theory have nothing to do with each other. But as Wheeler pointed out, if you imagine probing far more deeply into the fabric of space-time, you reach a point where the fluctuations and uncertainties that characterize particles take hold of space-time itself. The smooth uniformity of the space around us and the ordered forward march of time give way to weird convolutions impossible to visualize. This is *Planck-scale physics* (from which Max Planck himself would no doubt have recoiled), now extensively explored by theoretical physicists.

4.39 Revisiting Heisenberg's uncertainty principle

Heisenberg took as his starting point the quantum state of the system under consideration (e.g. a single electron, an atom, a molecule etc.), and argued that the only sensible way to formulate a mechanics of the system was by modeling the act of observation on it. Here, by the word "observation" we mean any interaction experienced by the system, such as the "scattering" off of light on an electron. In the absence of any interaction, the system would be totally isolated from the outside world and so totally irrelevant. Only by some form of interaction or observation does the system exist in a "definite state."

Heisenberg represented observations on a system as mathematical operations on its quantum state. This allowed him to write equations governing the behavior of a quantum system and so led to results which were identical to the somewhat more accessible wave mechanics of Schrodinger (say in predicting the energy levels of the hydrogen atom). The equivalence of the two approaches can be appreciated by realizing that the expressions Heisenberg used to represent the observations are differential operators and that they act on the quantum state, which is represented by the *wave function* of the system. So this approach will result in a differential equation in the *wave function* ψ, identical to the wave equation which Schrodinger obtained by analogy with the wave equation for light.

Heisenberg's uncertainty principle results from the realization that any act of observation on the quantum system will "disturb" it thus denying perfect knowledge of the system to the observer. This is best illustrated by analysis of what would happen if we were to attempt to observe the position of the electron in an atomic orbit by scattering a photon of light off it. The photon's wavelength is related to its momentum by the same equation as for any other particle: $\lambda = h/p$.

So the greater the photon's momentum the shorter its wavelength is and vice versa. If then we wish to determine the position of the electron as accurately as possible, we should use the photon with the highest possible momentum (shortest wavelength), as it is not possible to resolve distances shorter than the wavelength of the light used. However, by using a high-momentum photon, although we will gain a good estimate of the electron's position at the instant of measurement, the electron will have been violently disturbed by the high momentum (energy) of the photon and so the electron's momentum will be very uncertain. This is the essence of Heisenberg's uncertainty principle. Knowledge of any one parameter implies uncertainty of some other so-called "conjugate" parameter.

A long-wavelength (low-momentum) photon can give only a rough estimate of the position of the electron, but does not disturb the atom very much. A short-wavelength (high-momentum) photon localizes the electron more accurately, but causes great disturbance. Note: a particle's wave function reflects its localization. The interpretation of the wave function is delta-x = *h*/(delta-*p*).

This spread in wavelengths (frequencies) causes the formation of a localized wave packet in the wave function reflecting the rough localization of the electron. When the electron is very specifically localized, say, in a *"quasi-point-like,"* high-energy collision with another particle, such as a photon, then the uncertainty in its momentum (and so the spread in the wavelength components of the wave function) is large, and the wave packet becomes very localized, in which case it is sensible to regard the electron as a particle. This picture of the electron wave makes rather a nonsense of the simple Bohr picture of the orbiting electrons. The dimensions of the electronic wave function are comparable to that of the atom itself. Until some act of measurement localizes the electron more closely, there is no meaning to ascribing any more detailed a position for the electron. However, this explanation is not altogether satisfactory as it stands. as we have left the electron with a rather poorly defined role in the atom.

In 1926 the German physicist Max Born ventured the suggestion that the square of the amplitude of the wave function at any point is related to the "probability" of finding the particle at that point. The wave function

itself is proposed to have no direct physical interpretation other than that of a "probability wave." When squared, it gives the chance of finding the particle at a particular point on the act of measurement. Hence, the probability density for finding the particle at the position x at time t is probability density = $|\psi(x, t)|^2$.

So the location of the electron in the atom is not wholly indeterminate. The solution to Schrodinger's equation for an electron in the electrical field of the proton will give a amplitude for the wave function as a function of distance from the proton (as well as the energy levels). When squared, the amplitude gives the probability of finding the electron at any particular point. Thus we can give only a probability for determining its position within the orbit and probabilities for finding it in the space between orbits. There is even a small probability of this so-called orbital electron existing actually inside the nucleus. Schrodinger's wavelength associates with every point in space (and time) two real numbers: the amplitude (or size) of the wave function, and its *phase.*

In general, the phase of a wave corresponds to the position in its cycle, with respect to an arbitrary reference point. In other words, it is a measure of how far away one is from a wave crest or trough. The phase is usually expressed as an angle. In contrast to the wave function's amplitude (which is related to the probability), its phase can never be directly observed – it is unobservable. Only differences in phase are observable (e.g. as interference patterns in optics).

4.40 Electron spin

Having just developed a rather sophisticated picture of the electronic wave function, we shall immediately retreat to the comfortingly familiar picture of Bohr's orbital atom to explain the next important development in quantum theory. By 1925, physicist attempting to explain the nature of atomic spectra had realized that all was not correct. Where, according to Bohr's model, just one spectral line should have existed, two were sometimes found very close together .

To explain this and other similar puzzles, the Dutch physicists Sam Goudsmit and George Uhlenbeck proposed that the electron spins on its axis as it orbits around the nucleus just as the Earth spins around the north-south axis as it orbits around the Sun. The splitting of the spectral is explained by the existence of magnetic effects inside the atom. The electron orbit around the nucleus forms a small loop of electric current and so sets up a magnetic field; the orbiting electron behaves like a small magnet. The spin of the electro also has a magnet associated with it which is referred to as the 'magnetic moment of the electron'. This interacts with the orbital magnetic moment, adding to or reducing the energy depending on the way in which the electron is spinning. This will lead to a slight difference in the energy for the different spins of the electron, and will result in the splitting of the spectral line associated with the Bohr orbit.

4.41 The Pauli exclusion principle

A straightforward look at the Bohr model of the atom tells us that some fundamental principle must be missing. There is seemingly nothing to prevent all the electrons of any atom from performing the same orbit. Yet we know that a typical atom will have its electrons speed over several different orbits. Otherwise, transitions between them would be rare, in contradiction to the observations of atomic spectra. So some rule must keep the electrons spread out across the orbits of the atom.

In 1925 the Austrian physicist Wolfgang Pauli derived the principle that no two electrons can simultaneously occupy precisely the same quantum state (i.e. have identical values of momentum charge and spin in the same region of space). He reached this conclusion after examining carefully the atomic spectra of helium. He found that transitions to certain states were always missing, implying that the quantum states themselves were forbidden.

For instance, the lowest orbit (or ground state) of helium in which the two electrons have the same value of spin is not present. But the state in which the two electron spins are opposite is observed.

The power of this principle in atomic physics can hardly be overstated. Because no two electrons can exist in the same state, the addition of extra orbital electrons will successively fill up the outer-lying electron orbits and will avoid over-crowding in the lowest one. Just two electrons are allowed in the ground state because the only difference can be the two values of spin available. More electrons are allowed in the higher orbits because their quantum states can differ by a wide range of orbital angular momenta around the nucleus (which also turns out to be quantized). It is the Pauli exclusion principle which is responsible for the chemical identities of all atoms of the save element as it is this principle which determines the allowed arrangements of the atomic electrons.

Although we have focused on the atom, the exclusion principle applies to any quantum system, the extent of which is defined principally by the wave functions of the component particles. In the case of totally isolated electrons of definite momentum whose wave functions extend over all space, the exclusion principle means that only two electrons with opposite spins can have the same momentum. In the case of electrons confined to a crystal (i.e. electrons whose wave functions extend over the dimensions of the crystal), the rule will apply to all electrons in the crystal.

Pauli's exclusion principle can be expressed alternatively in terms of the behavior of the wave function of a quantum system. Although we have talked so far only of the wave functions of individual particles, these can be aggregated for any quantum system to give a wave function describing the whole system. For example, the total wave function of the helium atom can describe the behavior of two electrons at the same time. Just as the wave function of a single electron is a wave packet reflecting the localization of the electron, a double-electron wave function will contain two wave function humps reflecting the localizations of the two electrons. The exclusion principle is a consequence of the fact that a multiple-electron wave function must change sign under the interchange of any two electrons. Wherever the wave function is positive it must become negative and vice versa. The wave function is said to be anti-symmetric under the interchange of two electrons.

This effect can be understood by considering the two-electron helium atom. Consider the wave function for one electron at position x_1 and the other at position x_2. Thus, the probability for the two electrons to be in the same place is zero and the exclusion principle follows. Note that since any two electrons are indistinguishable, all we are doing in interchanging them is relabeling the electrons and this should make no difference to the physical results (e.g. energy levels and probability densities). The *anti-symmetry* of the wave function allows just this. As all physical quantities are proportional to its square, changing only its sign will make no difference.

Particles such as the electron and the proton with spin $\frac{1}{2}\hbar$ (and other move exotic particles that we shall meet with other half-integral spins $^3/_2\,\hbar$, $^5/_2\,\hbar$, . . . obey the exclusion principle, have anti-symmetric wave functions under the interchange of two identical such particles and are referred to as *fermions,* this is because an ensemble of fermions obey statistics governing their dynamics, which were first formulated by the Italian physicist Enrico Fermi and the Englishman Paul Dirac. Fermi-Dirac statistics show how momentum is distributed amongst the particles of the ensemble. Because of the exclusion principle in any quantum system, there is a limit to the number of particles which can adopt any particular value of momentum and so this leads to a wide range of momentum carried by the particles. Particles such as the photon with spin $\hbar$ (and other particles we shall meet with integral spins such as 0, $\hbar$, $2\hbar$, $3\hbar$, . . . do not obey the exclusion principle and are called *bosons.* Their wave function does not alter under the interchange of two particles. An assembly of bosons obeys dynamical statistics first formulated by the Indian physicist Satiendranath Bose and Albert Einstein. In Bose-Einstein statistics there is no limit to the number of particles which can have the same value of momentum, and this allows the assembly of bosons to act coherently, as in the case of laser light.

Ch 5 Particle Physics

5.1 Point like entities

Over the past several decades many unresolved questions in particle physics have been answered, but our increasing sophisticated level of understanding has led to even deeper questions. In 1983, the discovery of the W and Z bosons provided firm evidence of the correctness of the Standard Model. This marked the beginning of the end of the phase of particle physics, which extended the methods of quantum electrodynamics, formulated at the end of the 1940s, to both the weak and strong nuclear forces. But a quantum theory of the other known force, gravity, was lacking. By the 199os, the increasingly, well observed structure of quarks, leptons and gauge bosons has established the Standard Model beyond reasonable doubt.

Since that time, another decade of experiments has confirmed the Standard Model and its generational structure with impressive accuracy. But the recently confirmed phenomenon of neutrino oscillations (creating neutrino mass) is definitely beyond its scope. Also, cosmological observations now indicate that as much as 96 percent of the Universe is made up of unknown sources of "dark matter" and "dark energy." Furthermore, yet another string revolution has resulted in a new understanding of string theories as the limit of "M-theory," whose exact structure is not yet known.

The last fifty years have seen an enormous advance in our understanding of the microscopic world. We now have a convincing picture of the fundamental structure of observable matter in terms of certain "point-like" elementary particles. We also have a comprehensive theory describing the behavior of and the forces between these elementary particles, which we believe provides a complete and correct description of nearly all non-gravitational physics.

The next few years will see the operation of the Large Hadron Collider at CERN, which promises an eventful decade of both confirmation and perhaps, surprise. The main goal of observing the Higgs boson would provide the final piece in the Standard Model jigsaw.

But there is also a very likely possibility of finding evidence of super-symmetric particles (the missing dark matter of the Universe) or other new physics beyond the Standard Model. Either of these will herald the dawn of a new era in particle physics.

Matter, so it seems, consists of just two types of elementary particles: quarks and leptons. These are the fundamental building blocks of the material world, out of which we ourselves are made. The theory describing the microscopic behavior of these particles has over the past decade or so, become known as the "Standard Model," providing as it does an accurate account of the force of electromagnetism, the weak nuclear force which is responsible for radioactive decay, and the strong nuclear force which holds atomic nuclei together. The Standard Model has been remarkably successful; up until a few years ago all experimental tests have verified the detailed predictions of the theory.

The Standard Model is based on the principle of "gauge symmetry," which asserts that the properties and interactions of elementary particles are governed by certain fundamental symmetries related to familiar conservation laws. Thus, the strong, weak and electromagnetic forces are all "gauge" forces. They are mediated by the exchange of certain particles, called gauge bosons, which are, for example, responsible for the interaction between two electric charges, and for the nuclear processes taking place within the Sun. Unsuccessful attempts have been made to fit the only other known force – gravity – into this gauge framework. However, despite our clear understanding of certain macroscopic aspects of gravity, a microscopic theory of gravity has so far proved elusive. Moreover, recent experiments in neutrino physics cannot be explained within the Standard Model, showing beyond doubt that there must be a theory beyond the Standard Model, and that the Standard Model itself is only an approximation, albeit a very good one, to the true theory of Nature.

The picture of the microworld has emerged slowly since the late 1960s, at which time only the electromagnetic force was well understood. At about the same time as the electroweak model was being developed physicists were using "deep inelastic scattering" experiments to probe the interior of the proton. The experiments provided the first indication that the proton was not truly elementary, but composed of point-like objects called "quarks." As the physical reality of quarks gained wider acceptance, a new gauge theory was formulated in an attempt to explain the strong forces between them. This theory is called "Quantum Chromodyamics" and attributes the strong force to the exchange of certain gauge bosons called gluons. Together, quantum chromodynamics and the Glashow – Weinberg – Salam electroweak theory constitute the "Standard model" of elementary particle physics. Early experiments involved collisions between electrons and positrons. These experiments were instrumental in confirming the physical reality of "quarks" and in testing many of the predictions of "quantum chromodynamics" and the "electroweak" theory.

In the next part of this chapter we begin by summarizing the "Standard Model" and describe the many tests of the model performed in electron-positron colliders over the past two decades. The recent neutrino experiments, which show that there must be a theory beyond the Standard Model, are then discussed. Finally, we address the question of what this story may be, using ideas from current research, such as grand unification, "supersmmetry" and "string theory."

5.2 Matter and Light

The physical world we see around us has two main components, matter and light, and it is the modern explanation of these things which is the purpose of many science books. During the course of the story, these concerns will be restated in terms of material particles and the forces which act between them, and we will most assuredly encounter new and exotic forms of both particles and forces.

The origin of the story, and the motivation for all that follows, is the explanation of everyday matter and visible light. Beginning as it does, with a laudable sense of history, at the turn of the last century, the story is one of twentieth-century achievement. For the background, we have only to appreciate the level of understanding of matter and light around 1900, and some of the problems in this understanding to prepare ourselves for the story of progress in which it follows.

5.3 The nature of matter

As mentioned in previous chapters, by 1900 most scientists were convinced that all matter is made up of a number of different sorts of atoms, as had been conjectured by the ancient Greeks millennia before and as had been indicated by chemistry experiments over the preceding two centuries. In the atomic picture, the different types of substance can be seen as arising from different arrangements of the atoms. In solids, the atoms are relatively immobile and in the case of crystals are arranged in set patterns of impressive precision. In liquids they roll loosely over one another and in gases they are widely separated and fly about at a velocity depending on the temperature of the gas. The application of heat to a substance can cause "phase transitions" in which the atoms change their mode of behavior as the heat energy is transferred into the kinetic energy of the atoms' motions.

Many familiar substances consist not of single atoms, but of definite combinations of certain atoms called molecules. In such cases it is these molecules which behave in the manner appropriate to the type of substance of concern. For instance, water consists of molecules, each made up of two hydrogen atoms and one oxygen atom (H2O). It is the molecules which are subject to a specific static arrangement in solid ice and the molecules which fly about in steam. The laws of chemistry, most of which were discovered "empirically" between 1700 and 1900, contain many deductions concerning the behavior of atoms and molecules.

At the risk of over-simplification, the most important of these can be summarized as follows:

1. Atoms can combine to form molecules, as indicated by chemical elements combining only in certain proportions (Richter and Dalton).
2. After a given temperature and pressure, equal volumes of gas contain equal numbers of molecules (Avogadro).
3. The relative weights of the atoms are approximately multiples of the weights of the hydrogen atom (Prout).
4. The mass of each atom is associated with a specific quantity of electrical charge (Faraday and Webber).
5. The elements can be arranged in families having common chemical properties but different atomic weights (Mendeleeff's periodic table).
6. An atom is approximately 10-10 m across, as implied by the internal friction of a gas (Loschmidt).

One of the philosophical motivations behind the atomic theory was the desire to explain the diversity of matter by assuming the existence of just a few fundamental and indivisible atoms. But by 1900 over 90 varieties of atoms were known, an uncomfortably large number for a supposedly fundamental entity. Also, there was evidence for the disintegration (divisibility) of atoms. At this breakdown 'ancient' atomic theory ended and modern physics began.

5.4 Two problems in classical theory

Just as these early atomic experiments revealed an unexpected richness in the structure of matter, so too, theoretical problems forced upon physicists more sophisticated descriptions of the natural world. The theories of special relativity and quantum mechanics arose as physicists realized that the classical physics of "mechanics," "thermodynamics" and "electromagnetism" were inadequate to account for apparent mysteries in the behavior of matter and light. Historically, the mysteries were contained in two problems, both under active investigation at the turn of the century.

5.5 The interaction of light with matter

All light, for instance sunlight, is a form of heat and so the description of the emission and absorption of radiation by matter was approached as a thermo-dynamical problem. In 1900 the German physicist Max Planck concluded that the classical thermo-dynamical theory was inadequate to describe the process correctly. The classical theory seemed to imply that if light of any one color (any one wavelength) could be emitted from matter in a continuous range of energy down to zero, then the total amount of energy radiated by the matter would be infinite. Much against his inclination, Planck was forced to conclude that light of any given color cannot be emitted in a continuous band of energy down to zero, but only in multiples of a fundamental quantum of energy representing the minimum negotiable bundle of energy at any particular wave-length. This is the starting point of quantum mechanics, which is the necessary description of anything very small (i.e. all atoms and elementary particles). As the elementary particles are both fast moving and small, it follows that their description must incorporate the rules of both special relativity and quantum mechanics. The synthesis of the two is known as "relativistic quantum" theory.

5.6 Relativistic Quantum Theory

Quantum mechanics, just like ordinary mechanics and electrodynamics, must be made to obey the principles of special relativity. Because the entities (particles, atoms, etc.) described by quantum theory quite often travel at speeds at or near c, this becomes an essential requirement. Special relativity will not just give corrections to conventional Newtonian mechanics, but will dictate dominant, unconventional relativistic effects.

We will see that the synthesis of relativity with quantum theory predicts wholly new and unfamiliar physical consequences (e.g. antimatter). This requires us to develop a new way of looking at matter via "quantum" fields. If we can then go in to develop the mechanics of interacting quantum fields, this will provide us with the most satisfactory description of the behavior of matter (both the conventional matter we have discussed so far, and the unconventional antimatter we will introduce along the way).

5.7 The Dirac equation

In 1927, the English physicist Paul Dirac combined special relativity, James Clerk Maxwell's electromagnetism, and quantum physics, and he came up with "quantum field theory." At the same time as Schrodinger and Heisenberg were formulating their respective versions of the quantum theory, Paul Dirac was attempting the same task. But, in addition, he was concerned that the quantum theory should manifestly respect Einstein's special relativity. This implies two distinct requirements: firstly, that the theory must predict the correct energy-momentum relation for relativistic particles, and secondly, that the theory must incorporate the phenomenon of electron spin in a Lorentz covariant fashion.

In one of the most celebrated brainstorming sessions of theoretical physics, Dirac simply wrote down the correct equation. He was guided in this task by realizing that Schrodinger's equation for the electronic wavefunction cannot possibly satisfy the requirements of special relativity because time and space enter the equation in different ways (as 1st and 2nd order derivatives respectively). Schrodinger's equation is perfectly adequate for particles moving with velocities much less than c, and it predicts the correct Newtonian energy-momentum relationship for particles. But because space and time are not treated correctly, it does not predict the correct relativistic relationships or incorporate energy-mass equivalence.

In the spirit of special relativity, Dirac sought an equation treating space and time on an equal basis. In this he succeeded, but found that in doing so the electron wave function contained two separate components which in the non-relativistic limit correspond to the probabilities that the electron is spin up (with spin quantum + /2) or spin down (with spin quantum – /2). Thus, is written as a two-component spinor. In fact, in the full theory it is a four-component object, for reasons which will become clear in the next section.

So in attempting to incorporate special relativity into quantum mechanics it was necessary to invent "electron spin." It is fascinating to wonder whether, if electron spin had not been proposed and discovered experimentally, it would have been proposed theoretically on this basis.

Dirac's equation can be used for exactly the same purposes as Schrodinger's but with much greater effect. Earlier we saw that the spin of the electron gives rise to a splitting in the energy levels of the hydrogen atom. This is because the magnetic moment of the electron may either be aligned with, or against, the magnetic field set up by the electron's orbital angular momentum. It was noticed in experiments that the half-integral unit of spin angular momentum /2 produced as big a magnetic moment as a whole integral unit of orbital angular momentum (i.e. spin is twice as effective in producing a magnetic moment as is orbital angular momentum). This is quantified by ascribing the value of 2 to the gyromagnetic ratio (the g-factor) of the electron. This is effectively the constant of proportionality between the electron spin and the magnetic moment resulting. In non-relativistic quantum mechanics, g = 2 is an empirical fact and with the Dirac equation, it is an exact prediction. The Dirac equation can also explain the fine splitting and hyperfine splitting of energy levels within the hydrogen atom. These result from the magnetic interactions between the electron's orbital angular momentum, the electron spin and the proton spin.

5.8 Antiparticles

One immediate consequence of predicting the relativistic relationship between energy and momentum for the electron wave function is that the Dirac equation seems to allow the existence of both positive-and-negative- energy particles. In an amazing feat of intellectual boldness, Dirac suggested that this prediction of "negative-energy" particles was not nonsense but, instead, the first glimpse of a hidden universe of antimatter. The concept of negative-energy entities is wholly alien to our knowledge of the Universe. All things of physical significance are associated with varying amounts of positive energy. So Dirac did not ascribe a straightforward physical existence to these negative-energy electrons.

Instead, he proposed an energy spectrum containing all electrons in the Universe. This spectrum consists of all positive-energy electrons which inhabit a band of energies stretching from the rest mass, up to arbitrarily high energies. These are the normal electrons which we observe in the laboratory and whose distribution over the energy spectrum is determined by the Pauli exclusion principle.

Dirac then went on to suggest that the spectrum also contains the negative-energy electrons which span the spectrum from arbitrarily high energies down to arbitrarily large negative energies. He proposed that these negative-energy electrons are unobservable in the real world. To prevent the real, positive-energy electrons from simply collapsing down into negative-energy states, it is necessary to assume that the entire negative-energy spectrum is full and that double occupancy of any energy state in the continuum is prevented by the Pauli exclusion principle. No electrons inhabit the energy gap. Viewed picturesquely, it is as if the world of physical reality conducts itself whilst (although) hovering over an unseen sea of negative-energy electrons.

But if this sea of negative-energy electrons is to remain unseen, what is its effect on the everyday world? The answer to this is that elementary particle interactions of various sorts can occasionally transfer enough energy to a negative-energy electron to boost it across the energy gap into the real world. For instance, a photon with energy may collide with the negative-energy electron and so "promote" it to reality. But this cannot be the end of the story, as we seem to have created a unit of electrical charge whereas we are convinced that this is a quantity in which is conserved absolutely.

Also, we started out with a photon of energy and have created an electron with a energy just over (a minimal energy value). Where has the energy difference of (the minimal energy value). We believe that positive energy is also conserved absolutely; it does not disappear into some negative energy sea.

These problems of interpretation are resolved by proposing that the hole left in the negative-energy sea represents a perceptible, positive- energy particle with an electrical charge opposite to that on the electron. (The absence of a negative-energy particle). This particle is referred to as the antiparticle of the electron or the positron, and is denoted e+.The positron was first discovered in 1931 by the American physicist Carl Anderson in a cloud chamber photograph of cosmic rays. Although the arguments given here concentrated specifically on the electron and the positron, it is important to appreciate that the Dirac equation applies to any relativistic spin-1/2 particle, and so too do the ideas of a "negative- energy sea" and "antiparticles." Both the proton p and neutron n can be described by the Dirac equation and seas of negative-energy protons and neutrons may be proposed as coexisting with those of the electrons. The holes in those seas, the antiprotons denoted p-bar, and antineutrons denoted n-bar, took somewhat longer to discover than the positron because the energy is larger. It requires high-energy accelerators to provide probes which are energetic enough to boost the antiprotons into existence. These were not available until the mid-1950s.

The electron wave function which is described in the Dirac equation can now be appreciated in its full four-component form. In the Newtonian limit, these components describe, respectively, the spin-up and spin-down states of both the electron and the positron. The development of the net concept in the micro-world is contained in the behavior of particles and antiparticles. We suggested that an energetic photon can promote a negative-energy electron from the sea, thus leaving a hole. So the photon can create an electron-positron pair from the vacuum. (In fact, this must take place in the presence of another particle to ensure conservation of energy and momentum.

Similarly, an electron and a positron can "annihilate" each other and give rise to energetic photons. The upshot of this is that particles such as the electron can no longer be regarded as immutable, fundamental entities. They can be created and destroyed just like photons, the quanta of the electromagnetic field.

Twenty years later (several years later – about 20 years), U.S. physicists Richard Feynman and Julian Schwinger and, independently, Japanese physicist Sin-Itiro Tomonaga developed Dirac's theory into a beautiful new quantum theory of the electron that is called "quantum electrodynamics" or QED. This theory tells us what a force is and how an electron knows that there is another electron nearby.

According to Maxwell, when an electron accelerates, it radiates an electromagnetic wave. That's how your cell phone works: The base unit accelerates electrons back and forth along the antenna, and the accelerated electrons generate the electromagnetic wave that travels to your receiver at the speed of light. Quantum physics says that this electromagnetic wave consists of photons, Newton's corpuscular description of light. Accelerating electrons produce photons. That's how electrons know of each other's presence: They send photons back and forth. If two traveling electrons meet each other, they exchange photons.

The next questions you may ask are –

- Where do these photons come from?
- Do the electrons carry photons around in case they meet otherelectrons?
- How do they know how many electrons they are going toencounter?

5.9 Quantum field theory (QFT)

In the most sophisticated form of quantum theory all entities are described by fields. Just as the photon is most obviously a manifestation of the electromagnetic field so too is an electron taken to be a manifestation of an electron field and a proton of a proton field.

Once we have learned to accept the idea of an "electron wave function" extending throughout space and time (by virtue of Heisenberg's uncertainty principle for a particle of definite momentum), it is not too great a leap to the idea of an electron field extending throughout space-time.

Any one individual electron wave function may be thought of as a particular frequency excitation of the field and may be localized to a greater or lesser extent dependent on its interactions.

The electrons field variable is, then, the (Fourier) sum over the individual wave functions, where coefficients multiplying each of the individual wave functions represent the probability of the creation or destruction of a quantum of that particular wavelength (momentum). The representation of a field as the summation over its quanta, with coefficients specifying the probabilities of the creation and destruction of those quanta, is referred to as second "quantization."

First quantization is the recognition of the particle nature of a wave or the wave nature of a. Second quantization is the incorporation of the ability to create and destroy the quanta in various reactions. There is a relatively simple picture which should help us to appreciate the nature of a quantum field and its connection with the notion of a particle. A quantum field is equivalent, at least mathematically, to an infinite collection of harmonic oscillators. These oscillators can be thought of as a series of springs with masses attached. When some of the oscillators become excited, they oscillate (or vibrate) at particular frequencies. These oscillations correspond to a particular excitation of the quantum field and hence to the presence of particles, i.e. field quanta.

We are familiar with electromagnetic and gravitational fields because their quanta are bosons, there are no restrictions on the number of quanta in any one energy state and so large assemblies of quanta may act together coherently to produce macroscopic effects. Electron and proton fields are not all evident because, being fermions, the quanta must obey Pauli's exclusion principle and this prevents them from acting together in a macroscopically observable fashion. So, although we can have concentrated beams of coherent photons (laser beams), we cannot produce similar beams of electrons. These instead must resemble ordinary incoherent lights (i.e. torchlights) with a wide spread of energies in the beam.

5.10 Perturbation theory

To describe elementary particle reactions in which quanta can be created and destroyed, it is necessary to propose an expression for the "Lagrangian" of the interacting quantum fields.

Let us concentrate on interacting electron and photon fields only. The Lagrangian will contain parts which represent free electrons and free photons, where A denotes a four-vector representing the electromagnetic field. It will also contain a part which represent the interactions between and photons, whose form will be dictated by general principles. These will include, for instance, Lorentz invariance and various conservation laws which the interactions are observed such as the conservation of electrical charge. The total Lagrangian is then the sum of all its parts.

This is the top-level specification of the fields being described and the way in which they interact. We can proceed to predict the values of physical quantities by following a method developed in the late 1940s by the American physicist Richard Feynman who derived a set of rules which specifies the propagation of the interacting field quanta as the sum of a set of increasingly complicated sub-processes involving the propagation of the "free" field quanta. Each sub-process in the sum can be represented in a convenient diagram referred to as a "Feynman diagram." The rules associate with each diagram a mathematical expression. To calculate the probability of occurrence P of any physical event involving the quanta of the field, it is first necessary to specify the initial and final states beings observed, denoted (<), the ket-symbol and (>), the bra-symbol respectively, and then to select all the Feynman diagrams which can connect the two. The mathematical expression for each diagram is then worked out to give the quantum-mechanical amplitude m for the sub- process. The amplitude for a number of the individual sub-processes may then be added to give the total amplitude m which is then squared to give the required probability of occurrence where,

y/ x=m. (=)

The probability of occurrence (i.e. of the transformation between initial and final states) may then be restated as the cross-sectional area of two colliding particles, as the mean (average) lifetime for a particle to decay, or as some other appropriate measureable parameter.

This is achieved by adopting the kinematical prescriptions which take into account factors like the initial flux of colliding particles, the density of targets available in a stationary target and so on. The reason why this approach can be adopted is that only the first of the simplest Feynman diagrams from the infinite series need be considered. This is because the strength of the interaction between electrons and photons (the strength of the electromagnetic force) is small. It can be regarded as a perturbation of free-particle-type behavior. Another way of stating this is that the probability of the electron or positron interacting with a photon is small.

In fact, each photon-electron vertex multiplies the probability of occurrence of the diagram by

e (* c).

As each new order of diagram contains a new photon line with two vertices, the relative magnitude of successive orders is reduced by

e2 (* c) = 1/137.

So only the first few sub-processes need be calculated to achieve an acceptable approximation to the exact answer. Planck's constant =

6.626068 × 10-34 m2 kg/s.

Related to the Planck constant is the reduced Planck constant, sometimes called the Dirac constant after Paul Dirac. It is equal to the Planck constant divided by 2, and is denoted . The reduced Planck constant is used when frequency is expressed in terms of radians per second ("angular frequency") instead of cycles per second. The energy of a photon with angular frequency, where = 2, is given by E = .

18.11 Virtual Particles

It is important to understand that the dynamics of the individual field quanta within any sub-process of the perturbation expansion are not constrained by energy or momentum conservation, provided that the sub- process as a whole does conserve both. This microscopic anarchy is permitted by Heisenberg's uncertainty principle in which states that energy can be uncertain to within,

(E) -E for a time (t) -t,

such that

(E)* (t) =

So an electron may emit an energetic photon, or a photon may convert into an electron-positron pair over microscopic timescales, provided that energy conservation is preserved in the long run. These illicit processes are known as 'virtual processes'. They form the intermediate states of elementary particle reactions.

So although we do not see them, we must calculate the probabilities of their occurrence and add them all up to find the number of different ways for a particle reaction to get from its initial to ts final state. A good example of a virtual process is the annihilation of an e+ e- pair into a photon. 5.12 Renormalization

In writing down all the Feynman diagrams of the sub-processes we find some whose amplitude appears to be infinite. These diagrams are generally those with bubbles on either electron or photon wave functions or surrounding electron-photon vertices. These diagrams give infinite contributions owing to ambiguities in defining the electron and the photon. An ordinary electron propagating through space is constantly emitting and absorbing virtual photons. It is engaging in self-interaction with its own electromagnetic field (of which its own charge is the source). So the wave function of the electron is already dressed up with these virtual photons. Similarly, a photon propagating through space is free to exist as a virtual e+ e- pair, and the full photon wave function already contains the probabilities of this occurring. Also, the electric charge, which we denote e, already contains the quantum corrections.

In 1949, Feynman, Schwinger, Dyson and Tomonaga showed how the infinite contributions to the perturbation series can be removed by redefining the electron, photon and electric charge to include the "quantum corrections." When the real electrons, photons and charges appear, the infinite diagrams are included implicitly and should not be recounted. The mathematical proof of this demonstration is known as "renormalization". Renormalization is a necessary formal process which shows that the particles in the theory and their interactions are consistent with the principles of quantum theory.

5.13 The quantum vacuum

In classical (non-quantum) physics, empty space-time is called the vacuum. The "classical vacuum" is utterly featureless. However, in quantum mechanics, the vacuum is a much more complex entity: it is far from featureless and far from empty. Actually, the quantum vacuum is just one particular state of the quantum field. It is the quantum-mechanical state in which no field quanta are excited, that is, no particles are present. Hence, it is the 'ground state' of the quantum field, the state of minimum energy.

In the vacuum, every oscillator is in its ground state. For a classical oscillator, this means that it is motionless: the spring holds the mass in a fixed position. However, for a quantum oscillator, the uncertainty principle means that neither position nor momentum is precisely fixed and both are subject to random quantum fluctuations. These fluctuations are called zero-point oscillations, or zero-point vibrations. So, the quantum vacuum is full of fluctuating quantum fields. There are no "real" particles involved, only virtual ones. Virtual particle-antiparticle pairs continually materialize out of the vacuum, propagate for a short time (allowed by the uncertainty principle) and then annihilate.

These zero-point vibrations mean that, in the vacuum – the state of minimum energy – there is a zero-point energy associated with any quantum field. Since, there is an infinite number of harmonic oscillators per unit volume the total zero-point energy density is in fact infinite. We have already seen that some sense can be made of infinite quantities through the process or renormalization. As it is usually implemented, this yields a zero energy density for the standard quantum vacuum.

5.14 The vacuum and the Casmir effect

In 1948, the Dutch physicist Hendrik Casimir came up with a clever idea to demonstrate that space is filled with "virtual particles." He proposed placing two uncharged metal plates in an empty chamber. When the plates are moved very close to each but not touching, the spacing between the plates excludes virtual particles larger than a particular wavelength. Because particles with longer wavelengths can still appear outside the plates, there are more particles outside than inside. The imbalance should push the plates together. Hendrik Casimir predicted that two clean, neutral, parallel, microscopically flat metal plates attract each other by a very weak force that varies inversely as the fourth power of the distance between them.

When Casimir first proposed it, his idea was just an idea – a thought experiment - because the technology of the 1940s and early 1950s wasn't up to the task. But in a few years, several scientists attempted the experiment, getting encouraging results and The "Casimir effect" was experimentally verified in 1958. In 1977, an experiment done at the University of Washington confirmed beyond all doubt the Casimir effect, as it's now called. It is extremely difficult to observe these vacuum fluctuations, since there is no state of lower energy with which the vacuum can be compared. However, there is one situation in which its effects can be seen indirectly. It can be understood in the following way. The zero- point energy filling the vacuum exerts pressure on everything.

In most circumstances, this pressure is not noticeable, since it acts in all directions and the effect cancels. However, the quantum vacuum has different properties between the two metal plates. Some of the zero-point vibrations of the electromagnetic field are suppressed namely, those with wavelengths too long to fit between the plates. So, the zero-point energy density between the plates is less than that of the standard vacuum, i.e. it is negative. From this it follows that the pressure outside is greater and hence the plates feel an attractive force.

5.15 Quantum electrodynamics (QED)

This is the name (often abbreviated to QED) given to the relativistic quantum field theory describing the interactions of electrically charged particles via photons. The discovery of the perturbation expansion revealed the existence of an infinite number of ever decreasing quantum corrections to any electromagnetic process. The renormalisability of QED means that we can avoid apparently infinite contributions to the perturbation expansion by careful definition of the electron and photon. Therefore we can calculate the value of observable parameters of electromagnetic processes to any desired degree of accuracy, being limited only by the computational effort required to evaluate the many hundreds of Feynman diagrams which are generated within the first few orders (first few powers of e2/(*c) of the perturbation expansion. This has led to some spectacular agreements between theoretical calculations and very accurate experimental measurements.

The g-factor of the electron is not, in fact, exactly equal to 2 (as predicted by the Dirac equation). Its value is affected by the quantum corrections to the electron. Essentially, the virtual photons of the quantum corrections carry off some of the mass of the electron while leaving its charge unaltered. This can then affect the magnetic moment generated by the electron during interactions. The measure of agreement between QED and experimental measurement is given by the figure for the modified g- factor. There are several other such amazing testaments to the success of QED, including numbers similar to the above for the g-factor of the muon (a heavy brother of the electron which we will meet soon), and yet more subtle shifting of the exact values of the energy levels within the hydrogen atom, the so-called Lamb shift.

5.16 Postscript

The realization that relativity and quantum mechanics must be made mutually consistent led to the discovery of "antiparticles," which led in turn to the concept of quantum fields. The theory of interacting quantum fields is the most satisfactory description of elementary particle behavior. All calculations in quantum field theory follow from the specification of the correct interaction "Lagrangian," which is determined by the conservation laws obeyed by the force under study.

The success of QED makes the most precise picture we have of the physical world or at least the electromagnetic phenomena in it. We have developed this picture of the world almost exclusively in terms of the particles interacting by the electromagnetic force. It is now time to turn our attention to the other particles and forces in nature to see if they are amenable to a similar treatment.

In what follows, we shall often use the language of "particle wave functions" rather than that of "quantum fields." Although somewhat imprecise, a particle wave function is a slightly more convenient and intuitive concept in most situations. However, there will be occasions in later chapters in which a proper understanding of certain phenomena demands that we consider the quantum fields themselves rather than wave functions.

5.17 The fundamental forces

It is an impressive demonstration of the unifying power of physics to realize that all the phenomena observed in the natural world can be attributed to the effects of just four fundamental forces. These are the familiar forces of gravity and electromagnetism, and the not-so-familiar weak and strong nuclear forces (generally referred to as the 'weak' and 'strong' forces'). Still more impressive is the fact that the phenomena occurring in the everyday world can be attributed to just two: gravity and electromagnetism. This is because only these forces have significant effects at observable ranges. The effects of the weak and strong nuclear forces are confined to within, at most, 10-15 m of their sources.

With this in, mind, it is worthwhile summarizing a few key facts about each of the four forces before going on to look at the variety of phenomena they display in our laboratories. In each case we are interested in the sources of the force and the intrinsic strength of the interactions to which they give rise. We are interested also in the space-time properties of the force; how it propagates through space and how it affects the motions of particles under its influence.

Finally, we must consider both the macroscopic (classical) description of the forces where appropriate and the microscopic (quantum- mechanical) picture where possible.

5.18 Gravity

Gravity is by far the most familiar of the forces in human experience, governing phenomena as diverse as falling apples and collapsing galaxies. At the non-relativistic level, the source of the gravitational force is mass and, because there is no such thing as negative mass, this force is always attractive. Furthermore, it is independent of all other attributes of the bodies upon which it acts, such as electric charge, spin, direction of motion, etc. The gravitational force is described classically by Newton's famous inverse-square law, which states that the magnitude of the force between two particles is proportional to the product of their masses and inversely proportional to the product of their masses and inversely proportional to the square of the distance between them:

F = G M1*m2/r2.

The strength of the force is governed by Newton's constant, G, and is extremely feeble compared with the other forces. We notice the effects of gravity only because is it the only long-range force acting between electrically neutral matter. In the micro-world the effects of gravity are mainly negligible. Only in exotic situations, such as on the boundary of a black hole and at the beginning of the Universe, do the effects of gravity on the elementary particles become important. The mechanism which gives rise to this force in the classical picture is that of the gravitational field, which spreads out from each mass-source to infinity. A test mass will interact with the gravitational field. At each point in space; the interaction between the test mass and the gravitational field reproduces the gravitational force. However, according to Newton's theory when a mass source moves, the gravitational field is sets up changes instantaneously to accommodate its new position. This instantaneous change is fundamentally incompatible with the theory of special relativity, which requires that disturbances cannot propagate faster than light. This motivated Einstein to formulate a new theory of gravity and relativity called general relativity which he completed in 1915.

A feature of Newton's formula is that the quantity characterizing a source of gravity – its gravitational mass – is identical to the quantity – its inertial mass – which characterizes its acceleration in response to an applied force, as given by another of Newton's famous equations:

$F = ma$.

This equivalence between gravitational and inertial mass, which was known to many generations of physicists before him, led Einstein to speculate on the connection between gravity and acceleration. The principle of equivalence is the apotheosis of this connection, and formed the basis of his conceptual leap from the theory of special relativity to the theory of general relativity. Put simply, the equivalence principle declares that, at any point, gravitation and acceleration are indistinguishable phenomena.

5.19 General Relativity revisited

We saw how, in special relativity, observers' perceptions of time and space are modified by factors depending on their relative velocities. From this it follows that during acceleration (changing velocity) an observer's scales of time and space must become distorted. By the principle of equivalence, a acceleration is identical to the effects of a gravitational field and so this too must give rise to a distortion of space-time. Einstein's general relativity goes on to explain the somewhat tenuous reality of the gravitational field as the warping of space-time around a mass-source. Thus a mass will distort space-time rather like a bowling ball laid on a rubber sheet. And the effect of gravity on the trajectory of a passing particle will be analogous to rolling a marble across the curved rubber sheet.

So general relativity suggests that instead of thinking of bodies as moving under the influence of a gravitational force, one should think of them as moving freely through a warped, or curved, space-time. Hence, the force of gravity is reduced to the curved geometry of space-time. Geometry has, as we know, different rules on a curved surface. For example, on the curved surface of the Earth, two north-pointing lines which are parallel at the equator (lines of longitude) actually meet at the north pole; whereas on a flat surface, two parallel lines never meet. In fact, in curved space-time, straight lines must be replaced by geodesics as the shortest path between two points; free particles move along geodesics.

On the surface of the Earth, geodesics correspond to great circles. Einstein embodied this interpretation of gravity as geometry in his field equations of general relativity:

Guv=8*PI*GTuvw

which loosely translates as,(Geometry of space-time)= 8*PI*G*(mass and energy).

So mass and energy determine the curvature and geometry of space-time; and the curvature and geometry of space-time determine the motion of matter. In other words 'matter tells space-time how to bend and space- time tells matter how to move'. The theory goes on to predict the existence of "gravitational waves" propagating through space as the result of some changes in a mass-source (such as the collapse of a star into a neutron star or black hole). In this event, distortions in space-time will spread out spherically in space at the speed of light, rather like ripples spread out circularly across the surface of a still pond into which a stone is dropped. There has been considerable experimental effort from cosmic events out in space, but none has so far succeeded because of the smallness of the disturbances. However, detectors are becoming more and more sensitive and there are great hopes that gravitational waves will be observed within the next decade or so.

5.20 Quantum gravity

It is important to remember that Einstein's general relativity is still a "classical theory": it does not account for gravity in the quantum- mechanical regime. A successful quantum theory of gravity has not yet been formulated, and the reconciliation of general relativity with quantum mechanics is one of the major outstanding problems in theoretical physics. It is straightforward enough to take the first few steps towards such a theory, following an analogy with "quantum electrodynamics." We can propose that the gravitational field consist of microscopic quanta called "gravitons" which must be massless (to accommodate the infinite range of gravity) and of spin 2 (for consistency with general relativity). The gravitational force between any two masses can then be described as an exchange of gravitons between them.

Problems arise, however, because, unlike quantum electrodynamics, certain graviton sub-processes always seem to occur with an infinite probability. Probability-quantum gravity is not renormalisable.

5.21 Electromagnetism

This is the force of which we have the fullest understanding. This is possibly a reflection of its physical characteristics: it is of infinite range, allowing macroscopic phenomena to guide our understanding of classical electromagnetism, and it is a reasonably weak force, allowing its microscopic quantum phenomena to be understood using perturbation theory. The strength of the electromagnetic force is characterized by the "fine structure constant"

= e2/ * c = 1/137.

The source of the force is, of course, electric charge which can be either positive or negative, leading to an attractive force between unlike charges and a repulsive force between like charges. When two charges are at rest, the electrostatic force between them is given by Coulomb's law, which is very similar to Newton's law of gravity, namely that the magnitude of the force is proportional to the product of the magnitudes of the charges involved (empirically) observed to exist only as multiples of the charge on an electron and inversely proportional to the square of the separation between them.

New mysteries are introduced by the concept of electric charge. What is it, other than a label for the source of a force we observe to act? Why does it exist only in quanta? Why is the charge quantum on the electron exactly opposite to that on the proton? These are largely taken for granted in classical electromagnetism and are only now being addressed in the modern theories.

Unlike gravity, when electric charges start to move, qualitatively new phenomena are introduced. A moving charge has associated with it not only an electric field, but also a magnetic field. A "test charge" will always be attracted (or repelled) along the direction of the electrical field, i.e. along a line joining the centers of the two charges. But the effect of the magnetic field is that a test charge will be subjected to an additional force along a direction which is mutually perpendicular to the relative motion of the source charge and to the direction of the magnetic field.

These properties imply that the combined electromagnetic force on a particle cannot be described simply by a number representing the magnitude of the force but, instead, by a vector quantity describing the magnitude of the forces acting in each of the three directions.

When a charge is subject to an acceleration, a variation in electric and magnetic fields is propagated out through space to signal the event.

If it is subject to regular accelerations, as may occur when an alternating voltage is applied to a radio aerial, then the charge emits an electromagnetic wave which consists of variations in the electric and magnetic fields perpendicular to the direction of propagation of the wave. Such an electromagnetic wave is part of the electromagnetic spectrum which contains, according to the frequency of oscillation of the fields, radio waves, infrared waves, visible light, ultraviolet light, X-rays and gamma rays.

Electromagnetic phenomena are all described in the classical regime by Maxwell's equations, which allow us to calculate, say the electric field resulting from a particular configuration of charges, or the wave equation describing the propagation of electric and magnetic fields through space. One interesting feature of these equations is that they are "asymmetric" owing to the absence of a fundamental quantum of magnetic charge. It is possible to conceive of a source of a magnetic field which would give rise to an elementary "magneto-static" force. Such a magnetic charge would appear as a single magnetic pole, in contrast to all examples of terrestrial magnets which consist invariably of north-pole-south-pole pairs. These conventional magnets are "magnetic dipoles" which are the result of the motions of the atomic electrons. The possibility of the existence of truly fundamental "magnetic monopoles" has been a popular topic of research in recent decades following their emergence from the most modern theories. We have already seen how we can formulate the quantum theory of electrodynamics by describing the interactions of charged particles via the electromagnetic field as the exchange of the quanta of the field, the photons, between the particles involved. QED is the paradigm quantum theory towards which our descriptions of the other forces all aspire.

5.22 The strong nuclear force

When the neutron was discovered by James Chadwick in 1932, it became obvious that a new force of nature must exist to bind together the neutron and protons (referred to generically as nucleons) within the nucleus. Prior to this discovery physicists seriously entertained the idea that the nucleus might have consisted of protons and electrons bound together by the electromagnetic force. Several features of the new force are readily apparent.

Of positively charged protons and neutral neutrons confined within a very small volume (typically of diameter 10-15 m), the strong force must be very strongly attractive to overcome the intense mutual electro-static repulsions felt by the protons.

The binding energy of the strong force between two protons is measured in millions of electron-volts, MeV, as opposed to typical atomic binding energies which are measured in electron-volts.

The second fact concerning the strong nuclear force is that it is of extremely short range. We know this because Rutherford's early scattering experiments of particles by atomic nuclei could be described by the electromagnetic force alone. Only at higher energies, when the particles are able to approach the nuclei more closely, are any effects of the strong force found. In fact, the force may be thought of as acting between two protons only when they are actually touching, implying a range of the strong force similar to that of a nuclear diameter of about 10-15 m.

Finally, the last fact we shall mention is that the strong nuclear force is independent of electric charge in that it binds both protons and neutrons in a similar fashion within the nucleus. One consequence of the solely microscopic nature of the strong force is that we should expect it to be a uniquely quantum phenomenon. We can expect no accurate interpretation in terms of classical physics but only in the probabilistic laws of quantum theory.

One of the prime sources of early information on the strong force was the phenomenon of radioactivity and the question of nuclear stability. This involves the explanation of the neutron/ proton ratios of the stable or nearly stable nuclei. These can be displayed as a band of stability on the plane defined by the neutron number, N, and the proton number, Z, of the nucleus.

The fact that it is predominantly the heavier nuclei which decay confirms our picture of the short-range nature of the strong force. If we naively think of the nucleus as a bag full of touching spheres, then if the force due to any one nucleon source were able to act on all other nucleons to enjoy proportionately stronger binding and thus greater stability. (It is assumed that by adding the nth nuclear bonding would occur and so a binding energy would increase with n, but this is observation is not to be the case. It is the heavier nuclei which suffer radioactive decay, indicating an insufficient binding together of the nucleons. This is because the nuclear force acts only between touching, or "nearest-neighbor" nucleons. The addition of any extra binding energy, whereas the electric repulsion of the protons is a long-range force, does grow with the number of protons present.

The question of nuclear stability may be described in part by the balance of the repulsive electrical forces and the attractive strong forces affecting the nucleons. It is possible to calculate the sum the these two forces for each nucleus and so to calculate the average binding energy indicating that the more strongly bound are the nucleon in each case, the more negative the binding energy indicating that the more strongly bound are the nucleons within the nuclei.

The relatively small negative binding energy of the light atoms results from them not having enough nucleons to saturate fully the nearest- neighbor strong interactions available.

The most strongly bound nuclei are those in the mid-mass range, like iron, which more efficiently use the strong force without incurring undue electric repulsion. The heavier nuclei suffer because the electric repulsion grows by an amount proportional to the number of protons present.

As nature attempts to accommodate heavier and heavier nuclei, a point is reached where it becomes energetically more favorable for the large nuclei to split into two more-tightly bound, mid-mass-range nuclei. This gives rise to an upper limit on the weights of atoms found in nature, occupied by uranium-238 with 92 protons and 146 neutrons. By bombarding uranium with neutrons, it is possible to exceed nature's stability limit causing the uranium + neutron nucleus to split into two. This is nuclear fission.

Radioactive " decay" occurs when an element is not big enough to split into two, but would still like to shed some weight to move up the binding energy curve to a region of greater stability. The particle (which is a helium nucleus consisting of two protons and two neutrons) will have existed as a "nucleus within a nucleus" prior to the decay. By borrowing energy for a short time according to Heisenberg's uncertainty principle, it will be able to travel beyond the range of the strong attractive forces of the remaining nucleons to a region where is subject only to the electrical repulsion due to the protons. Thus the nucleus is seen to expel an particle.

Because the energy is borrowed according to the probabilistic laws of quantum theory, it is not possible to specify a particular time for decay, but only to specify the time by which there will be a, say, 50% probability of a given nucleus having undergone the decay (corresponding to the average time needed for 50% of a sample to decay). This is called the half- life of the element, denoted by 1/2.

Another feature of nuclear stability can be explained by the action quantum principle. Although we have explained why too many protons in a massive nucleus may cause it to break up, we have not explained why this cannot be countered by simply adding an arbitrary number of neutrons to gain extra attractive strong forces. The reason is due to Pauli's exclusion principle. Because both protons and neutrons are fermions, no two protons and no two neutrons can occupy the same quantum state. We cannot simply add an arbitrary number of neutrons to dilute the repulsive effects of the electric charges on the protons, as the exclusion principle forces the neutrons to stack up in increasingly energetic configurations leading to a reduction in the negative binding energy per nucleon and so to decreased stability.

Although we have now reviewed some facts about the strong force (its short range, charge independence and spin dependence via the exclusion principle, etc.), we have done nothing to explain the mechanism of its action apart from noting that only a quantum picture will be suitable for such a microscopic phenomenon. Yukawa's formulation of his meson theory of the strong force is the point of departure into particle physics proper from the inferences of nuclear physics.

5.23 The weak nuclear force

One of the most obvious features of the neutron is that it decays spontaneously into a proton and an electron with a half-life of about 10 minutes. This period is much longer than any of the phenomena associated with the strong force, and it is difficult to imagine how the electromagnetic force could be responsible for this comparative stability. So, we are led to the conclusion that neutron decay is due to some qualitatively new force of nature.

It is this "weak" force that causes neutron decay in which lies behind the phenomenon of the radioactive decay of nuclei. The decay of the neutron into a proton allows a nucleus to relieve a crucial neutron surplus which, because of the action of the Pauli exclusion principle, may be incurring a substantial energy penalty and eroding the binding energy of the nucleus. The same interaction may also allow the reverse reaction to occur in which a nuclear proton transfers into a neutron by absorbing an electron. (This may occur because of the very small but infinite chance that the electron may find itself actually inside the nucleus, according to the positional uncertainty respected by the electron wave function.)

This reaction will allow a proto-rich nucleus suffering from undue electric repulsion to dilute its proton content slightly, thereby strengthening its binding.

One problem soon encountered in attempts to explain radioactive decay is that the electrons which are emitted from decaying nuclei are seen to emerge with a range of energies up to some maximum which is equal the difference in the masses or the initial and final nuclei involved. When the electrons emerge with less than this maximum figure we see and also an accompanying apparent violation of angular momentum conservation), Pauli postulated in 1930 that another, invisible, particle was also emitted during the decay, which carries off the missing energy and angular momentum. As the original reactions do conserve electric charge, then this new particle must be neutral. On the strength of this, Fermi called it neutrino.

Several properties of the neutrino are apparent from the facts of decay. Because of conservation of energy, it is necessary that the neutrino be very light or indeed massless (because some electrons do emerge with the maximum energy allowed by the mass difference). Similarly, because of angular momentum conservation the neutrino must be spin 1/2.

Another interesting feature is that the neutrino interacts with other particles only by the weak force and gravity (because the strong interaction is obviously not present in neutron decay, and because the uncharged neutrino experiences no electromagnetic effects).

The apparent invisibility of the neutrino is due to the very "feebleness" of the weak force. This reluctance to interact allows it to pass through the entire mass of the Earth with only a minimal chance of interaction en route. Because of this, the neutrino was not observed (i.e. collisions attributable to its path were not identified) until large neutrino fluxes emerging from nuclear reactors because available. This was achieved in 1956 by Reines, some 26 years after Pauli's proposal. The weak force, like its strong counterpart, acts over microscopic distances only. In fact, to all intents and purposes it makes itself felt only when particles come together at a point (i.e. below any resolving power available to physics, say less than 10-18 m).

5.24 Interacting fields

Our ultimate objective must be to predict the values of physical quantities which can be measured in the laboratory, such as particle reaction cross- sections, particles lifetimes, energy levels in bound systems, etc. We hope to achieve this by using the idea of quantum fields to tell us the "probabilities" of the creation and destruction of their quanta in various reactions, and to provide us with descriptions of the behavior of the quanta between creation and destruction of the wavefunctions. This will then allow us to calculate the probabilities associated with physical processes.

Now the probabilities follow somehow from the dynamics of any system, whether it is governed by Newtonian mechanics, quantum mechanics or quantum field theory, can be derived from a single quantity describing the system, called its Lagrangian. The Lagrangian (L) for any system is the difference between its kinetic energy (KE) and its potential energy,

(PE), L = KE – PE.

For a classical particle, say a cube ball, moving through the gravitational field of the Earth, the potential energy is due to its height x above the Earth (PE = mgx), and its kinetic energy is due to its velocity (KE = 1/2 mv2).

In quantum mechanics (or QFT), we are dealing with wave functions (or fields) in which extend throughout space-time. Here, we do not deal with the total Lagrangian L directly, but with the Lagrangian density L. The total Lagrangian can then be found by integrating the Lagrangian density over all space.

It is straightforward to write down the expression for the Lagrangian density of a "free electron" in terms of the electron wave function (or field). For both the cricket ball and the free electron, it is a trivial exercise to go from the Lagrangian to the equations of motion to go from the Lagrangian to the equations of motion to go from the Lagrangian to the equations of motion (F = ma for the cricket ball and the Dirac equation for the electron). But in the case of elementary particles in interaction we do not know in general the equations of motion and, where we do, we cannot solve them. We cannot therefore proceed immediately to calculate the quantities of physical interactions resulting from the motions of particles so a more subtle approach is required.

5.25 Weak Interaction

Next in the hierarchy of interactions is the weak interaction, which is responsible for the emission of electrons (beta rays) in radioactivity and for various other transformations that involve neutrinos. As its name implies, it is weak (relative to the electromagnetic and strong interactions), although it is much stronger than gravity. The W and Z particles have more than eighty times the mass of a proton (the original "heavy" particle). When Enrico Fermi developed the first theory of beta decay in 1943, he imagined direct weak interaction among a quarter of particles: the proton, neutron, electron, and neutrino. For many years thereafter, physicists speculated that one or more exchange particles or force carriers, as we would say now) might be involved in the process, living their brief existence between the time a neutron was a neutron and the moment it vanished to become a proton, electron, and antineutrino.

Not until 1983, however, did physicists discover the W and Z particles, with the aid of a large proton synchrotron at CERN in Geneva. they are really three close sibling, with positive, negative, and zero charge, much like the triad of positive, negative, and neutral pions, which were themselves once believed to be the force carriers of the strong nuclear force.

But there are a couple of big differences between the W-Z triad and the pion triad. The W and Z particles are, we believe, fundamental, lacking any physical size and not made of anything smaller. Moreover, the W and Z particles are enormously massive relative to pions.

5.26 Electromagnetic Interaction

The photon, has had an interesting history, from its "invention" by Albert Einstein in 1905 through its somewhat shadowy existence in the 1920s as a "corpuscle," not quite a real particle; on through its central role in the 1930s and 1940s, when physicists linked it with the electron and positron to build the powerful theory called quantum electrodynamics; up to the present day, when we see it as a fundamental particle of truly zero mass and truly zero size serving as the force carrier for electromagnetism. Indeed, you literally "see" protons nearly every waking moment, day and night. And there are lots of photons that you don't see –those that carry radio and television and wireless-phone signals, heat from warm walls, and X-rays through your body.

The Universe is filled with low-energy photons, the so-called "cosmic background radiation," left over from the Big Bang. All in all, there are about a billion photons in the Universe for every material particle.

It's a wonderful unification of two of the four kinds of force – weak and electromagnetic. Yukawa realized long ago that the more massive an exchange particle, the less its reach, the less the "range' of the force. If you imagine yourself as a god who can change the mass of an exchange particle to suit your whim, you find that as you make that particle more and more massive, leaving everything else unchanged, the force gets weaker and weaker (in addition to having shorter and shorter range). In the 1970s, three eminent theoretical physicists – Abdus Salam, Steven Weinberg, and Shelden Glashow – boldly advanced the idea that the weak and electromagnetic interactions are different faces of a single underlying interaction.

In effect, they said that the essential difference between these two kinds of interaction is only the difference in the nature of the force carriers. The electromagnetic force is of long range (we can sense it over meters or even miles) and is relatively strong because the carrier of this force is a particle with no mass, the photon. The weak interaction is of short range (reaching a distance smaller that the diameter of a proton) and is relatively weak, so it requires vary massive force carriers. Not many years later, the discovery of the W and Z particles provided the needed confirmation of this "electroweak" theory. In some sense, then, the weak and electromagnetic interactions are one and the same force. However, the weak interaction is universal, affecting all particles of all kinds, while the electromagnetic interaction affects only particles with electric charge.

5.27 Fermi's theory of the weak interactions

Prior to the early 1960s, just three different leptons were recognized: the electron, the muon and the neutrino (together with their antiparticles). The best place to study the weak interaction is in the processes involving these leptons only. This ensures that there are no unwanted strong interaction effects spoiling the picture. Unfortunately early opportunities to study purely leptonic reactions were limited, being restricted to the muon decay into an electron and neutrino. The most common ones decay into an electron and neutrino.

The most common weak interactions available for study are the radioactive beta-particle of nuclei and the decay of the pions and kaons (which are described generically as the weak decay of hadrons), and it was predominantly these reactions which formed the basis of the first description of the weak interactions formulated by Fermi in 1933.

5.28 Strong Interaction

The last entry in the table of particles is a set of eight particles (sixteen if you count the antiparticles). These are the aptly named gluons, which provide the strong-interaction "glue." Although not electrically charged, they carry odd mixtures of color charge, such as red-anti-green or blue- anti-red. There are eight independent combinations of color-anti-color that define the eight gluons. Every time a quark interacts with a gluon, the quark's color changes.

Unlike photons, which can interact with one another only indirectly via charged particles, gluons exert a direct force on one another, in addition to exerting forces on quarks.

5.29 Symmetry in the microworld

In the everyday world, symmetries in both space and time have a universal fascination for the human observer. In nature, the symmetry exhibited in a snowflake crystal or on a butterfly's wings might be taken to indicate some divine guiding hand. In the world of physics, and especially in the micro-world, symmetries are linked closely to the actual dynamics of the systems under study. Symmetries are the most fundamental explanation for the way things behave (the laws of physics).

This is of course, the case that physicists notice natural phenomena and write down equations of motion to describe them (notably Newton, Einstein and Dirac, to name a few).

But in describing the micro-world it is generally far too difficult to write down the equations of motion straight away; the forces are unfamiliar and our experiments provide only ground-floor windows into the skyscraper of the high-energy domain.

So we are forced to consider first the symmetries governing the phenomena under study, generally indicated by the action of conservation rules of one sort or another (e.g. energy, momentum and electric charge). The symmetries may then guide our investigation of the nature of the forces to which they give rise. Symmetry is described by a branch of mathematics called

"Group theory." A group is simply a set of symmetry transformations, changes under which a system stays the same. The action of these transformations on a specific set of objects is called a representation.

The notion is made non-trivial by the demand that repeated transformations are equivalent to another transformation. When a physical system has such a symmetry, the Lagrangian governing the system does not change under the group transformations.

This then implies the existence of a conserved quantity. This connection is due to a remarkable mathematical theorem by Emmy Noether which states that, for every continuous symmetry of a "Lagrangian," there is a quantity which is conserved by its dynamics.

We can proceed to put some flesh on this theoretical skeleton by giving examples of four kinds of symmetry used extensively in particle physics: continuous space-time symmetries, discrete symmetries, dynamical symmetries, and internal symmetries. After a look at each of these, we will mention in closing how even broken symmetries can provide a useful guide to the formulation of physical laws.

5.30 Charge conjugation symmetry - Broken symmetries

Symmetries are so sufficiently valuable that even broken ones can be useful. For many purposes a broken mirror is as good as a whole one. We have mentioned already how the individual reflection symmetries P, C and T might be broken in some classes of reaction (which turn out to be those governed by the weak nuclear force). But for other forces which do respect them, they are still a valuable guide for indicating which reactions are possible and which are forbidden.

5.31 Particles – The fundamental constituents of nature

At this point it is worth both introducing some of the generic names which are used for these particles and defining their essential features. A few of the most often used words are:

- Nucleons: neutrons and protons.
- Hadrons: all particles affected by the strong nuclear force.
- Baryons: hadrons which are fermions (half-integral spin particles) such as the
- Mesons: hadrons which are bosons (integral spin particles) such as the pion.
- Leptons: all particles not affected by the strong nuclear force, such as the electron and the muon.Particles which are baryons are assigned a baryon number B which takes the value B = 1 for the nucleons, and leptons. In all particle reactions, the total baryon number is found to be conserved (i.e. the total to the baryon numbers of all the ingoing particles must equal that of all outgoing particles). Similarly, particles which are leptons are assigned a lepton number which is also conserved in particle reactions. The Meson Modern particle physics can be thought of as starting with the advent of mesons. For these are not constituents of everyday matter, as are the protons and the electrons, but were first proposed to provide a description of nuclear forces.

The subsequent discoveries of a bewildering variety of mesons heralded an unexpected richness in the structure of matter, which took many decades to understand.

Yukawa's proposal

In attempting to describe the features of the strong nuclear force, physicists in the 1930s had to satisfy two basic requirements. Firstly, as the force acts in the same way on both protons and neutrons, it must be independent of electric charges and, secondly, as the force is felt only within the atomic nucleus, it must be of very short range.

In 1935 the Japanese physicist H. Yukawa suggested that the nuclear force between protons is mediated by a massive particle, now called the - meson or pion, denoted by, in contrast to the massless photon which mediates the range of the electro-magnetic force. It is the mass of the mediating particle which ensures that the force it carries extends over only a finite range. This is indicated by Heisenberg's uncertainty principle which allows the violation of energy conservation for a brief period.

If the proton emits a "pion" of finite mass, then energy conservation is violated by the amount equal to this mass energy. The time for which this situation can obtain places an upper limit on the distance which the pion can travel, and this distance is a guide to the maximum effective range of the force. From the alpha-particle scattering experiments, we know that the effective range of the strong force is about 10-15 m, which gives a pion mass of about 300 times that of the electron, or about 150 MeV. To account for all the possible interactions between nucleons, the pions must come in three charge states. For instance, the proton may transform into a neutron by the emission of a positively charged pion or, equivalently, by the absorption of a negatively charged pion. But the proton may also remain unchanged during a nuclear reaction, which can be explained only by the existence of an unchanged pion. So the pion must exist in three charge states: positive, neutral and negative.

5.32 The Quantum World

The Photon – the particle of light – was also known in 1926, but it wasn't considered a "real" particle. It had no mass; it couldn't be slowed down; and it was all too easily created and annihilated (emitted and absorbed). It wasn't a reliable, stable chunk of matter like the electron or the proton. So, even though the photon surely behaved like a particle in some respects, physicists fudged by calling it a "corpuscle" of light.

Only a few years later it gained full and equal status as a real particle, when physicists realized that electrons could be created and annihilated every bit as easily as photons, that the wave properties of electrons and the wave properties of photons were similar, and that a particle with no mass, such as a photon, was a perfectly ordinary particle.

In 1976, the known subatomic particles numbered in the hundreds. A few were added in the 1930s and the 1940s, and then a flood of particles were discovered in the 1950s and 1960s. Physicists had stopped calling the particles "elementary" or "fundamental" because there were just too many of them for that. Yet just as the number of particles seemed to be getting out of hand, physicists were coming up with a simplifying scheme. A manageably small number of particles appeared to be truly fundamental (including quarks, which, to this day, no one has seen directly). Most of the known particles, including the old familiar proto, were composite – that is, built from combinations of the fundamental particles. We can see an analogy here to what had happened decades earlier with our understanding of atoms and nuclei. By the time the neutron (an uncharged, or neutral, sibling of the proton) was discovered in 1932, the number of known atomic nuclei had grown to several hundred. Each was characterized by a mass and a positive charge. The charge determined the atomic number, or place in the periodic table. In other words, it defined the element (an element is a substance with unique chemical properties). The hydrogen nucleus had a charge of one unit, the helium nucleus a charge of two units, the oxygen nucleus a charge of eight units, the uranium nucleus a charge of ninety-two units, and so on. Some nuclei with the same charge (therefore nuclei of the same element had different masses.

Atoms built around these nuclei were called isotopes. Scientists felt sure that these several hundred nuclear types, of ninety or so elements averaging two or three isotopes each, were built of a smaller number of more fundamental constituents; but prior to the discovery of the neutron, they couldn't be sure exactly what those constituents were. The neutron made it all clear (though later it, too, was discovered to be composite). Nuclei were constructed of just two particles, the proton and the neutron. The protons provided the charge, and the protons and neutrons together provided the mass. Whizzing around the nucleus in the much larger volume of the whole atom were electrons. So, just three basic particles accounted for the structure of hundreds of distinct atoms.

For the subatomic particles, the "discovery" of quarks played a role similar to the discovery of the neutron for atoms. I put "discovery" in quotation marks because what Murray Gell-Mann and George Zweig, both at Caltech, did independently in 1964 was to postulate the existence of quarks, not prove their existence through observation (the name "quark" we owe to Gell-Mann).

The evidence for quarks, though still indirect, is by now overwhelming. Today, quarks are recognized as the constituents of protons, neutrons, and a whole host of other particles.

Physicists then invented the standard model of the subatomic particles. In this model there are twenty-four fundamental particles, including the electron, the photon, and half a dozen quarks, accounting for all observed particles and their interactions. Twenty-four is not as pleasingly small a number as three (the number of particles known in 1926), but so far these twenty-four remain stubbornly "fundamental." No one has found any of them to be made of other, more fundamental entities. But if the superstring theories have it right, there may be smaller, simpler structures that await discovery. Some of the fundamental particles are called leptons, some are called quarks, and some are called force carriers.

Leptons

There are six leptons, two each of three flavors. The word "lepton" comes from a Greek word meaning "small" or "light"; reasonable for the electron and the neutrinos. Discoveries, however, have a way of overtaking nomenclature; the tau lepton is anything but light. Some of the leptons are unstable. (Instability is the same thing as radioactivity. It means that after living a short – and not exactly predictable – time, the particle undergoes a sudden transformation, or decay, into other particles.) The leptons are all weakly interacting. This sets them apart from the strongly interacting quarks (which you'll meet in the next chapter). Like quarks, leptons have no known constituents (thus, we call them fundamental) and no known size. In all experiments to date (and in successful theories describing them), the leptons act as point particles.

The Electron

The electron is not only the best-known lepton – it has the distinction of being the first fundamental particle ever discovered. Before leaving the electron, I want to mention its antiparticle, the positron. In 1928 the English physicist Paul Dirac, as taciturn as he was brilliant, wrote down an equation (we now call it, not surprisingly, the Dirac equation) that brought together the principles of both relativity and quantum mechanics in an effort to describe the electron. To Dirac's own astonishment, not to mention the astonishment of his colleagues around the world, this equation had two startling things to say.

First, it "predicted" that the electron should have spin one-half.

Of course, the electron was already known to have spin one-half, but no one knew why, or even knew that this property of the electron would flow from a mathematical theory.

Secondly, as mentioned earlier, Dirac's equation implied the existence of antimatter. It predicted that the electron should have a companion particle – an anti-electron – with the same mass and spin as the electron but with opposite electric charge. According to Dirac's theory, when a "positron" and an electron met, there would be a mini-explosion. Leaving only a pair of photons, created in the encounter.

This prediction was a bit hard for physicists to swallow at the time, for no such lightweight positive particle had ever been observed, nor had an "annihilation" event been seen. Dirac himself briefly lost faith in his own equation's prediction. He toyed with the idea that in some way the proton might be the electron's antiparticle, but soon realized that this was hardly possible. Moreover, the idea of an antiparticle with a mass different from its counterpart's was quite "inelegant." A dozen years later, gamma rays were identified as electromagnetic radiation. They were eventually recognized as photons. In responding to Niels Bohr's comment that Dirac rarely spoke, Ernest Rutherford told Bohr the story of a disappointed pet-store customer who returned a parrot because it didn't talk. "Oh," said the store owner, "I'm sorry. You wanted a parrot who talks. I gave you the parrot who thinks."

Dirac, like Einstein before him, believed in equations that passed the tests of simplicity, generality, and "beauty." Is this faith-based physics? Yes, in a way it is faith in what sits firmly on a rock of prior knowledge – a faith that has successfully driven major advances in physics, going back to Kepler, Galileo, and Newton. Dirac argued, in effect, that his theory was too perfect to be wrong, that it was now up to the experimenters to prove it was right. And this is just what happened. In 1932 Carl Anderson at Caltech found the telltale track of an anti-electron – now commonly called a positron – in a cloud chamber exposed to cosmic rays.

Before long, more confirmation was provided by Frederic and Irene Joliot-Curie in France.

They created new radioactive elements, some of which decayed with the emission of positrons rather than electrons. Dirac, Anderson, and the Joliot-Curies were all awarded Nobel prizes in the 1930s (Frederic met Irene in 1925, when she worked as an assistant to her mother, Marie Curie, herself a Nobelist. When Frederic and Irene were married in 1926, they adopted the hyphenated name.) We now know that every particle has an antiparticle.

For a few neutral particles, the particles and antiparticle are the same. The photon, for example, is its own antiparticle. But most particles have a companion antiparticle that is distinct. All six of the leptons have distinct antiparticles.

The Muon

In the early 1930s, peace prevailed in the world of elementary particles. Matter was constructed of protons, neutrons, and electrons. Processes of change – interactions – involved photons and the still hypothetical but quite credible neutrinos. It contained only five particles. This peace was not to last. In Europe and the United States, experimenters studying cosmic radiation were finding evidence for charged particles about two hundred times more massive than electrons (nine times less massive than protons and neutrons). At first, this seemed like good news.

The Japanese theorist Hideki Yukawa, not long before, had postulated the existence of new strongly interacting particles of just about that mass whose exchange among neutrons and protons within the nuclear, he argued, could give rise to the strong nuclear force. A wonderful accord between theory and experiment or so it seemed. But experiment, the final arbiter, said no. Not long after 1945, when physicists in many countries returned from war work and took up pure research again, cosmic-ray experiments showed that the Yukawa particles, though indeed real, were not the same as most of the cosmic-ray particles that had been leaving tracks in cloud chambers and photographic emulsions.

Yukawa's particles came to be called pions. The particles more abundant in cosmic radiation were (eventually) named muons. Leptons are the fundamental spin-one-half particles that experience no strong interactions (and contain no quarks).

Baryons are composite, strongly interacting particles made of quarks and having spin one-half (or possibly 3/2 or 5/2). The word "baryon" comes from a Greek word meaning "large" or "heavy." The proton and the neutron and many heavier particles are baryons. But there are some heavy particles that are not baryons.

5.33 The muon puzzle

In 1937, five years after his discovery of the positron, Anderson observed in his cloud chamber yet another new particle originating from cosmic rays. The particle was found to exist in both positive-charge and negative- charge states with a mass some 200 times that of the electron, about 106 MeV. At first, the particle was thought to be Yukawa's pion and only gradually was this proved not to be the case. Most importantly, the new 'mesons' seemed very reluctant to interact with atomic nuclei, as indicated by the fact that they are able to penetrate the Earth's atmosphere to reach the cloud chamber at ground level. For particles which were expected to be carrying the strong nuclear force such behavior was unlikely. Also, there was no sign of the neutral meson.

Theorists eventually accepted that this new particle was not the pion; instead it was named the muon, and is denoted by (). The muon was a baffling discovery as it seemed to have no purpose in the scheme of things. It behaves exactly like a heavy electron and it decays into an electron in 2 x 10-6 s; and so is not found in ordinary matter. Although we shall see later how the muon can fit into a second generation of heavy elementary particles, the reason for this repetition is still by no means obvious. Thus, the muon is not a meson at all, but a lepton like the electron.

Mesons, like baryons, are composite, strongly interacting particles made of quarks. But instead of half-odd-integer spin (1/2, 3/2, and so on), they have spin 0, 1, or some other integer value. "Meson" comes from a Greek word meaning "middle" or "intermediate." The word (or an alternate choice, "mesotron") was applied originally to the particle intermediate in mass between the electron and the proton. What we now call the muon was originally called the mu meson. What we now call the pion (the Yukawa particle) was originally called the pi meson. Now we know that the muon and pion have almost nothing in common. The pion is a meson; the muon is not. The pion is, if fact, the least massive meson. Many others are heavier – some much heavier, even, then the proton.

Quarks are the fundamental, strongly interacting particles that never appear singly. They are constituents of baryons and mesons and are themselves "baryonic," meaning that they possess a property called baryonic charge. When a quark and an antiquark unite to form a meson, the baryonic charge is zero.

When three quarks unite to form a baryon, the baryonic charge is one. Force carriers are particles whose creation, annihilation, and exchange give rise to forces. We believe that force carriers, like the leptons and quarks, are fundamental particles with no substructure.

There are other names that I'll mention here only in passing. Hadrons include the baryons and mesons that interact strongly; nucleons encompass the neutrons and protons that reside in nuclei; fermions "(after Enrico Fermi) are particles like leptons and quarks and nucleons that half- integer spin; and bosons (after the Indian physicist Satyendra Nath Bose) are particles like the force carries and mesons that have integer spin.

By the late 1940s, experimenters had established that this component of cosmic radiation had no appreciable strong interaction, for it penetrated matter too readily. The muon behaved in every way like an electron, except that it was two hundred times heavier. This was a major mystery at the time, for large mass seemed to be associated only with strong interactions. Particles that did not interact strongly – electrons, neutrinos, and photons – were lightweight. "Who ordered that?" the eminent Columbia University physicist IU. I. Rabi reportedly asked.

At Caltech, the brilliant theorist Richard Feynman is said to have kept a curiously phrased question on his blackboard: "Why does the muon weigh?" For every measurement that was made only confirmed that the muon and electron were, except for mass, seemingly identical.

So whatever the reason might be for the muon's existence, physicists had to accept it as a corpulent close cousin of the electron. The fact that the muon lives a mere two millionths of a second on average and the electron, if undisturbed, lives forever is not, by the way, a significant difference. Every particle would like to decay if it could. The conservation of charge, along with the conservation of energy, presents the electron from decaying, for there are no lighter charged particles into which it can decay.

The muon is not so constrained and eventually (two millionths of a second is a very long time in the subatomic realm) it does decay. The particular way in which it decays tells us that in fact it differs from the electron in more than mass. If mass were the only difference, we would find that the muon sometimes decays into an electron and a photon (or -ray), as represented by the formula :

Mu e +

This decay, if it occurred, would preserve lepton number (one lepton before the decay, and one after). But no such decay has been seen. Evidently the muon and electron differ in some characteristic which prevents one of them from turning into the other. That characteristic is what came to be called "flavor." And flavor, like electric charge and baryonic charge, is conserver: it's the same after an interaction as it was before. If you look at the muon-decay event you'll see that before the decay occurs, there is one particle present with the muon's flavor – namely, in this example, the negative muon itself. After the decay, one particle with the muon's flavor, a muon neutrino, comes into being. So the muon flavor is preserved. What about electron flavor? The particles created in the decay include an electron and an anti-neutrino of the electron type.

So, as physicists reckon these numbers, there an electron "flavor numbers" of +I (for the electron) and –I (for the anti-neutrino) after the decay: a total of zero, the same as before the decay. To be sure, assigning a numerical value to "flavor" seems a bit far-fetched, but it works. Each of the three lepton flavor is conserved.

The Muon Neutrino

Among the experiments that established the reality of the muon in the late 1940s was then carried out by Cecil Powell and his collaborators at the University if Bristol in Great Britain.

Using a special photographic emulsion as a detector, they observed the track of what we now call a emulsion as a detector, they observed the track of what we now call a pion, followed by the track of a muon, followed in turn by the track of an electron. It appeared that the pion had decayed into a muon and one or more unseen neutral particles, and that the muon had then decayed into an unseen particles were neutrinos, but the question was: One kind of neutrino or two?

Physicists had reason to believe that the muon, like the electron, would have an associated neutrino, but they could not at first tell whether the muon's neutrino was the same as the electron's or different.

Since the muon had some property other than mass that distinguished it from the electron (the property we now call flavor), it was rather natural to suppose that the muon world have its own neutrino, distinct from the electron's neutrino. But physicists like to assume that the simplest explanation is probably the correct one. Why invent a new particle if you don't need it? Perhaps the electron's neutrino could do double duty and pair up with the muon as well as with the electron. Only experiments could decide.

The critical experiment was carried out in 1962 (half a dozen years after Reines and Cowan detected the electron's antineutrino) by a Columbia University group headed by Leon Lederman, Melvin Schwartz, and Jack Steinberger. Working at the then-preeminent 33-GeV accelerator at Brookhaven Lab and Long Island, they duplicated in the laboratory what cosmic rays do high in the atmosphere. Protons strike nuclei, creating pions. The pions fly some distance and then decay into muons and neutrinos. Given enough time and distance, the muons also decay, but these experimenters were interested in the neutrinos produced by the decaying pions. The positive pion decays into a positive muon, which is an antimuon, and a neutrino. The negative pion decays into a negative muon and an antineutrino.

The Tau

In the family of the leptons we have the electron, the muon, their neutrinos, and the anti-particles of these four leptons. That should wrap it up. Especially since, at the time, there was evidence for three quarks and theoretical reasons to believe in a fourth. (The fourth quark was discovered in 1974.) In short, physicists knew of four leptons and suspected that there were four quarks.

At the Stanford Linear Accelerator Center in California, Martin Perl decided to explore unknown territory, looking for a charged lepton more massive – possibly much more massive – than the muon.

Perl was an activist in both politics and science. In the 1960s, he joined protests against the Vietnam War and worked for social justice at home. In the early 1970a, he led a successful struggle to create within the staid American Physical Society a unit dedicated to issues of science and society. At Stanford, some of Perl's colleagues considered his determination to look for a massive lepton unwise. He was, after all, guided only by curiosity and hope, not by theory. No theorist could tell him he would probably succeed. His experimental colleagues told him he would probably fail.

Indeed, Perl's search, culminating in the discovery of the tau, took many years. He started looking in the late 1960s and announced the first evidence for this new lepton in 1975. The first evidence was a bit shaky. Not until 1978 after Perl and others had completed additional experiments, did the physics community fully accept the reality of the tau. (By that time, two more quarks had been found, bringing their total to five.) Perl's 1995 Nobel prize vindicated his determination and persistence

Here's how Perl found the tau. He worked with colliding beams of electrons and positrons (antielectrons) at a "storage ring" attached to the Stanford Linear Accelerator. In this device, electrons and positrons are collected and "stored" in a large doughnut-shaped enclosure, guided in opposite directions around their underground racetrack by magnetic fields. Each time an electron collided with a positron in Perl's experiment, a total energy of up to 5 GeV was available, some of which could be transformed into the mass of new particles. We are accustomed to thinking that when a pair of particles annihilate, their mass energy (or "rest energy") is released.

But in the Stanford Linear Accelerator, the kinetic energies of the particles vastly exceed their mass energies (which amount to only 1 MeV). When particles annihilate not just their mass energy goes but also their kinetic energy can go into making new mass.

The Tau Neutrino

The summer of 2000 brought evidence for the last lepton – believed to be the last fundamental particle that hadn't been seen before. Researchers at Fermi lab in Illinois (at that time the world's most powerful accelerator) fired a beam of 800-GeV protons at a tungsten target, and then did all they could to get rid of the emerging debris, except for neutrinos.

Magnets deflected charged particles out of the way. Thick absorbers captured most neutral particles. What got through the obstacle course to reach the photographic-emulsion detectors included lots of neutrinos of all kinds. A few of the tau neutrinos (about one in a million million, according to the estimates of the researches) interacted in the emulsion to create tau leptons, which, because of their charge, left tiny tracks in the emulsion, each about a millimeter long.

These tracks, along with additional tracks of particles created by the decaying taus, established the reality of the tau neutrinos.

By the time of this sighting of the tau neutrino, no physicist had any doubt of its existence. Yet all breathed a sigh of relief when their well- founded faith in the nature of the lepton family was rewarded with solid evidence. Leptons clearly have three flavors and we have a mature theory of their behavior, covering beta radioactivity and a whole host of other observed phenomena.

Neutrino Mass

When Pauli first suggested the neutrino in 1930 the "neutron," as he called it then), he said that it ought to have about the same mass as an electron and in any case not more than 1 percent of the mass of a proton. The idea that the neutrino might have no mass at all came a little later, as the result of beta-decay experiments. in a particular beta-decay event, there is a certain definite quantity of energy carried away by the electron and neutrino combined.

The electron (the observed particle) takes some of this energy; the neutrino (the unobserved particle) takes the rest. The neutrino, if it has mass, has to take away at least its own rest energy, mc2. Even if it dribbled out of the radioactive nucleus with no appreciable speed, it would still carry away that much energy. So the electron can take away at most the total available energy minus the neutrino's mc2.

Physicists in the 1930s, completing successively more precise experiments, found that in beta decay the most energetic electrons shooting from the nucleus were carrying away just about all of the available energy. The mass of the neutrino, if any, had to be very small. (by now, the upper limit on the electron neutrino's mass has been pushed down to less than one hundred-thousandth the mass of an electron.)

The possibility of mass exactly zero was consistent with Fermi's theory and was b then a perfectly ordinary idea, given that the massless photon was already a familiar friend. Theorists, in fact, soon developed a version of Fermi's theory that required neutrinos to be massless. So for some decades it was generally assumed that neutrinos were massless particles. A massless neutrino is an appealing idea, as simple as can be. We now have good evidence that neutrinos have some small mass.

One way to describe this state of affairs is to say that at a certain level of approximation – often a very good approximation – nature reveals to us laws of striking simplicity, yet shows us laws of greater complexity when we look deeper. To give one other example: according to the theories of John Wheeler and other researchers, when we look at tiny enough regions of space over short enough spans of time, the bland smoothness of space and time in our ordinary world gives way to quantum foam.

Much of the ordinary physical world around us is not simple. You need look no further than your local weather forecast to know that scientists are far from describing our physical environment in simple terms. Ripples on water, leaves trembling on trees, smoke rising from a campfire – all defy and simple description.

So, roughly speaking, we have three layers of complexity. There is, in the top layer – the visible layer – great complexity (rippling water, trembling leaves, the weather). Underlying this complexity is a layer of startling simplicity uncovered by scientists over the past few centuries (Newton's gravity, Maxwell's electromagnetism, Dirac's quantum electron). In the deepest layer, complexity rears its head again. Tiny deviations from simplicity appear. But these are not like the complexities of our immediate environment. They reflect what may be a still deeper, equations of general relativity, which can be written concisely in a few lines, have a stunning simplicity "(they explain gravity in terms of space and time only, and show for the first time why gravity imparts equal acceleration to all falling objects, whatever their size or composition). Yet these are the equations that tell us Newton's inverse-square law is not quite right.

Modern quantum theory, with its beautiful and simple basic equations and its few concepts, nevertheless tells us that what we call a vacuum – nothingness – is in fact a lively place in which particles are constantly being annihilated and created. Simplicity is subtle, and so is the beauty that goes with it.

5.36 Back to Neutrino Mass

So, based on what modern physics has taught us about layers of complexity and simplicity, we might say that it would be "nice" if neutrinos were massless, but that we should not be surprised if, based on some deeper – perhaps ultimately simple – theory, they have a small mass after all. Scientists recently concluded that neutrinos do have mass, that the mass is small, and that different neutrinos have different masses. The evidence for neutrino mass first came in the 1990s from measurements in an enormous underground detector in Japan called Super Kamiokande, and was reinforced in 2001 and 2002 by measurements at SNO, the Sudbury Neutrino Observatory (in Canada), also located deep underground.

A quantum system can exist in two "states" at once. It's further possible that a "muon neutrino" may not be a "pure" particle at all and may not have a single identifiable mass. It may be a mixture of two other particles, each of definite mass. The "tau neutrino," in turn, may also be a mixture, but a different mixture, of the same two other particles.

If the two other particles had the same mass, all of this speculation about mixing would be just so much mathematical manipulation with no observable consequences. But if the two other particles have different mass, the wave nature of particles enters the picture.

The quantum waves associated with the two mixed particles oscillate at different frequency (because frequency is related to energy, and energy, in turn, is related to mass). This makes the muon neutrino turn gradually (over a flight distance of perhaps hundreds of miles) into a tau neutrino (or perhaps an electron neutrino), then back to a muon neutrino, and so on – a phenomenon called "neutrino oscillation."

5.37 The three Flavors

No one knows why there are three "flavors" of particles. String theorists think that their specialty might someday provide and answer. Members of this hardy band of theoretical physicists are working on the intriguing but still speculative idea that each fundamental particle consists of a "vibrating" or "oscillating" string of a size unimaginably small compared with the size of a proton or an atomic nucleus.

If the sting theorists are right, gravity will be incorporated into the theory of particles; the number, masses, and other properties of the fundamental particles may be predicted; and leptons and quarks will have some size after all. All of that will provide a revolution in our view of nature in the small.

Surprisingly, physicists feel confident that the third flavor marks the end of the trail – that no more lay ahead. The fact that the total number of neutrinos arriving from the Sun is just three times the number of electron neutrinos that arrive is one argument that three is the limit. Another argument has to do with what in quantum theory are called "virtual processes," which might roughly be called the influence of the invisible on the visible. Between "before" (say, the impending collision of two particles) and "after" (say, the emergence of newly created particles from the collision), quantum theory says that every imaginable thing happens, limited only by certain conservation laws. The intermediate things that happen include the creation and annihilation of particles of all kinds. Even though these intermediate "virtual particles" are not seen, they influence the actual outcome, the "after" of the "before".

If there were one or more additional flavors of particles, they would participate in this virtual dance and their existence could be inferred.

There is no evidence for such participation of particles of additional flavor. Perhaps the most convincing argument against additional flavors comes from the decay of a heavy neutral particle called the Z0 (pronounced "zee-naught"), which is nearly one hundred times more massive than a proton. Enter another quantum effect.

The measured mass of the Z0 has a range of values. Each time the mass of a Z0 is measured, a different value is obtained. When the experimenters put together all of the masses they have measured for the same particle, they see a clustering of mass values about some average, with a spread that is called the "uncertainty of the mass." (The uncertainty is meaningful because it is larger than the possible error in any single measurement. There really is no single mass for the particle.) According to Heisenberg's uncertainty principle, this mass uncertainty is related to a time uncertainty: more of one means less of the other. So if the mass uncertainty is big, the Z0 must live only a very short time. If the mass uncertainty is small, the Z0 must live longer. In fact, the lifetime of the Z0 is so short that there is no way to measure it with a clock. The mass uncertainty serves as a surrogate clock. Just by finding out how the mass measurements are spread out, researchers can determine the particle's lifetime.

Now comes the rest of the argument. Certain modes of Z0 decay – those producing charged particles – can be observed. Among the invisible modes are decay into neutrino-antineutrino pairs. Physicists, knowing the rate of decay into visible (charged) particles and knowing the mean lifetime for all modes of decay combined, can infer the rate of decay into invisible particles. They calculate that the Z0 can decay into exactly three kinds of neutrino-antineutrino pairs (the electron, muon, and tau varieties), not two or four or five or any other number. The only way that there could be more than three flavors is if the neutrinos of the fourth or later flavors were extremely massive – say, many times the mass of a proton – which is hardly expected, given that the masses of the three known neutrinos are far less than the mass of a proton. So the flavors stop at three.

Quarks

Quarks are odd creatures indeed. They are among the limited number of particles we call fundamental, yet we have never seen one by itself – and hardly expect to. We have only a vague idea of the masses of the quarks (since we can't get hold of one to weigh it), and no idea at all why the heaviest of them is tens of thousands of times more massive than the lightest one. Yet there they are, six of them in three groups, an array that matches the array of leptons. Like the leptons, they have one-half unit of spin. And like the leptons, they have no measurable physical extension; so far as we can tell, they exit at points. Otherwise, they differ quite dramatically from the leptons, for they interact strongly (which the leptons do not), and they link up in twosomes or threesomes to make particles, such as pions, protons, and neutrons, which we can see where leptons do not team up to make other particles. Physicists can only learn about quarks by studying protons, neutrons, and other composite particles (particles made of more fundamental entities). Mass, like so many things, behaves differently in the subatomic world because of the e1uivalence of mass and energy. When three quarks team up to form a proton, only a small part of the proton's mass comes from the masses of the quarks. Most of it comes from pure energy trapped in the proton.

Those in the first group of the six quarks have the prosaic names "up" and "down" (these are merely names, with no relation to any directions). For the second group, physicists became more whimsical, calling the quarks strange and "charm." In the late 1940s, some unexpectedly long- lived particles heavier than protons showed up in the cosmic radiation. These particles became the "strange" particles, and now we understand why they hang around for as long as a ten-billionth of a second or so. It's because they contain a kind of quark not found in neutrons and protons: the strange quark. Later, when some quite heavy mesons and baryons were found to live about a trillionth of a second (when they "should" have decayed much faster, in a billionth of a trillionth of a second or less), that was "charming." Needless to say, these particles' lease on life is attributable to yet another quark that they contain: the charm quark.

When it came to the quite heavy third group of quarks, discovered in 1977 and 1995, the physicists got cold feet. For a time, these last two quarks were called "truth" and "beauty." But conservatism prevailed. "Truth" became "top," and "beauty" became "bottom." (There was some theoretical rationale for the new names. Still, it's a pity that truth and beauty got lost.)

Although property "fractionated" by the quark is baryon number sometimes called baryonic charge, analogous to electric charge). The proton and neutron are baryons (recall that this originally meant "heavy particle"). The critically important feature of baryon number is that it is conserved. A heavier baryon can decay into a lighter one. When it does so, the number of barons number is not lost. (Similarly, electric charge is conserved. There I always as much of it after a reaction as before.) Fortunately for the structure of the Universe and for us humans, the lightest baryon has nowhere to go.

It is stable, because there is no lighter baryon into which it can decay. That lightest baryon is the proton. It seems to live forever. Nearest neighbor to the proton is the neutron, just a bit heavier. This means that the neutron is unstable: it can decay into the lower-mass proton (and an electron and antineutrino) without violating the law of baryon conservation or the law of energy conservation. Left alone, the neutron lives, on average, a whole fifteen minutes before it vanishes in a puff of three other particles. Fortunately again for us humans,, the neutron is stabilized within atomic nuclei, so certain combinations of up to 209 protons and neutrons can bundle together and live forever.

This means that our world is made of scores of different elements, not just the single element hydrogen. And it's all because mass s energy and energy is mass. In effect, the neutron within a stable nucleus, thanks to its potential energy, is sufficiently reduced in mass that it cannot decay. (In some unstable nuclei, neutrons can decay.

This process gives rise to beta radioactivity). Which brings us back to quarks. Each carries a baron number of 1/3. So three quarks, such as combine to form a proton or neutron, have baryon number 1. Since an antiquark has baryon number -1/3, a quark and antiquark together have zero baryon number. Such quark-antiquark combinations from mesons. So the antiquark is not just some curiosity. It lies within a great array of particles (none of which, however, is stable.

The quarks have a wide range of masses, from a few MeV for the up and down quarks to more than 170,000 MeV for the top quark. As mentioned earlier, no one has any idea why this is so. Another important property of quarks: color has nothing to do with the colors that we see). Color is much like electric charge (and, in fact, is sometimes called color charge); it is a property carried by a particle that is never lost or destroyed. A quark can be "red," "green," or "blue."

Antiquarks are anti-red, anti- green, or anti-blue. Red, green, and blue combined equally are "colorless." An equal combination of anti-red, anti-green, and anti-blue is likewise colorless.

The lightest baryons are the proton and the neutron, which lie at the heart of every atom we know. Then comes barons with Greek names – lambda, sigma, omega (and many more). The proton and neutron are composed of u (up) and d (down) quarks. The other baryons in the table contain one or more s (strange quarks. Still heavier baryons not listed) contain c (charm) and b (bottom) quarks. No top-containing baryons have yet been found. The lightest charm-containing baryon is about two-and-a- half times the mass of the proton. The so-far only known bottom-containing baryon has a mass about six times the proton's mass. Note that all of the baryons except the proton are unstable (radioactive).

5.38 SU(3) and quarks

By the early 1960s it became clear that many hundreds of so-called "elementary" resonance particles exist, each with well-defined values of the various quantum numbers such as spin, iso-spin, strangeness and baryon number and with widths which are generally seen to increase (or lifetimes which are generally seen to decrease) as teir masses become larger.

At that time, the most urgent task for physicists was to discover the correct classification scheme for the particles, which would do for the elementary particles what Mendeleeff's periodic table had done for the variety of chemical elements known as the 19th century.

Quarks closely examined

In 1964, Gell-Mann and George Zweig pointed out that the representations of SU(3) which were occupied by particles could be chosen from amongst all those mathematically possible by assuming them to be generated by just two combinations of the fundamental representation. Gell-Mann called the entities in the fundamental representation quarks. The three varieties of quark, or flavors as they are now called, have since come to be known as the up, down and strange quarks, the up and down labels referring to the orientation of the quarks' iso-spin. One significant consequence of this scheme is that if three quarks are to make up each baryon with a baryon number 1, then the quarks themselves must have baryon number 1/3.

The "quarks" are referred to as "entities" rather than "particles" for good reason. It is not necessary to assume their existence as observable particles to enjoy to successes of the SU(3) flavor scheme. They may be thought of as the mathematical elements only for such a scheme, devoid of physical reality. This was a fortunate escape clause at the beginning of the quarks' career because their fractional electronic charges and the failure to observe them in experiments encouraged skepticism in the naturally conservative world of physics. As we shall see, indirect evidence for the physical reality of quarks is now very convincing – despite the fact that they have never been seen directly in isolation.

But this evidence has mounted only rather slowly since in 1968 the beginning of the 'deep inelastic' experiments at the Stanford Linear Accelerator Center (SLAC) in California. Prior to this, most physicists preferred to reserve judgment on the reality of quarks, content to rely on the mathematics or SU(3) only.

5.39 Quark properties and dimensions

Quarks are points. They are "entities" that have no size whatsoever, yet possess mass, color, spin, electric charge, and baron number. As quarks flit about, they are incessantly emitting and absorbing gluons. So the space that was the inside of the spherical container contains not three but dozens of particles, all in a mad dance f motion, creation, and annihilation.

The gluons, like the quarks, are points – points with color, anti-color, and other properties, but without mass. Miraculously, or so it seems, the swarm as a whole maintains color neutrality. It also maintains a total electric charge of one unit (this is a proton), a total baryon number of one unit, and a total spin of one-half unit. If one quark starts to stray outside of the boundaries of the original surface, it gets tugged strongly back by the gluons. This strong force is quite remarkable. Unlike gravity (which gets weaker with increasing distance) or the electric force (which gets weaker with increasing distance) or the electric force (which also gets weaker with increasing distance), the pull of gluons increases as the distance gets greater. So the space occupied by the quarks and gluons, the space that is the interior of a proton, needs no "skin". The gluons police the quarks and one another relentlessly at the boundaries, making sure through stronger and stronger force that no particle wanders away. There is evidence that right in the middle of the proton, each quark menders relatively freely (like a child in the middle of the play area who can safely be ignored by the teacher). So we have never detected a quark or gluon, it is impossible to tear one of the particles loose from the other. Yet if you draw an analogy to a rubber band, which pulls more strongly the more it is stretched, you might wonder if it isn't possible to pour so much energy into a proton that even the strong gluon bond snaps, popping a quark free, just as a rubber band, with a sufficiently energetic pull, will break and release whatever is attached to its ends. As so often happens, the analogy from the ordinary world around us doesn't translate well to the subatomic world. Once again, the equivalence of mass and energy makes itself felt.

5.40 The formulation of QED

QED seeks to explain the interaction of charged particles, say electrons, in such a way that total electronic charge is always conserved. Hence, the gauge theory of QED successfully explains the interactions of electrons (and other charged particles). In perturbation theory, the probability of an event occurring is given by the sum of contributions from a series of increasingly more complex Feynman diagrams.

5.41 Particle physics and Cosmology

Recently, physics has witnessed the convergence of two of its most fascinating and most fundamental branches: elementary particle physics and cosmology. These two subjects, dealing with the Universe on the smallest and largest possible scales respectively, are now thought to be inextricably intertwined within the framework of the Big-Bang theory of the origin of the Universe. This intimate interrelationship between particle physics and cosmology is revealed in the profound implications each discipline holds for the other. According to the Big-Bang theory, the Universe began some 1010 million years ago from a space-time singularity, a single point of infinite energy-density and infinite space-time curvature. The act of creation – the Big Bang – was an enormous explosion from which an extremely hot and dense, rapidly expanding Universe came into being. The early Universe was a thick, hot primordial 'soup', filled with a great abundance of elementary particles of every kind, its evolution governed by the fundamental forces between them. Consequently the early Universe was also the ultimate particle accelerator: Its extremely high temperature and high density offer an unrivalled opportunity to probe physics beyond the reach of terrestrial accelerators and test ideas such as grand unified theories, supersymmetry and superstrings.

Quantum Chromodynamics - Colored Quarks

Just as Bohr's early quantum theory of the atom had been advanced to describe Rutherford's discoveries, so Quantum Chromodynamics (QCD) was put forward as description of the behavior of the quarks inside the proton. Pressing the analogy further, just as Bohr's description of the atom was an extension of the quantum theory propounded earlier by Planck, so QCD is an application of the ideas of "gauge field theory" developed in the 1960s.

QCD was proposed in 1973 by Fritzsch, Leutwyler and Gell-Mann (the last of whom, appropriately enough, was one of the original proponents of quarks in 1963), although a similar idea had been put forward in 1966 by Nambu. The basic idea is to use a new charge called color as the source of the inter-quark forces, just as electric charge is the source of electromagnetic forces between charged particles.

Quark confinement

The fact that a single quark has never been observed has for years the single greatest puzzle of elementary particle physics. No matter how energetically protons are collided together in the enormous accelerators at CERN and elsewhere, no quarks are seen to emerge in the debris. Many other varieties of particles are produced, but never fractionally charged particles which may be identified with the quarks.

This means that the forces which bind the quarks together are much stronger than the forces of the collision – which means that they are enormously strong. As an indication, we may note that the energies which bind the electrons into their atomic orbits are of the order of a few electron-volts. The energies binding the protons and neutrons in the nucleus are of the order of a few million electron-volts. Pairs of protons have been collided a energies of hundreds of thousands of millions electron-volts and still no quarks are observed which means the chromo-dynamic force between them must be at least that strong.

5.43 Probing the vacuum

The primary means of studying the fundamental constituents of matter and their interactions is through performing scattering experiments. This has been the case since the very beginning of particle physics. One class of scattering experiments which has, over the past three decades, been extremely fruitful involves collisions between electrons and positrons. These e+ e- experiments have yielded a great deal of information about the nature of the strong, weak and electromagnetic forces, and have played a major role in establishing the "Standard Model" of particles: QCD and the Glashow-Weinberg-Salem electroweak theory.

5.44 The Standard Model of Particle physics

The previous sections have told the long story of the discovery of the various particles and interactions which form what we now call the Standard Model of Particle Physics. The key theoretical elements of the Standard Model were in place by the early 1970s, and the discovery of the W and Z bosons in 1983 convinced physicists that the Standard Model was correct. In the years since then, the Standard Model has been subjected to intense experimental scrutiny. All the constituent particles have been found and a great number of precision tests of the model have been performed. At the time of writing, not one laboratory experiment (barring the discovery of neutrino oscillations has been found to be inconsistent with the predictions of the Standard Model.

The model has passed all tests with flying colors. The Standard model is a gauge quantum field theory based on the three sacred principles of relative quantum mechanics and gauge invariance.

5.45 The hunt for the Higgs boson

We saw that there is now a wealth of evidence pointing to the Standard Model as a correct and consistent theory of particle physics. Indeed, no experiment we have discussed so far has been found to be inconsistent with the Standard Model. However, a key element of the Standard Model is missing. No experiment has yet detected the "Higgs" particle, which is believed to be responsible for the spontaneous breakdown of gauge symmetry, and which gives masses to the gauge bosons and fermions. The Higgs boson is in a sense the key-stone of the Standard Model, and its discovery would be a triumphant confirmation of the Standard model in its entirety. The existence of the Higgs boson has now been predicted for over three decades, so why has no experiment found it? The answer must be simply that the Higgs boson (or bosons0 is too massive to have been produced in previous experiments. In order to have a chance of seeing the Higgs, we must look to higher and higher energies. This requires a new generation of particle physics experiments.

5.46 Constraints on the Higgs boson mass

The fact that a Higgs boson has not yet been detected indicates that it is very massive. The best constraint comes from the LEP experiment at CERN, which in its final results of 2001 excluded a Higgs boson of mass below 114.4 GeV. So if we want to detect the Higgs, we must built machines which are capable of producing Higgs bosons of this mass or above.

But there is an immediate problem: what if the Higgs boson is much heavier than 114.4 GeV, say a TeV or more? Then we would still have no chance of detecting the Higgs, even with the next generation of experiments. Fortunately, there is a way in which we can estimate an upper bound for the mass of the Higgs boson. As we have seen throughout this book, physical processes which we observe in experiments (such as scattering amplitudes) receive "quantum corrections," corresponding to Feynman diagrams with loops of virtual particles. The contribution of such diagrams decreases as the masses of the virtual particles increase. In particular, all Standard Model processes have quantum corrections with loops containing virtual "Higgs bosons." The sizes of these quantum corrections due to Higgs bosons are determined by the mass of the Higgs. Thus, by precision measurements of Standard Model processes, one can estimate the Higgs mass. A combined estimate suggests that the most likely value for the Higgs is around 117 GeV, which is just above the lower bound of 114.4 GeV, obtained from the LEP data.

The combined Standard Model data suggest that LEP really was close to finding the Higgs, they may have actually seen it, and that the Higgs is probably just around the corner. This gives great confidence that the Higgs boson will be found by the next generation particle physics experiment, the Large Hadron Collider. The Large Hadron Collider (or LHC) was completed in 2007 at CERN. Commissioned in 1996 and construction began in 2001. See figure 18.1 The LHC was built in the tunnel which formerly housed LEP, but in contrast to LEP, LHC will collide not electrons and positrons, but protons and heavy ions. The energy losses due to synchrotron radiation are much lower for these heavier particles and this will enable much higher collision energies to be replaces. The beam energy of LHC will be around 7 teV, corresponding to a center-of-mass energy of 14 TeV. The construction of the LHC involved massive engineering and computing challenges, as well as financial ones. The machine and detectors alone cost around 6 billion Swiss francs (or 4.5 billion euro). The magnetic field used to guide the protons around the tunnel; will have a strength of 8.3 tesla. Huge electric currents (2000 amps) are required to generate these magnetic fields. In order to carry such currents without power loss, superconducting cables are used, which must be cooled to temperatures just a few degrees above absolute zero with liquid helium. The computing challenges are particularly pressing. High-energy collisions between protons typically generate hundreds of secondary particles. Consequently, vast amounts of data must be stored and analyzed. It is estimated that around 15 petabytes of data will be generated per year at the LHC. If stored on compact discs, one year's worth of data would result in a stack twice the size of Mount Everest and it would require the equivalent of 105 of today's highest-performance personal computers to process data at this rate.

In order to deal with such a huge amount of data, a new concept in scientific computing was needed to be employed known as the Grid. The Grid is a global computing infrastructure, based on 60 major computing sites spread around Europe, North America and Asia, and connected by a super-high-bandwidth telecommunications networks which will process and store the LHC data. The Grid will allow computing power around the world to be pooled for the first time, making it possible to perform computations that no single existing computer could ever hope to perform. There are still enormous challenges facing to perform. There are still enormous challenges facing the Grid and issues concerning the storage, security and accessibility of data are still being addressed. Once they are, the Grid could well result in another revolution in computing and telecommunications, just like the World Wide Web. Electrons don't carry photons with them. According to QED, the photons are created out of nothing. Their existences are allowed by Heisenberg's uncertainty principle. The uncertainty principle says that you can't determine with complete accuracy the position of an electron and at the same time measure what direction and how fast it's going. It also says that you can't measure simultaneously and with complete accuracy the energy of a particle and for how long the particle has that energy. And this uncertainty is the clue to everything.

Virtual Particles

It turns out that Heisenberg's uncertainty principle allows things to happen under the radar. You can borrow as much energy as you need and, as long as its paid back within the short period of time allowed, you can use it for anything. This illegal scheme is allowed in the quantum world. You "borrow" energy to create a particle, for example. The energy to create the particle comes out of nowhere, out of the blue. (Actually, it comes out of the black, the vacuum.) You borrow it, use it, and create your particle. The particle does its thing, lives out its life, and dies. You recoup your energy and give it back to the vacuum. If all of that happens within the time allotted, everything is fine.

5.47 Filling up the vacuum

Particles that are created out of the vacuum are called "virtual particles." A virtual particle is one that's created out of nothing and that exists only for the duration allowed by Heisenberg's uncertainty principle. Virtual particles exist for some fleeting moment and disappear. Real particles, on the other hand, exist for a long time or forever. Real photons are eternal. The conclusions that QED arrived at are very strange. Yet, the theory works – better than any other theory that humans have ever created. QED is the most accurate theory ever. Its predictions have been found valid to one part in a billion.

QED tells us that the vacuum isn't really empty. The vacuum is filled with these virtual particles that suddenly appear, only to disappear almost immediately. The vacuum, teeming with these virtual particles, has energy. In 1967, the Russian physicist Yakov Zel'dovich showed that the energy of the vacuum is actually the energy that Einstein needed to create antigravity. The vacuum energy is the same as the "cosmological constant," the "dark energy" that permeates the Universe and is now driving its accelerating expansion.

Ch 6 The Structure of Particle Physics – The Standard Model

Subatomic Physics

In this chapter you will get an overview of the objects that are treated by particle physics, namely matter, forces and space-time. We will also discuss how one goes about to characterize the strength of a reaction between particles with the concept of the cross section. Which is a central issue of our subject. We will take a quick tour of the elements of matter at the subatomic level. After having read this chapter, you should know quarks and leptons by name and a few of their fundamental properties. You should be able to assign these constituents to families and generations. And you should know the quantum numbers which are conserved in interactions among these particles.

Of course it is not to scale otherwise, the nucleus would be completely invisible. The atom consists of a positively charged nucleus of small size of the order of ten to the minus fourteen meters in the surrounding electron cloud which is about ten thousand times larger that is to say, ten to the minus ten meters in radius and negatively charged. For all we know, the electron is an elementary particle, without internal structure and probably without size. It belongs to the family of leptons. Electrons and the nucleus are bound together by the electromagnetic force. The nucleus on the other hand, is not elementary. It contains protons and neutrons bound together by the nuclear force. We will talk about the physics of the atomic nucleus in the next section. Protons and neutrons are not elementary either. They contain quarks bound together by the strong force.

As far as we can tell today, quarks are equally elementary in the same sense as electrons are. But be careful, all of this describes just 5% of the universe, the rest is dark matter and dark energy, and we will introduce you to these two forms of matter and energy in module number eight. One can assign quarks and leptons to families according to their properties, and to generations according to their mass as shown in this table.

The lepton families are shown on the left of the diagram. All charged leptons have similar properties as the electron, which is the most well-known member of this family. The electric charge is minus one elementary charge. The neutral leptons, the neutrinos, are produced in radioactive decays. They have zero electric charge and they're very small mass.

The quarks families are shown on the right of this table. These are the constituents of proton and neutron and of all particles which we call hadrons. There are also two types.

The up-type quarks have an electric charge of 2/3 of the elementary charge, the down-type quarks have a charge of -1/3e. Every family has three generations, almost identical copies of the first one, but a lot heavier than the first one.

In the first generation, we find the familiar electron, the light neutrino nu_L, the quarks up and down u and d, these are the constituents of the ordinary matter around us today.

The second generation, has the muon μ, the medium mass neutrino nu_M, the charm quark and the strange quark, c and s. And finally, in the third generation we find the tau lepton, the heavy neutrino nu_H, the top and bottom quarks, t and b.

There is a quantum number which is called flavor which distinguishes between the generations inside each family. It is conserved by all interactions except the weak interactions.

The range of masses in this table is very, very large. Neutrinos are the lightest matter particles. Their masses are found experimentally to be below two electron volts but they are probably rather in the range of milli electron volts meV. We will talk about them in an upcoming section, which covers weak interactions. The top quark has a mass as large as a Hafnium nucleus. The Hafnium nucleus has an atomic number 178 and 72 positive electric charges. But as far as we know the top quark is a point-like particle like all the others in this table.

With these constituents we can implement an order scheme, a descriptive nomenclature, à la Liné. We call hadrons all particles which contain quarks. They are sensitive to strong and nuclear reactions, leptons are not. Baryons are bound states formed by three quarks. Mesons contain a quark and an anti-quark. And the two together form what we call hadrons. The nucleons p and n are the lightest baryons. They form the atomic nuclei we find around us. And particles like the pion are the lightest mesons formed by quarks and anti-quarks.

There are no hadrons including a top quark because of it's short lifetime. This quark decays before it can ever form a bound state with any others. Leptons on the right of this table are elementary in the sense that they are not composed of other particles for all we know. Observations are compatible with them having no size, they are thus point-like and so are the quarks. But not the hadrons.

We will often see the neutrinos, nu_e, nu_mu and nu_tau in tables of this sort. I don't adhere to this. These particles, nu_e, nu_mu and nu_tau are mixtures of the true particles with a specific mass which we denote by nu_L, nu_M and nu_H in this course.

Ww thus respect their mass hierarchy and assign them to their respective generations according to that property like we do for all other particles. The enormous strength of the strong force binds together, quarks to form very stable bound states. To be sensitive to the strong force, particles need to have the necessary charge, which is called color charge.

Quarks do have color charge, leptons do not.

This leads to a whole zoology of hadrons as shown in this summary table. The Particle Data Group is responsible to collect all the data concerning their properties, an exhaustive list of states as well as a summary of the known properties is available on the site of

the Particle Data Group quoted in red on the bottom of the slide. On the left you see a list of baryons containing three quarks. Like the proton which is a uud bound state. And the neutron which is a dud bound state.

On the right is a list of meson formed by quarks, quark anti-quark pairs, like the positive pion which is a u-dbar bound state. Leptons are not sensitive to the strong force because they do not carry the necessary color charge. They intreact only, via the electromagnetic and weak forces and do not form long lasting bound states among themselves.

6-2 A list summarizing some important properties of elementary particles.

In the upper half of the table you find the constituents of matter. Leptons and quarks are of spin one half as you can see. In the lower half you find particles that transmit forces.

These are bosons of integer spin. The charges are indicated in units of an elementary charge for each one of them, like the one of the electron in the case of the electric charge. Here we say that its charge is -1 which means that it is -1 times the elementary charge, e.

The weak charge is called weak isospin and it has two components in contrast to the electromagnetic charge which has only one component. We denote them by T and T3,

the total length of the weak isospin and its third component. This charge depends on the orientation of the spin of the particle with respect to its direction of motion, which means it depends on the helicity of the particle. The strong charge has three components.

It's called color, and we use the abbreviation r for red, g for green, and b for blue to denote them. Color is a property of quarks and of gluons where the latter even carry a color and an anti-color simultaneously. For every particle, there's an anti-particle which has the same mass but all charges opposite, as indicated in this red table. Charges are additive quantum numbers. For a system of particles the total charge is the sum

of the charges of its constituents. But there are other quantum numbers that are conserved.

The first one quoted here is the total number of baryons, i.e. the number of baryons minus the number of anti baryons in a system. This number is conserved, this applies in particular to quarks which have a baryon number of 1/3, and anti-quarks which have a baryon number of minus 1/3. The total lepton number, the number of leptons minus the number of antileptons is also constant in a closed system. To a certain extent this is even true generation by generation but the neutrinos which show up here, nu_e, nu_mu and nu_tau are mixtures of the true particles, nu_L, nu_M and nu_H. So, there's some mixture between generations in weak interactions. But more important than that, the charges of all types are rigorously conserved as far we know.

This evidently concerns the electric charge Q that you are familiar with but also the weak isospin T and T3, and the color charges R, G, and B.

The flavor which distinguishes between generations is a special case. It is conserved by the strong and electromagnetic interactions but can be altered by the weak one. We will define what it exactly means to conserve a quantum number when we discuss **Feynman diagrams,** later in this section. In the next section, we will introduce in more detail the forces which interact between matter particles.

Everything is made up of atoms. This is what you were hopefully taught at school and it was the historical starting point for particle physics. The atom was decomposed and its structure studied in the early 20th century. Beginning to understand the atom was a significant achievement but in the realm of modern physics the atom is too large (about $1x10^{-10}$ metres wide) to be described as particle physics.

The standard model is our current theory that best describes particle physics, having been developed throughout the latter half of the 20th century. It is a well established theory, having stood up to significant testing. Indeed it has predicted many particles, to a good precision, before their experimental discovery. The latest confirmation being the discovery of the Higgs boson at the large hadron collider (LHC). The standard model describes the types of particles that make up matter and the possible interactions between these particles.

Types of Particle

All particles belong to one of two categories: **fermions** or **bosons**. The classification of which category a particle belongs to is based on the particle's spin. Spin is an important quantum mechanical property that is intrinsic to the particle and nothing to do with the particle actually physically spinning.

A fermion is any particle that has a half-integer spin (1/2, 3/2, 5/2, etc.). The consequence of this, called the Pauli exclusion principle, is that fermions can't occupy the same quantum state; a quantum state simply being all the values of the particle's quantum properties (energy, spin etc.). A boson is any particle that has an integer spin (0, 1, 2, etc.). There is no restriction on bosons, any number of them can occupy the same quantum state.

The categorisation of boson or fermion determines the particle's behaviour. Elementary fermions are the particles that make up matter and elementary bosons carry forces between these matter particles.

Quarks and Leptons

Elementary fermions, the building blocks of matter, are grouped into two types: quarks and leptons. There are six different types, known as flavours, of quarks. The quark flavours are called up, down, strange, charm, top and bottom. Quarks are never observed individually, instead they bind together to form composite particles called hadrons. There are two possible compositions of hadron, known as baryons and mesons. Baryons are formed by three quarks binding together. Mesons are formed by a quark binding to an antiquark, more on antimatter later. In addition to its mass, electrical charge and spin, particles have additional quantum numbers (also confusingly referred to as charges) associated with them. The charges associated with quarks are baryon number, isospin, charm, strangeness, topness and bottomness.

There are also six different types of leptons. They are broken down into three different flavours of lepton: electron, muon and tau. Each flavour has an electrically charged particle (like the electron) and an associated neutrino. Neutrinos are electrically neutral particles with an almost negligible mass. To be clear, the six leptons are the electron, electron neutrino, muon, muon neutrino, tau and the tau neutrino. The charges associated with leptons are lepton number, electron lepton number, muon lepton number and tau lepton number.

Elementary fermions fit a convenient pattern, called generations. The fermions are grouped into three generations, with each generation containing two quarks and the two leptons associated with a specific lepton flavour. Higher generations contain significantly higher mass particles than the previous generation. A fermion can only decay down into lower generation fermions. Therefore, the ordinary matter we observe in everyday life (which must be stable) is entirely made up of first generation fermions.

The pattern of the twelve elementary fermions, how they are arranged into generations and then bind together to form matter, such as atoms. A table showing the properties of the different flavours of quark. Mass is given in energy style units and electrical charge is given in units of the magnitude of the electron's charge.

Antimatter

Every matter particle has a partner, called its antiparticle. Antiparticles have almost all their properties the same as their partner, such as mass and lifetime. However, their charges are the opposite sign to their partner's charges. For example, the up antiquark has the same mass and spin as an up quark but its electric charge is -2/3 ***e***, its baryon number is -1/3 and its weak isospin is -1/2. An antiparticle is symbolically denoted by a bar above the symbol for it's regular partner.

When a particle collides with its corresponding antiparticle they annihilate each other, the two particles are destroyed and converted into pure energy. This energy will then go on to create new particle-antiparticle pairs.

Apart from this strange behaviour, antiparticles behave pretty normally, in the sense that they seem to interact with gravity in the same way as standard particles and antiparticles can bind together to form antimatter such as antibaryons, antiatoms and so on.

An example of tracks left by particles in a bubble chamber. The first antiparticle discovered, the positron, was discovered by observing similar tracks. The positron track was identical to an electron's but spiralling in the opposite direction.

Exchange Forces

The fundamental interactions (also interchangeably called "forces") of nature are caused by elementary fermions exchanging force carrier particles. Precisely, the exchange of virtual bosons. But what is a virtual particle? The underlying mathematical framework describing interactions between particles is quantum field theory (QFT). In QFT, particles are well defined excitations of quantum fields. Virtual particles are fluctuations in a field, that only exists for a limited time. These fluctuations are allowed, through quantum mechanics, to break energy conservation but this limits the time that the particle is allowed to exist, as described by the uncertainty principle.

Heisenberg's uncertainty principle, a cornerstone of quantum mechanics, that relates the uncertainty in energy with the uncertainty in time.

Combining the uncertainty principle with Einstein's principle of mass-energy equivalence can then provide estimates for the lifetime of a virtual particle and its maximum possible range of motion in that time. If this virtual particle is a force carrier, these provide rough estimates of the range of an interaction and the characteristic time for it to take effect. For example, an interaction caused by a heavy exchange particle will have a shorter range than a light exchange particle.

The maximum range is estimated by assuming the force carrier moves at the speed of light.

There are only four fundamental interactions that have been observed within our universe: gravity, electromagnetism, the strong interaction and the weak interaction.

Gravity

Gravity is the only fundamental interaction unexplained within the standard model. Our current model of gravity is Einstein's theory of general relativity. However, general relativity hasn't been reconciled with QFT. This isn't a large practical problem because gravity is the weakest force by a significant margin and gravitational forces can be assumed as negligible for interactions between particles.

Electromagnetism

The electromagnetic interaction affects all particles that carry an electrical charge. It causes like charged particles to repel each other and oppositely charged particles to attract each other. The force carrier for electromagnetism is the photon (**γ**). The photon is electrically neutral and massless, hence the electromagnetic interaction has an infinite range (although it does get significantly weaker with distance). The characteristic time for electromagnetic interactions is somewhere between 10^{-20} and 10^{-15} seconds.

Examples of the electromagnetic interaction include atoms being held together, the force preventing stuff from falling through the floor, friction between objects and the interactions with electrons that form the basis of all electrical technology. All everyday phenomena are either produced by electromagnetic or gravitational interactions.

Strong Interaction

Analogous to how the electromagnetic interaction couples to electrical charge, the strong interaction couples to a different kind of charge, the colour charge. All quarks carry a colour, which is either red, blue or green. Antiquarks can carry anti-red, anti-blue or anti-green. This has nothing to do with the quark's visible colour, it is simply another quantum number. All observable particles (ie. hadrons) have to be colour neutral. Mesons can obviously meet this requirement e.g. red + anti-red is colour neutral. Baryons can also be colour neutral if each quark is a different colour (an RGB combination), this is a labelling concept that draws from white light being a sum of different coloured lights.

The force carrier for strong interactions is the gluon (***g***). Gluons are massless and they also carry colour charges, meaning gluons can strongly interact with each other, unlike photons. A gluon carries both a colour and an anti-colour, this leads to there being eight different gluon types. The strong interaction is attractive but it gets stronger with distance. This effect leads to the phenomena of colour confinement. When attempting to pull apart quarks, the increasing distance between them increases the energy of the colour field between them. Eventually, it becomes a high enough energy that a quark-antiquark pair is created. This phenomenon prevents individual quarks from ever being observed.

Even though gluons are massless, the effect of colour confinement limits the range of the strong interaction to about 10^{-15} metres, roughly the size of the nucleus within an atom. The characteristic time for strong interactions is less than 10^{-20} seconds. At the atomic scale, the strong interaction is the dominant interaction and hence it is responsible for binding together quarks into hadrons and holding the nucleus together.

Weak Interaction

The weak interaction causes massive particles to decay into lighter particles. Unlike the electromagnetic and strong interactions, the weak interaction is allowed to break the conservation of some quantum numbers such as strangeness.

Breaking these conservation laws allows the weak interaction to change the flavour of quarks and so it can cause quarks to decay between generations. However, electrical charge, baryon number and lepton numbers must still be conserved.

The weak interaction has three force carriers: the W^+, the W^- and the Z^0. All three particles have heavy masses (around 80 GeV/c^2) which limits the range of the interaction to about 10^{-18} metres, significantly smaller than the size of a proton. The characteristic time for weak interactions is typically longer than 10^{-15} seconds. The weak nuclear interaction is responsible for nuclear decays and hence it is the fundamental interaction behind radioactivity.

Within the standard model, the weak interaction has been successfully unified with the electromagnetic interaction. Essentially, electromagnetism and the weak interaction are actually manifestations of the same interaction, the electroweak interaction. At our universe's current low energy they appear to be very different but at a high enough energy (around 246 GeV) they merge to become the same interaction.

Higgs Boson

In the 1960s, Peter Higgs predicted a field permeating the universe to explain where the mass of standard model particles came from. Interactions with the Higgs field gives mass to all of the elementary particles in the standard model and as such is hugely important for the standard model to work. The higgs boson (***H***) is an excitation of the predicted Higgs field. Searching for the boson was the easiest way of confirming the Higgs field theory and hence a search for the Higgs boson was a long term ambition for particle physics.

In 2012, a particle matching the properties expected of the standard model Higgs boson was discovered at the LHC. This was the final discovery of a particle in the standard model and a confirmation of the model.

6-4 Particle mass does not all come from the Higgs Boson

About 21 grams is the mass of all of the electrons in your body. Now all of that mass comes from the Higgs mechanism. Which means as your electrons are traveling through space-time, they interact with the Higgs field and it is that that gives them their mass. The field slows them down and stops them from traveling at the speed of light. But most of your mass does not come from the Higgs mechanism. And neither does all the stuff you see around you. The mass is coming from something quit different. That is because most of your mass and most the mass around us comes from neutrons and protons. And they are not fundamental particles they are made from constituent particles called quarks. Now the theory that describes quarks and their interactions with each other through gluons is called Quantum Chromo-Dynamics.

And Chromo is the Greek word for color. So in some way these objects are meant to carry the color charge. But they are much smaller than the wavelength of visible light. So they are in no way actually colored. But it is a useful analogy that helps us think about how they interact and the particles they can make up. Now, the rules are pretty simple. In order for a particle to exist, it must be colorless or white. Now you can accomplish that in two different ways. You can make three quarks where each one is different color ie., red, green an blue, so overall they combine to the color white. Or you could use a quark and anti-quark where one is a color like green and the other is its "anti-color" for instance magenta.

Now, what I would like to do is through a thought experiment, simulate how quarks bind together to form different particles. Now, one must remember that empty space is not truly empty. There are fluxuations in the gluon field. You have to imagine that space is rippling with bumps the flow coming and gong.

Now to get rid of those fluxuations actually takes energy. And this is an important part of binding the quarks together. The existence of quarks actually suppresses the gluon fluxuations and creates what is called a flux tube. An area where there is really nothing in the vacuum of which is in between the quark and the anti-quark. That pairs them up and creates what is called a meson as which is the quark anti-quark pair.

What is interesting about thee flux-tube is that as these quarks become more separated the flux-tube remains the same diameter and the same sort of depth of suppression of the field, which means that the field does not actually increase. Its not like a spring, its not like an elastic band. The force is the same strength that is pulling the quarks back together. But you are putting more work in as you move these quarks further apart form each other. So, for a time people thought that these quarks are always going to be confined how ever far you move them. and as a result, you are really going to get a really long flux tube. But what actually happens is that if you put in enough energy, you can actually create a quark anti-quark pair. Never the less, the quarks are still confined. You can never see an individual quark. Because if you tried to pull one out, you put so much energy in the situation, another quark anti-quark pair will be created.

Now, to form a proton your going to need an UP-Quark another UP-quark and a Down-quark. Now the standard model of the proton that you have probably seen involves these quarks bounded together by little gluon springs that go between them. We know that that picture is completely wrong now. Even in the best situation, you might think that you would see *flux-tubes* around the edge of the triangle with three quarks. But we know that they don't do that. Rather, you get these Y-shaped flux-tubes.

The strange thing about a proton is that there maybe there are more then only three quarks there. You can have additional quark anti-quark pairs pop in and out of existence. So that any given time there could be five or seven or nine, any odd number of quarks that could make up the proton. In a proto, the quarks sit on the "lumps" or "bumps" of the gluon field. Along with the UP and Down quarks, there is a strange quark and an anti-strange quark. Which is strange because you don't normally think of these quarks being inside a proton, but they can be at any particular point in time. You can also see that these quarks have cleared out the vacuum. And you can also see that these kind of flux tubes which are the areas where the gluon field has been suppressed. And that's really what is binding these quarks together.

That's the strong force that binds the quarks in the heart of the proton. It is intrinsically related to the fact that clearing out the fluxuations has more energy then that bind them where they are. It cost energy to create a vacuum so where is the mass of the proton really coming from? Well of course, the constituent quarks do interact with the Higgs field and that gives them a small amount of mass. But if you add up the mass of all the quarks in the proton, it would only account for about one percent of its total mass. So where is the rest of the mass coming from? The answer is energy. We know Einstein's famous equation E=mc-squared. Well that says that we have got a lot of energy for just a little bit of mass.

But you rearrange the equation you can see that we can get an enormous amount of mass if there is lots of energy there. And that is really where most of the mass of the proton is coming from. It is from the fact that there is energy fluxuations in the gluon field and the quarks are interacting with those gluons. That is where most of your mass is coming from. It's coming from the energy that is in there. You know Einstein talked about, well if I had a hot cup of tea, it would actually have a slightly greater mass then when the same cup of tea when cold. So most of our mass we owe to *E=mc-squared*. You owe to the fact that your mass is packed with energy because of the interaction between the quarks and these gluon fluctuations in that gluon field. So what we think of as ordinary empty space, that turns out to be the thing that gives us all most of our mass.

Feynman Diagrams

Feynman diagrams are simple, visual representations of particle interactions. They are used by theoretical particle physicists to calculate the probability of a specific interaction between a set of initial particles that leads to a set of final particles. A feynman diagram only shows one possible arrangement of particles that can produce the interaction. To calculate the total probability of the interaction occurring, a sum must be performed over all the possible arrangements that lead to the interaction. The use of feynman diagrams make these extremely complicated calculations more tangible and easier to digest.

Within the diagram, the trajectory of particles are represented by lines (either straight, dotted or squiggly depending on the type of particle). Vertices are where the particles meet and interact. Antiparticles are represented as travelling backwards in time. Virtual particles are represented by internal lines, lines which connect two vertices (starting and ending within the diagram). Real particles are represented by lines that either point into the diagram or out of the diagram.

Below are some examples of Feynman diagrams for a variety of particle interactions.

A less probable arrangement that produces the same repulsion as above. In this case the photon creates a virtual electron-positron pair while travelling between the electrons.

The exchange of a gluon between two quarks, creating a strong force between them. Notice the conservation of overall colour. Colour is displayed in brackets next to each particle.

Limitations of the Standard Model

Despite being a successful, thoroughly tested theory we know that there are problems the theory can't explain. These gaps in our understanding indicate that the standard model is only an approximation (at low energies) of a more complete theory.

These are currently the major unsolved problems within the standard model:

- Neutrino mass - The standard model initially assumed neutrinos were massless, until it was experimentally determined that they had some tiny mass. Their exact masses and the mechanism behind is not yet understood.
- Interaction unification - Two of the four fundamental interactions have already been unified into the electroweak interaction. The next step is to try and unify this with the strong force, at an even higher unification energy. The aim would then be to unify this with gravity and have one fundamental unified interaction.
- Gravity - Gravity isn't explained within the standard model. The properties of a graviton, the name coined for a potential force carrier, are still debated let alone observed by experiment. Most likely, a new model is needed to reconcile quantum mechanics with gravity; one of many potential candidates is string theory.
- Dark matter and dark energy - Only about 5% of our universe consists of 'ordinary matter' as explained by the standard model. The rest is made up of completely unexplained kinds of matter and energy, which we currently call dark matter and dark energy.
- Matter-antimatter asymmetry - We observe that our universe is almost exclusively matter, where is all the antimatter? The standard model gives no reasons why antimatter and matter shouldn't be produced in equal amounts. Therefore, it is a question of cosmology and particle physics as to why our universe has developed into such an asymmetrical state.

Ch 7 Philosophies and Paradoxes in Physics

7.1 A complete break from classical physics

A complete break with classical physics comes with the realization that not just photons and electrons but all "particles" and all "waves" are in fact a mixture of wave and particle called particle/wave duality. It just happens that in our everyday world the particle component overwhelmingly dominates the mixture when interacting with material objects. The wave aspect is still there, in accordance with the relation $p*\lambda = h$, although it is totally insignificant from a macro-world perspective. However, in the world of the very small, where particle and wave aspects of reality are equally significant, things do not behave in any way that we can understand from our experience of the everyday world. It isn't just that the Bohr atom with its electron "orbits" is a false picture; all pictures are false, and there is no physical "analogy" we can make to understand what goes on inside of atoms. Atoms simply behave like atoms, and nothing else.

The idea of the "field" was introduced by Faraday, and was not generally used until Maxwell showed that all electric and magnetic phenomena could be described using only four equations involving electric and magnetic fields. These equations, known as Maxwell's equations, are the basic equations for all electromagnetism. They are fundamental in the since that Newton's three laws of motion and the law of universal gravitation are for mechanics. In a sense, they are even more fundamental, since they are consistent with the theory of relativity, whereas Newton's laws are not. Because all of electromagnetism is contained in this set of four equations, Maxwell's equations are considered one of the great triumphs of the human mind.

Electromagnetic waves, such as radio and TV signals, can be sent, and received, by means of antennas. At whatever frequency, all EM waves travel at the speed of light. And light itself is an EM wave. Although we will not present Maxwell's equations in mathematical form since they involve calculus, instead, we will summarize them here in words.

They are: (1) a generalized form of Coulomb's law known as Gauss's law that relates electric fields to its sources, (2) electric charge; a similar law for the magnetic field, except that since there seems to be no magnetic monopoles, there will be no single magnetic charges, and so magnetic field lines are always continuous – they do not begin nor end (as electric field lines do, on charges); (3) an electric field is produced by a changing magnetic field; (4) a magnetic field is produced by an electric current, or by a changing electric field.

Law number (3) is Faraday's law. The first part of (4), that a magnetic field is produced by an electric current, was discovered by Oersted, and the mathematical relation is given by Ampere's law. But the second part of (4) is an entirely new aspect predicted by Maxwell. Maxwell argued that if a changing magnetic field produces an electric field, as given by Faraday's law, then the reverse might be true as well: a changing electric field will produce a magnetic field. This was a hypothesis by Maxwell, based on the idea of "symmetry" in nature.

Indeed, the size of the effect in most cases is so small that Maxwell recognized it would be difficult to detect it experimentally. With the fore-mentioned scientific discoveries leading up to relativity theory and quantum theory, scientific philosophies and paradoxical interpretations have emerged as a consequence.

7.2 Revisiting the Twin Paradox

Not long after Einstein proposed the special theory of relativity, an apparent paradox was pointed out. According to this twin paradox, suppose one of a pair of 20-year-old twins takes off in a spaceship traveling at a very high speed to a distant star and back again, while the other twin remains on Earth. According to the Earth twin, the traveling twin will age less. Whereas 20 years might pass for the Earth twin, perhaps only 1 year (depending on the spacecraft's speed) would pass for the traveler twin. Thus, when the traveler returns, the Earth bound twin could expect to be 40 years old whereas the traveling twin would be only 21.

This is the "viewpoint" of the twin on the Earth. But what about the traveling twin? If all inertial reference frames are equally good, won't the traveling twin make all the claims the Earth twin does, only in reverse? Can't the astronaut twin claim that since the Earth is moving away at a high speed, time passes more slowly on Earth and the twin on Earth will age less? This is the opposite of what the Earth twin predicts.

They cannot both be right, for after all the spacecraft returns to Earth allowing for a direct comparison of ages and clocks to be made. This, at least on the surface, seems to be a paradox.

There is, however, not a paradox at all. The consequences of the special theory of relativity – in this case, time dilation – can be applied only by observers in inertial (non-accelerated) reference frames. The Earth is such a frame. During the acceleration periods, the twin on the spacecraft is not maintaining a steady inertial (non-accelerating) reference frame, so the spacecraft twin's predictions based on special relativity are not valid. The twin on Earth is in an inertial frame and can make valid predictions. Thus, there is no paradox. Consequently, the traveling twin's point of view expressed above is not correct. The predictions of the Earth twin are valid according to the theory of relativity, and the prediction that the traveling twin returns having aged less is the proper one. Einstein's general theory of relativity, which deals with accelerating reference frames, confirms this result.

7.3 Probability versus Determinism

The classical Newtonian view of the world is a deterministic one. One of its basic ideas is that once the position and velocity of an object are known at a particular time, its future position can be predicted if the forces on it are known. For example, if a stone is thrown a number of times with the same initial velocity and angle, and the forces on it remain the same, the path of the projectile will always be the same. If the forces are known (gravity and air resistance, if any) the stone's path can be precisely predicted. This "mechanistic" view implies that the future unfolding of the Universe, assumed to be made up of particulate bodies, is completely determined.

This classical deterministic view of the physical world has been radically altered by quantum mechanics. As we saw in the analysis of the double-slit experiment, electrons all prepared in the same way will not all end up in the same place. According to quantum mechanics, certain probabilities exist that an electron will arrive at different points. This is very different from the classical view, in which the path of a particle is precisely predictable from the initial position and velocity and the forces exerted on it. According to quantum mechanics, the position and velocity of an object cannot even be known accurately at the same time. This is expressed in the uncertainty principle, and arises because basic entities, such as electrons, are not considered simply as particles: they have wave properties as well.

Quantum mechanics allows us to calculate only the probability that, say, an electron (when thought of as a particle) will be observed at various places. Quantum mechanics says there is some inherent "unpredictability" in nature.

Since matter is considered to be made up of atoms, even ordinary-sized objects are expected to be governed by probability rather than by strict determinism. For example, there is a finite (but very small) probability that when you throw a stone, its path will suddenly curve upward instead of following the downward-curved parabola of normal projectile motion. Quantum mechanics predicts with very high probability that ordinary objects will behave just as the classical laws of physics predict. But these predictions are considered probabilities, not certainties. The reason that macroscopic objects are behave in accordance with classical laws with very high numbers of objects are present in a statistical situation, deviations from the average (or most probable) are negligible. It is the average configuration of vast numbers of molecules that follows the so-called fixed laws of classical physics with such high probability, and gives rise to an apparent *"determinism."* Deviations from classical laws are observed when small numbers of molecules are dealt with. We can say, then, that although there are no precise "deterministic" laws in quantum mechanics, there are "statistical" laws based on probability.

It is important to note that there is a difference between the probability imposed by quantum mechanics and that used in the 19th century to understand thermodynamics and the behavior of gases in terms of molecules. In thermodynamics, probability is used because there are far too many molecules to be kept track of. But the molecules were still assumed to move and interact in a deterministic way according to Newton's laws. Probability in quantum mechanics is quite different. It is seen as *inherent* in nature, and not as a limitation on our abilities to calculate.

Although a few physicists have not given up the deterministic view of nature and have been reluctant to accept quantum mechanics as a complete theory – one was Einstein – nonetheless, the vast majority of physicists do accept quantum mechanics and the probabilistic view of nature. This view, which as presented here is the generally accepted one, is called the Copenhagen interpretation of quantum mechanics in honor of Niels Bohr's home, since it was largely developed there through discussions between Bohr and other prominent physicists.

Because electrons are not simply particles, they cannot be thought of as following particular paths in space and time. This suggests that a description of matter in space and time may not be completely correct. This deep and far-reaching conclusion has been a lively topic of discussion among philosophers. Perhaps the most important and influential philosopher of quantum mechanics was Bohr. He argued that a space-time description of experiments on atoms or electrons must be given in terms of space and time and other concepts familiar to ordinary experience, such as waves and particles. We must not let our descriptions of experiments lead us into believing that atoms or electrons themselves actually exit in space and time as particles. This distinction between our interpretations of experiments and what is "really" happening in nature is crucial.

Thus far, we have been discussing the position and velocity of an electron as if it were a particle. But it isn't simply a particle. Indeed, we have the uncertainty principle because an electron – matter in general – has wave as well as particle properties. What the uncertainty principle really tells us is that if we insist on thinking of an electron as a particle, then there are certain limitations on this simplified view – namely, that the position and velocity cannot both be known precisely at the same time; and that the energy can be uncertain (or non-conserved) in the amount Δ-E for a time Δ-t approx. $\hbar/\Delta$-E. Because Planck's constant, h, is so small, the uncertainties expressed in the uncertainty principle are usually negligible on the macroscopic level. But at the level of the atom, the uncertainties are significant. Because we consider ordinary "objects" to be made up of atoms containing nuclei and electrons, the uncertainty principle is relevant to our understanding of all of nature. The uncertainty principle expresses, perhaps most clearly, the probabilistic nature of quantum mechanics. It thus is often used as a basis for philosophic discussion. We still don't know exactly how gravity works, but Einstein's limit was finally proved experimentally at the beginning of the twenty-first century - gravity does travel at the speed of light. If the Sun suddenly vanished, we also wouldn't feel the catastrophic impact of the loss of its gravitational pull until eight minutes later. Locality reigns.

Or at least that seemed to be the case, until experiments based on the work of an obscure physicist from Northern Ireland John Bell proved the existence of entanglement. Entanglement is genuine action at a distance, something that even now troubles many scientists. Of course, today we have a more sophisticated view of the Universe - and have a face up to the fact that the concept of "distance" itself is perhaps not as clear and obvious as it once was. A theorist of Duke University has suggested that the quantum world has an extra unseen dimension through which apparently spatially separated objects can communicate as if they were side by side. Others imagine spatial separation to be invisible - in effect, nonexistent –

to entangled particles. Even, so, there is a powerful reluctance to allow that anything, however insubstantial and unable to carry information, could travel faster than light. Although Einstein's objections to quantum theory based on its dependence on probability are frequently repeated (usually in one of several quotes about God not throwing dice), it was the breach of locality that really seemed to wound Einstein's sense of what was right.

The phenomenon that challenges locality, that makes action at a distance a possibility once more, the phenomenon of entanglement, emerges from quantum theory, the modern science of the very small. To reach the conception of entanglement, we need to trace quantum theory's development from a useful fudge to fix a puzzling phenomenon, to a wide-ranging structure that would undermine all of classical physics.

7.4 Holding on tight to Determinism

Quantum physics was fully developed by 1925. This field started with Einstein's 1905 paper on the photoelectric effect where he showed that light is lumpy (quantized). Quantum theory would become extremely successful over the years, the most successful physical theory of our period, "according to Einstein himself. However, even with its success, Einstein never accepted that quantum physics was the final answer. What was it about quantum theory that Einstein didn't like? That it abolished determinism. Heisenberg's uncertainty principle which is at the heart of quantum physics, says that you no longer can know at once everything you want to know about the behavior of objects. Although the principle applies to everything in the Universe, from electrons to stars, the effects are noticeable only in the realm of the atom, when you work with subatomic particles.

If the Universe behaves according to Heisenberg's principle, you can never know for certain the "*past*" or "*present*" of anything in the Universe. And you can't face the future with any certainty about what you'll encounter. The outcome of everything became probabilistic. "God doesn't play dice with the Universe," Einstein once said. He wasn't ready to abandon classical determinism from physics. And he put up a big fight. He and Niels Bohr, who was the main champion on the new physics argued back and forth about the meaning of reality. In many cases, Bohr convinced Einstein of the flaws of a particular argument against quantum physics. But Bohr couldn't answer Einstein's argument about the *EPR* problem, which is explain in this chapter. Einstein found a way to sidestep Heisenberg's uncertainty principle an determine, with complete accuracy, the outcome of the collision of two particles and their entire future history.

At the time, technology hadn't advanced enough for the extremely delicate *EPR* experiment to be performed. Now it has. Since 1982, several real *EPR* experiments have been performed, and the results show that Einstein was wrong. The basic premise of quantum physics is correct. The world is not deterministic! Quantum theory is widely held to resist any realist interpretation and to mark the advent of a 'postmodern' science characterized by paradox, uncertainty, and the limits of precise measurement.

The realm of macro-physical objects and the subatomic or the macro-physical domain. These confusions took place at an early stage in the history of quantum physics (more specifically, in the well-known series of debates between Einstein and Bohr). They emerge most clearly in subsequent discussions of the 1932 Einstein-Podolsky-Rosen (*EPR*) paper, which laid down criteria for a realist interpretation compatible with the known laws of physics, among them those of relativity theory.

The *EPR* argument gave rise to J.S. Bell's equally famous theorem to the effect that any such interpretation – one that entailed the existence of *"hidden variables"* – would also entail some highly problematic consequences, including (what Einstein refused to accept) nonlocal effects of quantum *"entanglement"* at arbitrary space-time distances. However, David Bohm's theory is able to accommodate this problem while also maintaining a realist ontology and producing results in accordance with the well-established *QM* observational results and predictions. Moreover, it avoids the kinds of extravagant conjecture – such as the 'many-worlds' interpretation currently championed by David Deutsch – which take orthodox QM as their basis for proposing a massive (scarcely thinkable) revision to our grasp of what constitutes a "realist" worldview.

7.5 The Clock in the Box

The great debate between Bohr and Einstein on the interpretation of quantum theory began in 1927 at the fifth Solvay Congress, and continued until Einstein's death in 1955. Einstein also corresponded with Born on the subject, and a flavor of the debate can be gleaned from The Born-Einstein Letters. The debate centered on the series of imaginary tests of the predictions of the Copenhagen interpretation – not real experiments." The game was that Einstein would try to think up an experiment in which it would be theoretically possible to measure two complementary things at once, the position and mass of a particle, or its precise energy at a precise time, and so on. Bohr, and Born, would then try to show how Einstein's thought experiment simply could not be carried out in the way required to pull the rug from under the theory. One example, the "clock in the box" experiment, will serve to show how the game was played.

Imagine a box, said Einstein, that has a hole in one wall covered by a shutter that can be opened and closed again under the control of a clock inside the box. Apart from the clock and shutter mechanism, the box is filled with radiation. Set up the apparatus so that at some precise, predetermined time on the clock the shutter will open and allow one photon to escape before it closed again.

Now weigh the box, wait for the photon to escape, and weigh the box again. Because mass is energy, the difference in the two weights tells us the energy of the photon that escaped. So we know, in principle, the exact energy of the photon and the exact time that it passed through the hole, refuting the uncertainty principle. Bohr, as always in these arguments, won the day by looking at the practical details of how the measurements could be carried out.

The box must be weighted, so it must be suspended by a string, for example, in a gravitational field. Before the photon escapes from the box, the mythical experimenter notes the position of a pointer, firmly fixed to the box, against a scale. After the photon escapes, the experimenter can, in principle, add weights to the box to restore the pointer to the same place. But this in itself involves the uncertainty relations. The position of the pointer can only be determined to within the limits set by Heisenberg's relation, and there is an uncertainty in the momentum of the box associated with that uncertainty in the position of the pointer. The greater the accuracy of the measurement of the weight of the box, the greater is the uncertainty in the all-important knowledge of its momentum.

Even if you try to restore the original situation by adding a small weight to the box to stretch the spring back to its original position, and measure the extra weigh to determine the energy of the escaping photon, you can never do better than reducing the uncertainty to the limits allowed by the Heisenberg relation, in this case

$\Lambda\text{-}E\ \Lambda\text{—}t > \hbar.$

The details of this and the other thought experiments involved in the Einstein-Bohr debate can be found in Abrahams Paris' '*Subtle is the Lord . . .*' Pais stresses that there is nothing fanciful in Bohr's insistence on a full and detailed description of the mythical experiments – in this case, heavy bolts that fix the framework of the balance into place, the spring that allows mass to be measured but must thereby allow the box to move, the little weight that has to be added, and so on.

The results of all experiments have to be interpreted in terms of classical language, the language of everyday reality. We could fix the box rigidly into place so that we had no uncertainty about its position, but then it would be impossible to measure the change in mass. The dilemma of quantum uncertainty arises because we try to express quantum ideas in everyday language, and that is why Bohr stressed the nuts and bolts of the experiments.

7.6 Einstein Forgets Relativity

As fore-mentioned, over the past three years, Bohr had re-examined the imaginary experiments Einstein had proposed at the Solvay conference in October 1927. Each was designed to show that quantum mechanics was inconsistent, but he had found the flaw in Einstein's analysis in every case. Not content to rest on his laurels, Bohr devised some thought experiments of his own involving an assortment of slits, shutters, clocks and the like as he probed his interpretation for any weakness. He found none. But Bohr never conjured up anything as simple and ingenious as the thought experiment that Einstein had just finished describing to him in Brussels at the sixth Solvay conference.

The theme of the six-day meeting that began on 20 October 1930 was the magnetic properties of matter. The format remained the same: a series of commissioned reports on various topics related to magnetism, each followed by a discussion. Bohr had joined Einstein as a member of the nine-strong scientific committee and both were therefore automatically invited to the conference. After the death of Lorentz, the Frenchman Paul Langevin had agreed to take on the demanding dual responsibilities of presiding over the committee and the conference. Dirac, Heisenberg, Kramers, Pauli and Sommerfeld were among the 34 participants.

The Newtonian Universe is purely deterministic with no room for chance. In it, a particle has a definite momentum and position at any given time. The forces that act on the particle determine the way its momentum and position vary in time. The only way that physicists such as James Clerk Maxwell and Ludwig Boltzmann could account for the properties of a gas that consists of many such particles was to use probability and settle for a statistical description. The forced retreat into a statistical analysis was due to the difficulties in tracking the motion of such an enormous number of particles. Probability was a consequence of human ignorance in a deterministic Universe where everything unfolded according to the laws of nature. If the present state of any system and the forces acting upon it are known, then what happens to it in the future is already determined. In classical physics, determinism is bound by an *umbilical cord* to *causality* – the notion that every effect has a cause. It was impossible to determine exactly where the electron was after a collision. The best that physics could do, he said, was to calculate the probability that the electron would be scattered through a certain angle. This was Born's 'new physical content', and it all hinged on his interpretation of the wave function.

The wave function itself has no physical reality; it exists in the mysterious ghost-like realm of the possible. It deals with abstract possibilities, like all the angles by which an electron could be scattered following a collision with an atom. There is a real world of difference between the possible and the probable. Born argued that the square of the wave function, a real rather than a complex number, inhabits the world of the probable. Squaring the wave function, for example, does not give the actual position of an electron, only the probability, the odds that it will found here rather than there. For example, if the value of the wave function of an electron at *X* is double its value at *Y*, the probability of it being found at *X* is four times greater than the probability of finding it at *Y*. The electron could be found at *X*, *Y* or somewhere else.

Niels Bohr would soon argue that until an observation or measurement is made, a microphysical object like an electron does not exist anywhere. Between one measurement and the next it has no existence outside the abstract possibilities of the wave function. It is only when an observation or measurement is made that the 'wave function collapses' as one of the 'possible' states of the electron becomes the 'actual' state of the probability of all the other possibilities becomes zero.

Schrodinger dismissed Born's probability interpretation. He did not accept that a collision of an electron or an alpha particle with an atom as "absolutely accidental," i.e. "completely undetermined." Otherwise, if Born was right, then there was no way to avoid quantum jumps and causality was once again threatened.

As a meeting of minds it was a close second to Solvay 1927, with twelve current and future Nobel laureates present. It was the backdrop to the "second round" if the ongoing struggle between Einstein and Bohr over the meaning of quantum mechanics and the nature of reality. Einstein had travelled to Brussels armed with a new thought experiment designed to deliver a fatal blow to the uncertainty principle and the Copenhagen interpretation. An unsuspecting Bohr was ambushed after one of the formal sessions.

As mentioned earlier, Einstein quizzed Bohr by saying, "Imagine a box full of light. In one of its walls is a hole with a shutter that can be opened and closed by a mechanism connected to a clock inside the box." This clock is synchronized with another in the laboratory. Weigh the box. Set the clock to open the shutter at a certain time for the briefest of moments, but long enough for a single photon to escape. We now know, explained Einstein, precisely the time at which the photon left the box. Bohr listened unconcerned; everything Einstein had proposed appeared straightforward and beyond contention. The uncertainty principle applied only to pairs of complementary variables – position and momentum or energy and time. It did not impose and limit on the degree of accuracy with which any one of the pair could be measured. Just then, with a hint of smile, Einstein uttered the deadly words: weigh the box again. In a flash, Bohr realized that he and the Copenhagen interpretation were in deep trouble. To work out how much light had escaped locked up in a single photon, Einstein used a remarkable discovery he had made while still a clerk at the Patent Office in Bern: energy is mass and mass is energy. This astonishing spin-off from his work on relativity was captured by Einstein in his simplest and most famous equation: $E=mc^2$, where E is energy, m is mass, and c is the speed of light.

By weighing the box of light before and after the photon escapes, it is easy to work out the differences in mass. Although such a staggeringly small change was impossible to measure using equipment available in 1930, in the realm of the thought experiment it was child's play. Using $E=mc^2$ to convert the quantity of missing mass into an equivalent amount of energy, it was possible to calculate precisely the energy of the escaped photon. The time of the photon's escape was known via the laboratory clock being synchronized with the one inside the light box controlling the shutter.

It appeared that Einstein had conceived an experiment capable of measuring simultaneously the energy of the photon and the time of its escape with a degree of accuracy proscribed by Heisenberg's uncertainty principle.

'It was quite a shock for Bohr', recalled the Belgian physicist Leon Rosenfeld, who had recently begun what turned into a long-term collaboration with the Dane. 'He did not see the solution at once.' While Bohr was desperately worried by Einstein's latest challenge, Pauli and Heisenberg were dismissive. 'Ah, well, it will be all right, it will be all right', they told him. 'During the whole evening he was extremely unhappy, going form on to the other and trying to persuade them that it couldn't be true, that it would be the end of physics if Einstein were right,' recalled Rosenfeld, 'but he couldn't produce any refutation.'

Rosenfeld was not invited to Solvay 1930, but had travelled to Brussels to meet Bohr. He never forgot the sight of the two quantum adversaries heading back to the Hotel Metropole that evening: 'Einstein, a tall majestic figure, walking quietly, with a somewhat ironical smile on his face, and Bohr trotting near him very excited ineffectually pleading that if Einstein's device would work, it would mean the end of physics. For Einstein it was neither an end nor a beginning. It was nothing more than a demonstration that quantum mechanics was inconsistent and therefore not the closed and complete theory that Bohr claimed. His latest thought experiment was simply an attempt to rescue the kind of physics that aimed to understand and observer-independent reality.

Bohr spent a sleepless night examining every facet of Einstein's thought experiment. He took the imaginary box of light apart to find the flaw that he hoped existed. Einstein did not picture, even in his mind's eye, either the details of the inner workings of the light box or how to weigh it. Bohr, desperate to get to grips with the device and the measurements that would have to be made, drew what he called a 'pseudo-realistic' diagram of the experimental set-up to help him.

Given the need to weigh the light box before the shutter is opened at a pre-set time and after the photon has escaped, Bohr decided to focus on the weighting process. With mounting anxiety and little time, he chose the simplest possible method. He suspended the light box from a spring fixed to a supporting frame. To turn it into a weighting scale, Bohr attached a pointer to the light box so its position could be read on a scale attached to the vertical arm of what resembled a hangman's gallows. To ensure that the pointer was positioned at zero on the scale, Bohr attached a small weight to the bottom of the box. There was nothing whimsical in the construction, as Bohr included even the nuts and bolts used to fix the frame to a base, and drew the clockwork mechanism controlling the opening and closing of the hole through which the photon was to escape.

The initial weighting of the light box is simply the configuration with the attached weight chosen to ensure that the pointer is at zero. After the photon escapes, the light box is lighter and is pulled upwards by the spring. To reposition the pointer at zero, the attached weight has to be replaced by a slightly heavier one. There is no time limit on how long the experimenter can take to change the weights. The difference in the weights is the mass lost due to the escaped photon, and from $E=mc^2$ the energy of the photon can be calculated precisely.

From the arguments he deployed at Solvay 1927, Bohr held that any measurement of the position of the light box would lead to an inherent uncertainty in its momentum, because to read the scale would require it to be illuminated. The very act of measuring its weight would cause an uncontrollable transfer of momentum to the light box because of the exchange of photons between the pointer and the observer causing it to move. The only way to improve the accuracy of the position measurement was to carry out the balancing of the light box, the positioning of the pointer to zero, over a comparatively long time.

However, Bohr argued that this would lead to a corresponding uncertainty in the momentum of the box. The more accurately the position of the box was measured, the greater the uncertainty attached to any measurement of its momentum.

Unlike at Solvay 1927, Einstein was attacking the "*energy-time*" uncertainty relation, not the "*position-momentum*" incarnation. It was now, in the early hours of the morning, that a tired Bohr suddenly saw the flaw in Einstein's experiment. He reconstructed the analysis little by little until he was satisfied that Einstein had indeed made an almost unbelievable mistake. Relieved, Bohr went to sleep for a few hours, knowing that when he awoke it would be to relish his triumph over breakfast.

In his desperation to destroy the Copenhagen view of quantum reality, Einstein had forgotten to take into account his own theory of general relativity. He had ignored the effects of gravity on the measurement of time by the clock inside the light box. General relativity was Einstein's greatest achievement. 'The theory appeared to me then, and it still does, the greatest feat of human thinking about Nature, the most amazing combination of philosophic penetration, physical intuition, and mathematical skill,' said Max Born. He called it 'a great work of art, to be enjoyed and admired from a distance.' When the bending of light predicted by general relativity was confirmed in 1919, it made headlines around the world. J.J. Thomson told one British newspaper that Einstein's theory was 'a whole new continent of new scientific ideas.'

One of these new ideas was gravitational "time dilation." Two identical and synchronized clocks in a room with one fixed to the ceiling and the other on the floor would be out of step by 300 parts in a billion billion, because time flows more slowly at the floor than at the ceiling. The reason was gravity. According to general relativity, Einstein's theory of gravity, the rate at which a clock ticks depends upon its position in a gravitational field. Also, a clock moving in a gravitational field ticks slower than one that is stationary. Bohr realized that this implied that weighting the light box affected the time-keeping of the clock inside.

The position of the light box in the Earth's gravitational field is altered by the act of measuring the pointer against the scale. This change in position would alter the rate of the clock and it would no longer be synchronized with the clock in the laboratory, making it impossible to measure as accurately as Einstein presumed the precise time the shutter opened and the photon escaped from the box. The greater the accuracy in measuring the energy of the photon, via $E=mc^2$, the greater the uncertainty in the position of the light box within the gravitational field. This uncertainty of position prevents, due to gravity's ability to affect the flow of time, the determination could not simultaneously measure exactly both the energy of the photon and the time of its escape. Heisenberg's uncertainty principle remained intact, and with it the Copenhagen interpretation of quantum mechanics.

When Bohr came down to breakfast he was no longer looking 'like a dog who has received a thrashing' the night before. Now it was Einstein who was stunned into silence as he listened to Bohr explain why his latest challenge, like those of three years earlier, had failed. Later there would be those who questioned Bohr's refutation because he had treated macroscopic elements such as the pointer, the scale, and the light box as if they were quantum objects and therefore subject to limitations imposed by the uncertainty principle. To handle macroscopic objects in this way ran counter to his insistence that laboratory equipment be treated classically. But Bohr had never been particularly clear about where to draw the line between the micro and macro, since in the end every classical object is nothing but a collection of atoms.

Whatever reservation there was, Einstein accepted Bohr's counter-arguments, as did the physics community at the time. As a result he ceased his attempts to circumvent the uncertainty principle to demonstrate that quantum mechanics was logically inconsistent. Instead Einstein would henceforth focus on exposing the theory as incomplete.

In November 1930 Einstein lectured in Leiden on the light box. Afterwards a member of the audience argued that there was no conflict within quantum mechanics. 'I know, this business is free of contradictions,' replied Einstein, 'yet in my view it contains a certain unreasonableness.' In spite of this, in September 1931, he once again nominated Heisenberg and Schrodinger for a Nobel Prize. But after going two rounds with Bohr at the Solvay conferences, one sentence in Einstein's letter of nomination was telling: 'In my opinion, this theory contains without doubt a piece of the ultimate truth.' 'His inner voice continued to whisper that quantum mechanics was incomplete, that it was not the whole truth as Bohr would have everyone believe.'

In the years following Solvay 1930, there was little direct contact between Bohr and Einstein. A valuable channel of communication ceased with Paul Ehrenfest's suicide in September 1933. In a moving tribute, Einstein wrote of his friend's inner struggle to understand quantum mechanics and 'the increasing difficulty of adaptation to new thoughts which always confronts the man past fifty.

There were many who read Einstein's words and mistook them as a lament at his own plight. Now in his mid-fifties, he knew he was regarded as a relic from a bygone age, refusing, or unable, to live with quantum mechanics. But he also knew what separated him and Schrodinger from most of their colleagues: 'Almost all the other fellows do not look from the facts to the theory but from the theory to the facts.'

7.7 In search of a complete unified theory

But there is a fundamental paradox in the search for such a complete unified theory. The ideas about scientific theories outlines above assume we are rational beings who are free to observe the Universe as we want and to draw logical deductions from what we see. In such a scheme it is reasonable to suppose that we might progress ever closer toward the laws that govern our Universe. Yet if there really were a complete unified theory, it would also presumably "determine" our actions – so the theory itself would determine the outcome of our search for it. And why should it determine that we come to the right conclusions from the evidence? Might it not equally (as) well determine that we draw the wrong conclusion? Or no conclusion at all?

The only answer that we can give to this problem is based on Darwin's principle of natural selection. The idea is that in any population of self-reproducing organisms, there will be variations in the genetic material and upbringing that different individuals have. These differences will mean that some individuals are better able than others to draw the right conclusions about the world around them and to reproduce, so their pattern of behavior and thought will come to dominate. It has certainty been true in the past that what we call intelligence and scientific discovery has conveyed a survival advantage. It is not so clear that this is still the case: our scientific discovery may well destroy us all, and even if they don't, a complete unified theory may not make much difference to our chances of survival.

However, provided the Universe had evolved in a regular way, we might expect that the reasoning abilities that natural selection has given us would also be valid in our search for a complete unified theory and so would not lead us to the wrong conclusions.

7.8 The EPR paradox

Einstein accepted Bohr's criticisms of other thought experiments, but by the early 1930s he had turned to a new kind of imaginary test of the quantum rules. The basic idea behind this new approach was to use experimental information about one particle to deduce the properties such as position and momentum, of a second particle. This version of the debate was never resolved in Einstein's lifetime, but it has now been successfully tested, not by an improved thought experiment but a real experiment in the lab. Once again, Bohr wins and Einstein loses.

During the early 1930s, Einstein's personal life was in a turmoil. He had to leave Germany because of the threat of persecution by the Nazi regime. By 1935 he was settled in Princeton, and in December 1936 his second wife, Elsa, died after a long illness. Amid all this turmoil, he continued to puzzle over the interpretation of quantum theory, defeated by Bohr's arguments but not convinced in his heart that the Copenhagen interpretation, with its inherent uncertainty and lack of strict causality, could be the last word as a valid description of the real world. Max Jammer has described in exhaustive detail the various twists and turns of Einstein's mind on this subject at that time, in "The Philosophy of Quantum Mechanics." Several threads came together in 1934 and 1935, when Einstein worked in Princeton with Boris Podolsky and Nathan Rosen on a paper presenting what has become known as the "*EPR Paradox*," even though it does not really describe a paradox at all.

A. Einstein, B. Podolsky, and N. Rosen, "Can quantum-mechanical description of physical reality be considered complete?" "Physical Review," volume 47, pp. 777-780, 1935. The paper is among those reprinted in the volume Physical Reality, edited by S. Toulmin, Harper & Row, 1970. The point of the argument was that, according to Einstein and his collaborators, the Copenhagen interpretation had to be incomplete – that there really is some underlying clockwork that keeps the Universe running, and that only gives the appearance of uncertainty and unpredictability at the quantum level, through statistical variations.

Imagine two particles, said Einstein, Podolsky, and Rosen, that interact with one another and then fly apart, not interacting with anything else at all until the experimenter decides to investigate one of them. Each particle has its own momentum, each is located at some position in space. Even within the rules of quantum theory, we are allowed to measure precisely the total momentum of the two particles, added together, and the distance between them, at the time they were close together. When, much later, we decide to measure the momentum of one of the particles we know, automatically, what the momentum of the other one must be, because the total must be unchanged.

Alternatively, we could have measured the precise position of the first particle and, in the same way, deduced the position of the second particle. Now, it is one thing to argue that the physical measurement of the momentum of particle A destroys knowledge of its own position, so that we cannot know its precise position, and that similarly, the momentum, which remains unknown. But it seemed quite another thing entirely, to Einstein and his colleagues, to two measurements we choose to make on particle A. How can particle B "know" whether it should have a precisely defined momentum or a precisely defined position? It seemed as if, in the quantum world, the measurements we make on a particle here affect its partner there, in violation of causality, an instantaneous "communication" traveling across space, something called "action at a distance."

If you accepted the Copenhagen interpretation, the *EPR* paper concluded, it "makes the reality of position and momentum in the second system depend upon the process of measurement carried out on the first system which does not disturb the second system in any way. No reasonable definition of reality could be expected to permit this." This is where the team diverged from most of their colleagues and from all of the Copenhagen school. Nobody disagreed with the logic of the argument, but they did disagree on what constitutes a "reasonable" definition of reality. Bohr and his colleagues could live with a reality in which the position and momentum of the second particle had no objective meaning until they were measured regardless of what you did to the first particle.

A choice had to be made between a world of *"objective reality"* and the quantum world, of that there was no doubt. But Einstein remained in a very small minority in deciding that of the two options open he would cling to objective reality and reject the Copenhagen interpretation.

But Einstein was an honest man, always ready to accept sound experimental evidence. If he had lived to see it, he would certainly have been persuaded by the recent experimental tests of what is effectively a kind of *EPR* effect that he was wrong. Objective reality does not have any place in our fundamental description of the Universe, but action at a distance or *acausality* does have such a place. The experimental verification of this is so important that it deserves a chapter to itself. But first, for completeness, we ought to look at some of the other paradoxical possibilities inherent in the quantum rules – particles that travel backward in time and, at last, Schrodinger's famous half-dead cat.

7.9 EPR's impact on the scientific community

In spite of these mutual misgivings, there were always young physicists eager to work with Einstein. One was Nathan Rosen, a 25-year-old New Yorker who arrived from MIT in 1934 to serve as his assistant. A few months before Rosen, the 39-year-old Russian-born Boris Podolsky had joined the institute. He had first met Einstein at Caltech in 1931 and they had collaborated on a paper. Einstein had an idea for another paper. It would mark a new phase in his debate with Bohr, as it unleashed a fresh assault on the Copenhagen interpretation.

At Solvay 1927 and 1930, Einstein attempted to circumvent the uncertainty principle to show that quantum mechanics was inconsistent and therefore incomplete. Bohr, aided by Heisenberg and Pauli, had successfully dismantled each thought experiment and defended the Copenhagen interpretation. Afterwards, Einstein accepted that although quantum mechanics was logically consistent it was not the definitive theory that Bohr claimed Einstein knew he needed a new strategy to demonstrate that quantum mechanics is incomplete, that it does not fully capture physical reality. To this end he developed his most enduring thought experiment.

For several weeks early in 1935, Einstein met Podolsky and Rosen in his office to thrash out his ideas. Podolsky was assigned the task of writing the resulting paper, while Rosen did most of the necessary mathematical calculations. Einstein, as Rosen recalled later, 'contributed the general point of view and its implications'.

Only four pages long, the Einstein-Podolsky-Rosen paper, or the *EPR* paper as it became known, was completed and mailed by the end of March. 'Can Quantum Mechanical Description of Physical Reality Be Considered Complete?', with its missing 'the', was published on 15 May in the American journal *Physical Review.* The *EPR* answer to the question posed was a defiant 'No!'. Even before it appeared in print, Einstein's name ensured that the *EPR* paper generated the kind of publicity nobody wanted.

On Saturday, 4 May 1935, the New York Times carried an article on page eleven under the attention-grabbing headline 'Einstein Attacks Quantum Theory': 'Professor Einstein will attack science's important theory of quantum mechanics, a theory of which he was a sort of grandfather. He concluded that while quantum mechanics is "correct" it is not "complete".' Three days later, the New York Times carried a statement from a clearly disgruntled Einstein. Although no stranger to talking to the press, he pointed out that: 'It is my invariable practice to discuss scientific matters only in the appropriate forum and I deprecate advance publication of any announcement in regard to such matters in the secular press.'

In the published paper, Einstein, Podolsky and Rosen started by differentiating between reality as it is and the physicist's understanding of it: 'Any serious consideration of a physical theory must take into account the distinction between the objective reality, which is independent of any theory, and the physical concepts with the objective reality, and by means of these concepts we picture this reality to ourselves.' In gauging the success of any particular physical theory, *EPR* argued that two questions had to be answered with an unequivocal 'Yes': is the theory correct" Is the description given by the theory complete?

'The correctness of the theory is judged by the degree of agreement between the conclusions of the theory and human experience', said *EPR*. It was a statement that every physicist would accept when 'experience' in physics takes the form of experiment and measurement. To date there had been no conflict between the experiments performed in the laboratory and the theoretical predictions of quantum mechanics. It appeared to be a correct theory. Yet for Einstein it was not enough for a theory to be correct, in agreement with experiments; it also had to be complete.

Whatever the meaning of the term 'complete', *EPR* imposed a necessary condition for the completeness of a physical theory: 'every element of the physical reality must have a counterpart in the physical theory.' This completeness criterion required *EPR* to define a so-called 'element of reality' if they were to carry through their argument. Einstein did not want to get stuck in the philosophical quicksand, which had swallowed so many, of trying to define 'reality.' In the past, none had emerged unscathed from an attempt to pinpoint what constituted reality. Astutely avoiding a 'comprehensive definition of reality' as 'unnecessary' for their purpose, *EPR* adopted what they deemed to be a 'sufficient' and 'reasonable' criterion for designating an 'element of reality': 'If, without in any way disturbing a system, we can predict with certainty (i.e. with probability equal to unity) the value of a physical quantity, then there exists an element of physical reality corresponding to this physical quantity.' Einstein wanted to disapprove Bohr's claim that quantum mechanics was a complete, fundamental theory of nature demonstrating that there existed objective 'elements of reality' which the theory did not capture. Einstein had shifted the focus of the debate with Bohr and his supporters away from the internal consistency of quantum mechanics to the nature of reality and the role of theory.

EPR asserted that for a theory to be complete there had to be one-to-one correspondence between an element of the theory and an element of reality. A sufficient condition for the reality of a physical quantity, such as momentum, is the possibility of predicting it with certainty without disturbing the system. If there exists an element of physical reality that was unaccounted for by the theory then the theory is incomplete. The situation would be akin to a person finding a book in a library and when trying to check it out, being told by the librarian that according to the catalogue there was no record of the library having the book. With the book bearing all the necessary markings indicating that it was indeed a part of the collection, the only possible explanation would be that the library's catalogue was incomplete.

According to the uncertainty principle, a measurement that yields an exact value for the momentum of a microphysical object or system excludes even the possibility of simultaneously measuring its position. The question that Einstein wanted to answer was: Does the inability to measure its exact position directly mean that the electron does not have a definite position? The Copenhagen interpretation answered that in the absence of a measurement to determine its position, the electron has no position. *EPR* set out to demonstrate that there are elements of physical reality, such as an electron having a definite position that quantum mechanics cannot accommodate – and therefore, it is incomplete.

EPR attempted to clinch their argument with a thought experiment. Two particles, *A* and *B*, interact briefly and then move off in opposite directions. The uncertainty principle forbids the exact measurement, at any given instant, of both the position and the momentum of either particle. However, it does allow an exact and simultaneous measurement of.

The key to the *EPR* thought experiment is to leave particle *B* undisturbed by avoiding any direct observation of it. Even if *A* and *B* are light years apart, nothing within the mathematical structure of quantum mechanics prohibits a measurement of the momentum of *A* yielding information about the exact momentum of B without B being disturbed in the process. When the momentum of particle *A* is measured exactly, it indirectly but simultaneously allows, via the law of conservation of momentum, an exact determination of the momentum of B. Therefore, according to the *EPR* criterion of reality, the momentum of *B* must be an element of physical reality. Similarly, by measuring the exact position of *A*, it is possible, because the physical distance separating *A* and *B* is known, to deduce the position of B without directly measuring it. Hence, *EPR* argue, it too must be an element of physical reality. EPR appeared to have contrived a means to establish with certainty the exact values of either the momentum or the position of *B* due to measurements performed on particle *A*, without the slightest possibility of particle *B* being physically disturbed.

Given their reality criterion, *EPR* argued that they had thus proved that both the momentum and position of particle *B* are 'elements of reality', that *B* can have simultaneously exact values of position and momentum. Since quantum mechanics via the uncertainty principle rules out any possibility of a particle simultaneously possessing both these properties, these 'elements of reality' have no counterparts in the theory. Therefore the quantum mechanical description of physical reality, *EPR* conclude, is incomplete.

Einstein's thought experiment was not designed to simultaneously measure the position and momentum of particle *B*. He accepted that it was impossible to measure either of these properties of a particle directly without causing an irreducible physical disturbance. Instead, the two-particle thought experiment was constructed to show that such properties could have a definite simultaneous existence, that both the position and the momentum of a particle are 'elements of reality'. If these properties of particle *B* can be determined without *B* being observed (measured), then these properties of *B* exist as elements of physical reality independently of being observed (measured). Particle *B* has a position that is real and a momentum that is real.

EPR was aware of the possible counter-argument that 'two or more physical quantities can be regarded as simultaneous elements of reality only they can be simultaneously measured or predicted'. This, however, made the reality of the momentum and position of particle *B* dependent upon the process of measurement carried out on particle *A*, which could be light years away and which does not disturb particle *B* in any way. 'No reasonable definition of reality could be expected to permit this', said *EPR*.

Einstein around 1940

Courtesy of California Institute of Technology and the Hebrew University of Jerusalem

Central to the *EPR* argument was Einstein's assumption of locality – that some mysterious, instantaneous "action-*at-a-distance*" does not exist. Locality ruled out the possibility of an event in a certain region of space instantaneously, faster-than-light, influencing another event elsewhere. For Einstein, the speed of light was nature's unbreakable limit on how fast anything could travel from one place to another. For the discoverer of relativity it was inconceivable for a measurement on particle *A* to affect instantaneously, at a distance, the independent elements of physical reality possessed by particle *B*. As soon as the *EPR* paper appeared, the alarm was raised among the leading quantum pioneers throughout Europe. 'Einstein has once again made a public statement about quantum mechanics, and even in the issue of Physical Review of May 15 (together with Podolsky and Rosen, not good company by the way),' wrote a furious Pauli in Zurich to Heisenberg in Leipzig. 'As is well known,' he continued, 'that is a disaster whenever it happens.'

This problem of getting the interpretation proved to be rather more difficult than just working out the equations', said Paul Dirac 50 years after the 1927 Solvay conference. The American Nobel laureate Murray Gell-Mann believes part of the reason was that 'Niels Bohr brain-washed a whole generation of physicists into believing that the problem had been solved.'

A poll conducted in July 1999 during a conference on quantum physics held at Cambridge University revealed the answers of a new generation to the vexed question of interpretation.

Of the 90 physicists polled, only four voted for the Copenhagen interpretation, but 30 favored the modern version of Everett's many worlds. Significantly, 50 ticked the box labeled 'none of the above or undecided'. The unresolved conceptual difficulties, such as the measurement problem and the inability to say exactly where the quantum world ends and the classical world of the everyday begins, have led to an increasing number of physicists willing to look for something deeper than quantum mechanics. 'A theory that yields "maybe" as an answer,' says the Dutch Nobel Prize-winning theorist Gerard 't Hooft, 'should be recognized as an inaccurate theory.' He believes the Universe is deterministic, and is in search of a more fundamental theory that would account for all of the strange, counter-intuitive features of quantum mechanics.

Others like Nicolas Gisin, a leading experimenter exploring entanglement, 'have no problem thinking that quantum theory is incomplete.' The emergence of other interpretations and the claim to completeness of quantum mechanics being in serious doubt have led to a reconsideration of the long-standing verdict against Einstein in his long-running debate with Bohr. 'Can it really be true that Einstein, in any significant sense, was as profoundly "wrong" as the followers of Bohr might maintain?' asks the British mathematician and physicist Sir Roger Penrose. 'I do not believe so. I would, myself, side strongly with Einstein in his belief in a sub-microscopic reality, and with his conviction that present-day quantum mechanics is fundamentally incomplete.'

Although Einstein never managed to deliver a decisive blow in his encounters with Bohr, Einstein's challenge was sustained and thought-provoking. It encouraged men like Bohm, Bell and Everett to probe and evaluate Bohr's Copenhagen interpretation. After having been marginalized during the last 30 years of his life because of his criticism of the Copenhagen interpretation and his attempts to slay his quantum demon, Einstein has been vindicated, in part. Einstein versus Bohr had little to do with the equations and numbers generated by the mathematics and quantum mechanics. What does quantum mechanics mean? What does it say about the nature of reality? It was their answers to these types of questions that separated the two men. Einstein never put forward an interpretation of his own, because he was not trying to shape his philosophy to fit a physical theory. Instead he used his belief in an observer-independent reality to assess quantum mechanics and found the theory wanting.

What is the true nature of the electron? After many years of deep thought and numerous discussions with Einstein, Niels Bohr came to the conclusion that the question doesn't have any meaning. According to Bohr, *'it's meaningless to ask what an electron really is. Physics isn't about what is; it's about telling us something about the world.'*

Heisenberg's uncertainty principle says that you can't determine with complete accuracy the position of an electron and at the same time measure where and how fast it's going (its momentum). An electron occupies a place in space only when you measure it at that location. If you measure its location again and find it someplace else, that's all you can say. How it got from here to there is a question that has no meaning in physics. The uncertainty principle doesn't apply just to electrons and atoms. It applies to everything. However, because it involves Planck's constant, which is very small, you don't notice its effects when you are watching baseballs, cars, or planets move.

7.10 The Proof

The direct, experimental proof of the paradoxical reality of the quantum world comes from modern versions of the *EPR* thought experiment. The modern experiments don't involve measurements of the position and momentum of particles, but of spin and polarization – a property of light that is in some ways analogous to the spin of a material particle. David Bohm of Birkbeck College in London, introduced the idea of spin measurements in a new version of the EPR thought experiment in 1952, but it wasn't until the 1960s that anybody seriously considered actually performing experiments to test the predictions of quantum theory in such situations. The conceptual breakthrough came in a 1964 paper by John Bell, a physicist working at *CERN*, the European research center near Geneva. But to understand the experiments, we need first to step back a little from that key paper and make sure we have a clear idea what "spin" and "polarization" mean.

7.11 The strength of the Copenhagen Interpretation

Each attack on the Copenhagen interpretation has strengthened its position. When thinkers of the caliber of Einstein try to find flaws with a theory, but the defenders of the theory are able to refute all of the attackers' arguments, that theory must emerge the stronger for its trial.

The Copenhagen interpretation is definitely "right" in the sense that it works; any better interpretation of the quantum rules must include the Copenhagen interpretation as a working view that enables experimenters to predict the outcome of their experiments – at least in an statistical sense – and enables engineers to design working laser systems, computers, and so on.

There is no point in going over all of the groundwork that resulted in the refutation of all of the counterproposals to the Copenhagen interpretation, a job that has been well done by others. Perhaps, though, the most important point to notice is one made by Heisenberg in his book "*Physics and Philosophy*," back in 1958. All of the counterproposals, Heisenberg stressed, are 'compelled to sacrifice the essential symmetry of quantum theory (for instance, the symmetry between waves and particles or between position and velocity). Therefore, we may well suppose that the Copenhagen interpretation cannot be avoided if these symmetry properties are held to be a genuine feature of nature; and every experiment yet performed supports this view.'

There is an improvement on the Copenhagen interpretation (not a refutation or counterproposal) that still includes this essential symmetry, and that best-but picture of quantum reality will be described in Chapter Eleven. However, it is hardly surprising that Heisenberg failed to mention it in a book published in 1958, since the new picture was only just being developed at that time, by a PhD student in the United States. Before we get on to that, however, it is right to trace the path of a combination of theory and experiment that by 1982 had established beyond any doubt the accuracy of the Copenhagen interpretation as a working view of quantum reality. The story starts with Einstein, and ends in a physics laboratory in Paris more than fifty years later; it is one of the great stories of science.

7.12 Schrodinger's Cat

The famous cat paradox first appeared in print (*Naturwissenschaften*, vol 23 page 812) in 1935, the same year as the EPR paper. Einstein saw Schrodinger's proposal as the "prettiest way" to show that the wave representation of matter is an incomplete representation of reality, and together with the EPR argument the cat paradox is still discussed in quantum theory today. Unlike the EPR argument, however, it has not been resolved to everyone's satisfaction.

Yet the concept behind this thought experiment is very simple. Schrodinger suggested that should imagine a box that contains a radioactive source, a detector that records the presence of radioactive particles (a Geiger counter, perhaps), a glass bottle containing a poison such as cyanide, and a live cat. The apparatus in the box is arranged so that there is a fifty-fifty chance that one of the atoms in the radioactive material will decay and that the detector will record a particle. If the detector does record such an event, then the glass container is crushed and the cat dies; if not, the cat lives.

We have no way of knowing the outcome of this experiment until we open the box to look inside; radioactive decay occurs entirely by chance and is unpredictable except in a statistical sense. According to the strict Copenhagen interpretation, just as in the two-hole experiment there is an equal probability that the electron goes through either hole, and the two overlapping possibilities for radioactive decay and no radioactive decay should produce a superposition of states. The whole experiment, cat and all, is governed by the rule that the superposition is "real" until we look at the experiment, and that only at the instant of observation does the wave function collapse into one of the two states.

Until we look inside, there is a radioactive sample that has both decayed and not decayed, a glass vessel of poison that is neither broken nor unbroken, and a cat that is both dead and alive, neither alive nor dead. A neutron can briefly change into a proton plus a charged pion, provided the two quickly get back together.

And a pion can create a virtual neutron/antiproton pair for a similarly brief interval. It is one thing to imagine an elementary particle such as an electron being neither here nor there but in some "superposition" of states, but much harder to imagine a familiar thing like a cat in this form of suspended animation. Schrodinger thought up the example to establish that there is a flaw in the strict Copenhagen interpretation, since obviously the cat cannot be both alive and dead at the same time.

But is this any more "obvious" than the "fact" that an electron cannot be both a particle and a wave at the same time? *Common sense has already been tested as a guide to quantum reality and has been found wanting.* The one sure thing we know about the quantum world is not to trust our common sense and only to believe things we can see directly or detect unambiguously with our instruments. We don't know what goes on inside a box unless we look.

Arguments about the cat in the box have gone on for over seventy years. One school of thought says that there is no problem, because the cat is quite able to decide for itself whether it is alive or dead, and that the cat's consciousness is sufficient to trigger the collapse of the wave function. In that case, where do you draw the line? Would an ant be aware of what was going on, or a bacterium? Moving in the other direction, since this is only a thought experiment we can imagine a human volunteer taking the place of the cat in the box (the volunteer is sometimes referred to as "Wigner's friend," after Eugene Wigner, a physicist who thought deeply about variations on the *cat-in-the-box* experiment and who, incidentally, is Dirac's brother-in-law). The human occupant of the box is clearly a competent observer who has the quantum-mechanical ability to collapse wave functions.

When we open the box, assuming we are lucky enough to find him still living, we can be quite sure that he will not report any mystic experiences but simply that the radioactive source failed to produce any particles at the allotted time. Yet still, to us outside the box the only correct way to describe conditions inside the box is as a superposition of states, until we look. In Feynman (space-time) diagram of a genuine interaction of several particles revealed by a bubble-chamber. A single proton could be involved in a network of virtual interactions. Such interactions go on all the time. No particle is as lonely as it seems at first sight. A proton, antineutron, and pion can appear out of nothing at all, as a vacuum fluctuation, for a brief time before annihilating. The same interaction can be represented as a loop in time, with proton and neutron chasing each other around a time eddy linked by a pion. Both view are equally valid.

A proton can chase its own tail through time in the same way. The chain is endless. Imagine that we have announced the experiment in advance to an intrigued world, but to avoid press interference it has been performed behind locked doors. Even after we have opened the box and either greeted our friend or dragged the corpse out, the reporters outside don't know what is going on. To them, the whole building in which our laboratory is based is in a superposition of states. And so on, back out in an infinite regression.

But suppose we replace Wigner's friend by a computer. The computer can register the information about the radioactivity decay, or lack of it. Can a computer collapse the wave function (at least inside the box)? According to yet another point of view, what matters is not human awareness of the outcome of the experiment, or even the awareness of a living creature, but the fact that the outcome of an event at the quantum level has been recorded (interacted upon), or made an impact on the macro-world. The radioactive atom may be in a superposition of states, but as soon as the *Geiger counter* has "probed" the atom the properties of the atom is forced into one state or the other, either decayed or not decayed.

So, unlike the *EPR* thought experiment, the cat-in-the-box experiment really does have paradoxical overtones. It is impossible to reconcile with the strict Copenhagen interpretation without accepting the "reality" of a dead-alive cat, and it has led Wigner and John Wheeler to consider the possibility that, because of the infinite regression of "cause and effect," the whole Universe may only owe its "real" existence to the fact that it is observed by intelligent beings. The most paradoxical of all the possibilities inherent in quantum theory is a direct descendant of Schrodinger's cat experiment, and it jumps off from what Wheeler calls a delayed-choice experiment.

7.13 The participatory Universe

American physicist Wheeler has written many thousands of words on the meaning of quantum theory, in many different publications, over a span of four decades. Perhaps the clearest exposition of his concept of the "*participatory Universe*" is in his contribution to "*Some Strangeness in the Proportion*," the proceedings (edited by Harry Woolf) of a symposium held to celebrate the centenary of Einstein's birth.

The only things we know about the quantum world are the results of experiments. The cloud was, in a sense, created by the process of questioning, and in the same sense the electron is created by our process of experimental probing. This anecdote stresses the fundamental axiom of quantum theory that no elementary phenomenon is a phenomenon until it is a recorded phenomenon. And the process of recording can play strange tricks with our everyday concept of reality. To make his point, Wheeler has come up with a thought experiment, a variation on the two-slit experiment. In this version of the game, the two slits are combined with a lens to focus light passing through the system, and the standard screen is replaced by another lens that can cause photons coming from each of the two slits to diverge.

A photon that passes through one slit goes through the second screen and is deflected by the second lens off to a detector on the left; a photon passing through the other slit goes to a detector on the right. With this experimental setup, we know which slit each photon went through, as surely as in the version where we watch each slit to see if the photon passes.

Just as in that case, if we allow one photon at a time to pass through the apparatus we unambiguously identify the path it follows and there is no interference because there is no superposition of states. Now modify the apparatus again. Cover the second lens with a photographic film arranged in strips like a venetian blind. The strips can be closed to make a complete screen, preventing the photons from passing through the lens and being deflected. Or the strips can be opened, allowing the photons to pass as before. Now, when the strips are closed the photons arrive at a screen just as in the classic two-hole experiment. We have no way of telling which hole each one went through, and we have an interference pattern as if each individual photon went through both slits at once. Now comes the trick. With this setup, we don't have to decide whether to open or close the strips until the photon has already passed the two holes. We can wait until after the photon has passed the two slits, and then decide whether to create an experiment in which it has gone through one hole alone or through "both at once."

In this delayed-choice experiment, something we do now has an irretrievable influence on what we can say about the past. History, for one photon at least, depends upon how we choose to make a measurement.

Philosophers have long pondered the fact that history has no meaning – the past has no experience – except in the way it is recorded in the present. Wheeler's delayed-choice experiment fleshes out this abstract concept into solid, practical terms. "We have no more right to say 'what the photon is doing' – until it is registered – than we do to say 'what word is in the room' – until the game of question and response is terminated" (see *Some Strangeness*, page 358). How far can this concept be pushed? The happy quantum cooks, constructing their computers and manipulating genetic material, will tell you that it is all philosophical speculation, and that it doesn't mean anything in the everyday, macroscopic world. But everything in the macroscopic world is made of particles that obey the quantum rules. Everything we call real is made of things that cannot be regarded as real; "what choice do we have but to say that in some way, yet to be discovered, they all must be built upon the statistics of billions upon billions of such acts of observer-participation?"

Never afraid to make the grand intuitive leap (remember his vision of the single electron weaving its way through space and time). Wheeler goes on to consider the whole Universe as a participatory, self-excited circuit. Starting from the Big Bang, the Universe expands and cools; after thousands of millions of years it produces beings capable of observing the Universe, and "acts of *observer-participancy* – via the mechanism of the delayed-choice experiment – in turn give tangible 'reality' to the Universe not only now but back to the beginning." By observing the photons of the cosmic background radiation, the echo of the Big Bang, we may be creating the Big Bang and the Universe. If Wheeler is correct, Feynman was even closer to the truth than he realized when he said that the two-hole experiment "contains the only mystery."

Following Wheeler, we have wandered into the realms of *metaphysics*, and one can imagine many readers thinking that since all this depends upon hypothetical thought experiments, you can play any game you like and it doesn't really matter which interpretation of reality you subscribe to. What we need is some solid evidence from real experiments on which to base a judgment of the best choice of interpretation Aspect experiment in the early 1980s – the proof that quantum weirdness is not only real but observable and measureable. The whole Universe can be thought of as a delayed-choice experiment in which the existence of observers who notice what is going on is what imparts tangible reality to the origin of everything.

7.14 Quantum Uncertainty

All quantum systems are subject to inherent uncertainty. As a typical system evolves, there are many possible outcomes, many contending realities, on offer. For instance, in the various laser experiments described earlier, a photon had a choice of which path to follow through a piece of apparatus. In the case of a lab experiment, the observer will always see a single, specific, concrete reality selected from among the phantom contenders. Thus a measurement of the photon's path will always yield a result or either one path or the other, but never both. When it comes to the Universe as a whole, there is no outside observer, because the Universe is all that there is, so quantum cosmology runs up against a major problem of interpretation. The favored way out is to assume that all the contending quantum realities enjoy equal status. They are not mere "phantom worlds," or "potential realities," but "really real" - all of them. Each reality corresponds to an entire Universe, complete with its own space and time.

These many Universes are not connected through space and time but are somehow "parallel," existing alongside each other. In general there will be an infinity of them. You might be wondering, if there are so many Universes about, why do we see only one of them. This is explained by supposing that, when a Universe splits into, say, two alternative worlds, observers also split, with each copy perceiving her or his single respective world. In practice, quantum processes going on at the atomic level all the time are continually splitting the Universe, and the reader, into vast numbers of copies. Each version of you touchingly believes she or he is unique. Bizarre though this may seem it is entirely consistent with experience, so long as the various Universes remain separate. Trouble comes, however, if they start overlapping or interfering with each other.

This leads to a second question: is it possible to observe the other Universe? The usual answer is no, but there is no unanimous agreement about this. David Deutch, a physicist at Oxford University with a bent for the unusual, believes that microscopic experiments could in principle be done in which two or more worlds are temporarily joined, allowing physical influences to slip through. Under normal circumstances, a join between two quantum worlds would produce only atomic-level effects, rather than a sort of *"paranormal"* phenomena at the macro level. However, some scientists suspect there may be circumstances in which a mixing of quantum realities dramatically manifests itself on a human scale. Tachyons - hypothetical particles that would always travel faster than light - and mentioned that "faster than light" can mean "backwards in time."

Although Einstein's special theory of relativity unequivocally forbids ordinary matter, and by implications human beings, from traveling into the past, the general theory of relativity is questionable on the subject. The speed of light is a barrier to cause and effect. Steven Hawking has proposed a "chronology protection hypothesis," according to which nature will always find a way to prevent "wormholes" and other contrivances from permitting travel into the past. The toughest argument against visiting the past is undoubtedly the grandmother paradox.

During brain operations, the patient is usually kept conscious. Libet took the opportunity to attach electrodes to exposed brains. By stimulating the cortex electrically, he was able to produce the sensation of a tingle in the patient's hand. Descartes' dualism says that it is caused by a nonmaterial mind somehow priming the brain to do its bidding. Consciousness of time differs from consciousness of other physical qualities, such as spatial size or shape, in a significant respect.

When we see a shape such as a square, the electrical activity in our brains is not square-shaped. There is no "little square" inside our heads, projected on a movie screen for the subject to watch. Instead, a complicated pattern of electrical activity somehow produces the sensation "square." That is to say, the square is represented by an electrical pattern. Galileo, Newton and Einstein all chose time as the central conceptual pillar around which to build a scientific picture of physical reality.

7.15 Enters David Bohm

One of the most interesting and "offbeat" scientists of the postwar years was David Bohm, an American-born theoretical physicist who worked mainly in London. There was a concern about the paradox about the nature of time. We take for granted that, when a radio station transmits a signal, we receive the signal on our radio set at home after it was sent out by the transmitter. The delay is not very long - just a fraction of a second from point to point on Earth - so we are normally unaware of it. But a telephone conversation relayed via satellite can introduce a noticeable time lag. Anyway, the point is that we never get to hear the radio signal before it is sent.

Why should we? You may ask. After all, effects don't normally happen before their causes. The problem that lay at the root of my worries stems from the mid-19th century, when James Clerk Maxwell wrote down his famous equations that describe the propagation of electromagnetic waves such as light and radio. Maxwell's theory predicts that radio waves travel through empty space at the speed of light.

What Maxwell's equations do not tell us, however, is whether these waves arrive before or after they are transmitted. They are indifferent to the distinction between past and future. According to the equations, it is perfectly permissible for the radio waves to go backwards in time as well as forwards in time.

Given a pattern of electromagnetic activity, such as that corresponding to radio waves from a transmitter spreading out through space, the time-reversed pattern (in this case, converging waves) is equally permitted by the laws of electromagnetism.

In physics jargon, forward-in-time waves are called "retarded" (as they arrive late) and backward-in-time waves are called "advanced" (as they arrive early). Although Bohm was famed for his writings and philosophical works, particularly among readers of a "mystical bent," he was a curiously isolated figure among the physics community. He was perhaps best known for his 1950s textbook on quantum mechanics. But very early on he decided he didn't like quantum mechanics as conventionally formulated, *ala* Bohr. So it became Bohm versus Bohr. Bohm thus took up the lonely torch of quantum dissent where Einstein had left it on his deathbed. With the help of a small band of devotees, most notably his Birkbeck colleague Basil Hiely, Bohm sought a theory in which the apparently random and unpredictable aspects of quantum phenomena had their origin in some deeper-level deterministic processes.

Bohm had this fascinating idea that, although some features of the world might look complicated, or even random, beneath it all there lays a hidden order, somehow "folded up." In later years, he was to call it "the implicate order." 'In my opinion', replied Bohm, 'progress in science is usually made by dropping assumptions.' This seemed like a humiliating put-down at the time, but I have always remembered those words of David Bohm. History shows he is right.

So often, major progress in science comes when the orthodox paradigm clashes with a new set of ideas or some new piece of experimental evidence that won't fit into the prevailing theories. Then somebody discards a cherished assumption, perhaps one that has almost been taken for granted and not explicitly stated, and suddenly all is transformed.

7.16 Hidden Variables

An interpretation that rejects the positivism of Bohr in favor of realism (or "naïve realism" as some of its detractors prefer) is based on what are known as "hidden variables," by which is meant that a quantum object actually does possess attributes, even when these cannot be observed. The leading theory of this kind is known as the 'de-Broglie-Bohm model' *(DBB)* after Louis de Broglie, the first person to postulate matter waves, and David Bohm, who developed and expanded these ideas in the 1950s and 1960s. In *DBB* theory, both the particle position and the wave are assumed to be real attributes of a particle at all times. The wave evolves according to the laws of quantum physics and the particles are guided both by the wave and the classical forces acting on it. The path followed by any particular particle is then completely determined and there is no uncertainty at this level. However, different particles arrive at different places depending on where they start from, and the theory ensures that the number arriving at different points are consistent with the probabilities predicted by quantum physics.

As an example, consider the two-split experiment: according to *DBB* theory the form of the wave is determined by the wave so that most of them end up in positions where the interference pattern has high intensity, while none arrives at the points where the wave is zero.

As we have noted before, the emergence of apparently random, statistical outcomes from the behavior of determining systems is quite familiar in a classical context. For example, if we toss a large number of coins, we will find that close to half of them come down heads while the rest show tail, even though the behavior of any individual coin is controlled by the forces acting on it and the initial spin imparted when it is tossed. Similarly, the behavior of the atoms in a gas can be analyzed statistically, even when the motion of its individual atoms and the collisions between them are controlled by classical mechanical laws.

Whenever a measurement is made some uncertainty or error is always involved. For example, you cannot make an absolutely exact measurement of the length of a table. Even with a measuring stick that has markings 1 mm apart, there will be an inaccuracy of about 12 mm or so. More precise instruments will produce more precise measurements. But there is always some uncertainty involved in a measurement no matter how accurate the measuring device. We expect that by using more precise instruments, the uncertainty in a measurement can be made indefinitely small.

But according to quantum mechanics, there is actually a limit to the accuracy of certain measurements. This limit is not a restriction on how well instruments can be made; rather, it is inherent in nature. It is the result of two factors: the wave-particle duality, and the unavoidable interaction between the thing observed and the observing instrument. Let us look at this in more detail.

To make a measurement of an object without somehow disturbing it, at least a little, is not possible. Consider trying to locate a *ping-pong ball* in a completely dark room, you grope about to find its position; and just when you touch it with your finger, it bounces away. Whenever we measure the position of an object, whether it's a ping-pong ball or an electron, we always touch it with something else that gives us the information about its position. To locate a lost ping-pong ball in a dark room, you could probe about with your hand or a stick; or you could shine a light and detect the light reflecting off the ball.

When you search with your hand or a stick, you find the ball's position when you touch it. But when you touch the ball you unavoidably bump it, and give it some momentum.

Thus you won't know its future position. The same would be true, but to a much lesser extent, if you observe the ping-pong ball using light. In order to "see" the ball, at least one photon must scatter from it. And the reflected photon must enter your eye or some other detector. When a photon strikes an ordinary-sized object, it does not appreciably alter the motion or position of the object. But when a photon strikes a very tiny object like an electron, it can transfer momentum to the object and thus greatly change the object's motion and position in an unpredictable way.

The mere act of measuring the position of an object at one time makes our knowledge of its future position imprecise. Now let us see where the wave-particle duality comes in. Imagine a thought experiment in which we are trying to measure the position of an object, say an electron, with photons. (The arguments would be similar if we were using, instead, an electron microscope). As we saw in an earlier chapter objects can be seen to an accuracy at best of about the *wave-length* of the radiation used. If we want an accurate position measurement we must use a short wavelength. But a short wavelength corresponds to high frequency and high energy (since $E = hf$); and the more energy the photons have, the more momentum they can give the object when they strike it. If photons of longer wavelength and correspondingly lower energy are used, the object's motion when struck by the photons will not be affected as much. But the longer wavelength means lower resolution, so the object's position will be less accurately known.

Thus, the act of observing products creates a significant uncertainty in either the position or the momentum of the electron. This is the essence of the uncertainty principle first enunciated by Heisenberg in 1927. Heisenberg's uncertainty principle tells us that we cannot measure both the position and momentum of an object precisely at the same time.

The more accurately we try to measure the position, so that Δ-x is small, the greater will be the uncertainty in momentum, Δ-p. If we try to measure the momentum very precisely, then the uncertainty in the position becomes large.

The uncertainty principle does not forbid single exact measurements, however. For example, in principle we could measure the position of an object exactly. But then its momentum would be completely unknown. Thus, although we might know the position of the object exactly at one instant, we could have no idea at all where it would be a moment later.

7.17 David Bohm and the hidden variables

More constructively, this book presents various arguments in favor of one such alternative to the Copenhagen Interpretation, the *"hidden-variables"* theory developed since the early 1950s by David Bohm and consistently neglected or marginalized by proponents of the Copenhagen doctrine. This is a version of the pilot-wave hypothesis, first put forward by Louis de Broglie, according to which the particle is 'guided' by a wave whose probability amplitudes are exactly in accordance with the well-supported

QM predictions and measured results.

Where it challenges the orthodox theory is in Bohm's realist premise that the particle does have precise simultaneous values of position and momentum, and furthermore that these pertain to its objective state at any given time, whatever the restrictions impose upon our knowledge by that limits of achievable precision in measurement.

According to Deutch, the many-worlds (or multiverse) theory is the 'sole plausible,' i.e. the physically and logically consistent solution to the various well-known QM paradoxes of wave-particle dualism, remote simultaneous interaction, the observer-induced 'collapse of the wave-packet', and so forth. According to this hypothesis we must assume that all possible outcomes are realized in each such momentary 'collapse' since the observer splits off into so many parallel, coexisting, but epistemically non-inter accessible 'worlds' whose subsequent branching constitute the lifeline – or experiential world series – for each of those endlessly proliferating centers of consciousness. What must interested lay persons will have gathered from the current literature can perhaps be best summarized as follows.

(1) QM has given rise to a number of problems and paradoxes – among them the wave-particle dualism – as regards physical 'reality' and the kinds or degrees of exactitude dualism – as regards physical 'reality' and the kinds or degrees of exactitude in scientific knowledge that we can hope to gain concerning it.

(2) Those problem have to do with certain limits that apply to the detection or measurement of quantum phenomena, such as the impossibility of assigning precise simultaneous values of location and momentum, or the fact – famously enshrined in Heisenberg's uncertainty principle – that any observation of subatomic particles, for instance through an electron microscope, will involve their exposure to a stream of other such energy-bearing particles and will thus affect or in some sense determine what is 'actually' there to be observed.

(3) As for the quantum paradoxes, these take rise from the necessity, as it seems, of abandoning local realism (i.e. Einstein's rule that no casual influence can propagate faster than the speed of light) in favor or remote superluminal interaction between particles at no matter how great a distance. For there is now a large body of experimental evidence that such nonlocal effects can indeed be shown to exist and that any realist interpretation will consequently need to take them on board, thus creating additional (some would say insoluble) problems for its own case.

The notion that a body persists in a state of uniform motion unless acted upon by a resultant force would be counter-intuitive to Aristotle but natural for Galileo. Such were some of Einstein's chief objections in the famous series of debates with Niels Bohr, when he argued that the orthodox (Copenhagen) theory of quantum mechanics was necessarily 'incomplete' since it entailed the existence of unthinkable phenomena such as instantaneous remote correlation of 'spooky action-at-a-distance'. Although he had been among the chief contributors to the early development of quantum mechanics, Einstein was by now deeply dissatisfied with what he saw as its failure to provide any adequate realist or casual-explanatory account of *QM* phenomena. The debate between Einstein and Bohr was Quantum Mechanics and the orthodox Copenhagen Interpretation of QM being a 'Complete Theory'. Therefore it became an article of faith among upholders of the Orthodox view – Heisenberg especially – that Quantum mechanics was indeed 'complete' in the since that no alternative account (such as Bohm's hidden-variables theory) could possibly support a realist interpretation while also matching the well-proven QM formalisms and predictions.

Bohr disagreed so sharply with Einstein on the issue of whether the orthodox theory might yet turn out to be 'incomplete', or to leave room for some future advance that would reconcile quantum mechanics with the aims and methods of classical physics, including – most importantly in this context – the special and general theories of relativity. For one major problem with orthodox QM was that it seemed to entail the existence of nonlocal simultaneous (faster than light 'communication' between particles that had once interacted and then moved apart of whatever distance of space-time separation.

Preface: Niels Bohr solved the problem of interpreting quantum mechanics once and for all. That until recently was the orthodoxy in physics, especially among writers of textbooks on quantum mechanics, who mostly wanted to get off the philosophy as quickly as possible and on with the physics. Niels Bohr, incidentally, would not have agreed with them. Now the fashion is for physicists to say that quantum mechanics is so peculiar that no one understands it, least of all themselves, is quite appropriate. Perhaps the change was wrought by Bell's theorem in the 1960s; perhaps it really was just a matter of fashion. Of course, in saying that quantum mechanics is incomprehensible, one is not saying that it is false, only that the human mind is not tuned in to the way the world is. The philosophy of quantum mechanics deals with the question: What is the way the world is, if *QM* is true. Realism in the philosophy of quantum mechanics means the idea that quantum systems are like classical particles.

7.18 The Bell test

According to Einstein, no influence can propagate faster than the speed of light, which is called "locality." The first tests of Bell's inequality were carried out at the University of California, Berkeley, using photons, and were reported in 1972. In the mid-1970s, experimenters carried out the first measurements using another variation on the theme. In these experiments, the photons are gamma rays produced when an electron and a positron annihilate. Again, the polarizations of the two photons must be correlated, and again the balance of evidence is that, when you try to measure those polarizations, the results you get violate Bell's inequality.

So out of the first seven tests of Bell's inequality five came out in favor of quantum mechanics. In his *Scientific American* article d'Espagnat stresses that this is even stronger evidence in favor of quantum theory than it seems at first sight.

Because of the nature of the experiments and the difficulties of operating them, "a great variety of systematic flaws in the design of an experiment could destroy the evidence of a real correlation . . . on the other hand it is hard to imagine an experimental error that could create a false correlation in five independent experiments. What is more, the results of those experiments not only violate Bell's inequality but also violate it precisely as quantum mechanics predicts.

Since the mid-1970s, even move tests have been carried out, designed to remove any remaining loopholes in the experimental design. The bits of the apparatus need to be placed far enough apart so that any "signal" between the detectors that might produce a spurious correlation would have to travel faster than light. Hat is done, and still Bell's inequality was violated. Or perhaps the correlation occurs because the photons "know," even as they are being created, what kind of experimental apparatus has been set up to trap them. That could happen, without the need for faster than light signals, if the apparatus has been set up in advance, and has established an overall wave function that affects the photon at birth. The ultimate test, so far, of Bell's inequality therefore involves changing the structure of the experiment while the photons are in flight, very much in the way that the double-slit experiment can be changed while the photon is in flight in John Wheeler's thought experiment. This is the experiment in which Alain Aspect's team at the University of Paris-South closed the last major loophole for the local realistic theories in 1982.

Aspect and his colleagues had already carried out tests of the *inequality* using photons from a cascade process, and found that the inequality was violated. Their improvement involved the use of a switch that changes the direction of a beam of light passing through it. The beam can be directed toward either one of two polarizing filters, each measuring a different direction of polarization, and each with its own photon detector behind it. The direction of the light beam passing through this switch can be changed with extraordinary rapidity, every 10 nanoseconds (1o thousand-millionths of a second, 10×10^{-9} s), by an automatic device that generates a pseudorandom signal. Because it takes 20 nanoseconds for a photon to travel from the atom in which it is born in the heart of the experiment to the detector itself, there is no way in which any information about the experimental setup can travel from one part of the apparatus to the other and affect the outcome of any measurement – unless such an influence travels faster than light.

Even here we get a scent of the kind of problems that puzzled Bohr for so long. The only real things are the results of our experiments, and the way we make measurements influences what we measure. Here, in the 1980s, are physicists using as an everyday tool of their trade a laser beam whose job is simply to pump atoms into an excited state.

We can only use this tool because we know about excited states and have the quantum cookbook to hand, but the whole purpose of our experiment is to verify the accuracy of quantum mechanics, the theory we used to write the quantum cookbook. I'm not suggesting that the experiments are therefore wrong. We can imagine other ways to excite atoms before we make the measurements, and other versions of the experiment do give the same result. But just as the everyday conceptions of previous generations of physicists were colored by their use of, say, spring balances and meter rules, so the present generation is affected, far more then it sometimes realizes, by the quantum tools of the trade. Philosophers may care to take up the question of what the results of the Bell experiment really mean if we are using quantum processes in order to set up the experiment. With Bohr – what we see is what we get; nothing else is real.

By 1975 six such tests had been carried out, and four of them produced results that violated Bell's inequality. Whatever the doubts about the meaning of locality for photons, this is striking further evidence in favor of quantum mechanics, especially since the experiments used two fundamentally different techniques. In the earliest photon version of the experiment, the photons came from atoms of either calcium or mercury, which can be excited by laser light into a chosen energetic state. The route back from this excited state to the ground state involves an electron in two transitions, first to another, but lower, excited state, and then on to the ground state; and each transition produces a photon. For the transitions chosen in these experiments, the two photons are produced with correlated polarizations. The photons from the cascade can then be analyzed, using photon counters that are placed behind polarizing filters. The Bell test starts out from a "local" realistic view of the world. In terms of the proton spin experiment, although the experimenter can never know the three components of spin for the same particle, he can measure any one of them he likes. If the three components are called *X*, *Y*, and *Z*, he finds that every time he records a value +1 for the *X* spin of one proton, he finds a value -1 for the *X* spin if its counterpart, and so on. But he is allowed to measure the *X* spin of one proton and the *Y* (or *Z* but not both) spin of its counterpart, and in that way it ought to be possible to get information about both the *X* and *Y* spins of each of the pair.

Even in principle, this is far from easy, and involves measuring the spins of lots of pairs of protons at random and discarding the ones that just happened to end up measuring the same spin vector in both members of the pair. Particles with half-integer spin can only line up parallel or anti-parallel to a magnetic field. Particles with integer spin are also able to align across the field.

But it can be done, and this gives the experimenter, in principle sets of results in which pairs of spins have been identified for pairs of protons in sets that can be written *XY*, *XZ*, and *YZ*. What Bell showed in his classic 1964 paper was that if such an experiment is carried out, then according to the local realistic views of the world the number of pairs for which the *X* and *Y* components both have positive spin (*X*+*Y*+) must always be less than the combined total of the pairs in which the *XZ* and *YZ* measurements all show a positive value of spin (*X*-plus*Z-plus + *Y*-plus*Z-plus). The calculation follows directly from the obvious fact that if a measurement shows that a particular proton has spin *X*-plus and *Y*-minus, for example, then its total spin state must be either *X*-plus *Y*-minus *Z*-plus or *X*-plus *Y*-minus *Z*-minus. The rest follows from a mathematically simple argument based on the theory of sets. But in quantum mechanics the mathematical rules are different, and if they are carried through correctly they come up with the opposite prediction, that the number of *X*+ *Y*+ pairs is more, not less, than the number of *X*+*Z*+ and *Y*+*Z*+ pairs combined. Because the calculation was originally expressed starting out from the local realistic view of the world, the conventional phrasing is that the first inequality is called "Bell's inequality," and that if Bell's inequality is violated then the local realistic view of the world is false, but quantum theory has passed another test.

7.19 The measurement problem

The above may be difficult to accept, but it works, and if we apply the rules and use the map book properly, we will correctly calculate predictable outcomes of measurement: the energy levels of the hydrogen atom, the electrical properties of a semiconductor, the result of a calculation carried out by a quantum computer and so on.

However, this implies that we understand what is meant by *"measurement"* and this turns out to be the most difficult and controversial problem in the interpretation of quantum physics. A 45-degree photon passes through an *HV* polarizer, but instead of being detected, the two paths possible are brought together so that they can interfere in a manner similar to that in the two-slit experiment. Just as in that experiment, we do not know which path the photon passes along, so we cannot attribute reality to either.

The consequence is that the 45-degree polarization is reconstructed by the addition of the *H* and *V* components – as we can demonstrate by passing the photon through another +/- 45-degree polarizer and observing that all the photons emerge in the +/- 45-degree channel.

If, however, we had placed a detector in one of the paths and between the two polarizers, we would either have detected the particle or we would not, so we would know that its polarization was either *H* or *V*; it turns out that in such a case, it is impossible in practice to reconstruct the original state and the emerging photons are either *H* or *V*.

We are led to the conclusion that the act of detection is an essential part of the measuring process and is responsible for placing the photon into an *H* or *V* state. This is consistent with the positivist approach outlined earlier, because in the absence of detection we do not know that the photon possesses polarization so we should not assume that it does. We appear therefore to be able to divide the quantum world from the classical world by the presence or absence of a detector in the experimental arrangement.

However, this begs the question of what is special about a detector and why we cannot treat it as a quantum object. Suppose it were subject to the rules of quantum physics. To be consistent, we would have to say that the detector does not possess the attribute of having detected or of not having detected a photon until its state has been recorded by a further piece of apparatus – say a camera directed at the detector output.

But if we then treat the camera as a quantum object, we have the same problem. At some point we have to make a distinction between the quantum and the classical and abandon our hope of a single fundamental theory explaining both. This impasse is the basis of a now well-known example of the application (or misapplication) of quantum physics, known as Schrodinger's cat.

Erwin Schrodinger who was one of the pioneers of quantum physics and whose equation was referred to previously suggested the following arrangement. A 45-degree photon passes through an *HV* apparatus and interacts with a detector, but the detector is now connected to a gun (or other lethal device arranged so that when a photon is detected the cat is killed. We can then argue that if the *HV* attribute cannot be applied to the photon, then the attribute of detection cannot be applied to the detector, and that of life-or-death cannot be applied to the cat. The cat is neither alive nor dead: it is simultaneously alive and dead.

Light split into two components by an *HV* polarizer can be reunited by a second polarizer facing in the opposite direction (marked as *VH*). If the crystals are set up carefully so that the two paths through the apparatus are identical, the light emerging on the right has the same polarization as that incident on the left. This is also true for individual photons, a fact that is difficult to reconcile with the idea of measurement changing the photon's state of polarization.

The *quantum measurement problem* just described is the heart of the conceptual difficulties of quantum physics and the source of the controversies surrounding it. Detectors and cats seem to be different kinds of objects following different physical laws from those governing the behavior of polarized photons. The former exist in definite states (particle detected or not; cat dead or alive) while the latter exists in *super-positions* until measured by something with the properties of the former. It seems that our dream of a successful single theory is not going to be realized, and that quantum objects differ from classical objects not only in degree, but also in kind.

We may therefore be led to conclude that there is a real and essential distinction on between the experimental apparatus, particularly the detector, and a quantum object such as a photon or electron. In practice, this distinction seems pretty obvious: the measuring apparatus is large and made of a huge number of particles, and is nothing like an electron! However, it is quite hard to define this difference objectively and in principle: just how big does an object have to be in order to be classical? What about a water molecule composed of two hydrogen atoms and an oxygen atom, and containing ten electrons; or what about a speck of dust containing a few million atoms? It has been suggested that the laws of quantum physics are genuinely different in the case of an object made up from a large number of particles.

Philosophically, however, it would be appealing to have one theory for the physical world rather than separate theories for the quantum and classical regimes. Might there be a universal fundamental theory of the physical world that reduces to quantum physics when applied to a single particle or small number of particles and is the same as classical physics when the object we are dealing with is large enough? Physics has seen something like this before when the theory of relativity was developed.

This appears to predict that objects moving close to the speed of light should follow different laws from those of Newton, but in fact the new principles apply to all objects: even when moving slowly, they are subject to the rules of relativity; it is just that the relativistic effects are very small and unnoticeable at low speeds. Perhaps the same is true in the present case, with the role of the speed replaces by, say, the number of fundamental particles in an object.

To test this hypothesis, we could try to demonstrate the wave properties of a large scale object by performing an interference experiment: if the laws of quantum physics are different at the large scale. Then we should be unable to detect interference in a situation where it would be expected.

The largest object to have had its wave properties demonstrated by an interference experiment akin to the Young's slit apparatus is the 'buckminster fullerene' molecule which is composed of sixty carbon atoms in a 'football' configuration. However, this does not mean that larger objects do not have wave properties, but rather that no experiment has been devised that would demonstrate them. The practical difficulties of such experiments become rapidly greater as the size of the object increases, but no experiment has yet been performed in which predicted quantum properties were not observed when they should have been.

We turn now to consider how the Copenhagen interpretation deals with this problem? Consider the following quote from Niels Bohr: Every atomic phenomenon is closed in the sense that its observation is based on registrations obtained by means of suitable amplification devices with irreversible functions such as, for example, permanent marks on a photographic plate caused by the penetration of the electrons into the emulsion.

We may conclude from this that Bohr was content to make a distinction between quantum system and classical apparatus. As we saw above, for all practical purposes, this is not difficult, and our ability to do so reliably underlies the huge success quantum physics has had. However, the Copenhagen approach goes further and denies the reality of anything other than the changes that occur in the classical apparatus: only the life or death of the cat or the 'permanent marks on a photographic plate' are real. The polarization state of the photon is an idealistic concept extrapolated from the results of our observations and no greater reality should be attributed to it. From this point of view, the function of quantum physics is to make statistical predictions about the outcome of experiments and we should not attribute any truth-value to any conclusions we may draw about the nature of the quantum system itself.

Not all physicists and philosophers are content with positivism and considerable effort has been made to develop alternative interpretations that would overcome this problem. All of these have their followers, although none has been able to command the support that would be necessary to replace the Copenhagen interpretation as the consensus view of the scientific community. We shall now discuss some of them.

7.20 Some more evidence

The test ought to apply equally well for the spin measurements of material particles, which are very difficult to carry out, or for measurements of the polarization of photons, which are easier to carry out, though still difficult. Because photons have zero rest mass, move at the speed of light, and have no means of distinguishing time, however, some physicists are uneasy about the experiments that involve photons. It is not really clear what the concept of locality means for a photon. So although most of the tests of Bell's inequality carried out so far have involved measurements of the polarization of photons, it is crucially important that the only test so far carried out actually using measurements of proton spins does give results that violate Bell's inequality and therefore support the quantum view of the world.

This was not the first test of Bell's inequality, but it was reported in 1976 by a team at the Saclay Nuclear Research Centre in France. The experiment very closely follows the original thought experiment, and involves shooting low-energy protons at a target that contains a lot of hydrogen atoms. When a proton strikes the nucleus of a hydrogen atom – which is another proton – the two particles interact through the singlet state and their spin components can be measure. The difficulties of making the measurements are immense. Only some of the protons are recorded by the detectors, and, unlike the ideal world of the thought experiment, even when measurements are made it is not always possible to record the spin components unambiguously. Nevertheless, the results of this French experiment clearly demonstrate that local realistic views of the world are false.

7.21 What does it all mean

The experiment is very nearly perfect; even though the switching of the light beam is not quite at random, it does change independently for each of the two photon beams. The only real loophole that remains is that most of the photons produced are not detected at all, because the detectors themselves are very inefficient. It is still possible to argue that only the photons that violate Bell's inequality are detected, and that the others would obey the inequality if only we could detect them. But no experiments designed to test this unlikely possibility are yet contemplated, and it does seem like the height of desperation to make that argument. Following the announcement of the results from Aspect's team just before "Christmas 1982," nobody seriously doubts that the Bell test confirms the predictions of quantum theory.

In fact, the results of this experiment, the best that can be achieved with present-day techniques, violate the inequalities to a greater extent than those of any of the previous tests and agree very well with the predictions of quantum mechanics. As d'Espagnat has said, 'Experiments have recently been carried out that would have forced Einstein to change his conception of nature on a point he always considered essential . . . we may safely say the *non-separability* is now one of the most certain general concepts in physics.'

This does not mean that there is any likelihood of being able to send messages faster than the speed of light. There is no prospect of conveying useful information in this way, because there is no way of linking one event that causes another event to the event it causes through this process. It is an essential feature of the effect that it only applies to events that have a common cause – the annihilation of a positron/electron pair; the return of an electron to the ground state; the separation of a pair of protons from the singlet state. You may imagine two detectors situated far apart in space, with photons from such central source flying off to each of them, and you may imagine some subtle technique for changing the polarization of one beam of photons, so that an observer far away at the second detector sees changes in the polarization of the other beam. But what sort of signal is it that is being changed? The original polarizations, or spins, of the particles in the beam are a result of random quantum processes, and carry no information in themselves.

All that the observer will be seeing is a different random pattern from the random pattern that he would see without the cunning manipulations of the first polarizer! Since there is no information in a random pattern, it would be totally useless. The information is contained in the difference between the two random patterns, but the first pattern never existed in the real world, and there is no way to extract the information.

But don't be too disappointed, for the Aspect experiment and its predecessors do indeed make for a very different world view from that of our everyday common sense. They tell us that particles that were once together in an interaction remain in some sense parts of a single system, which responds together to further interactions. Virtually everything we see and touch and feel is made up of collections of particles that have been involved in interactions with other particles right back through time, to the Big Bang in which the Universe as we know it came into being. The atoms in my body are made of particles that once jostled in close proximity in the cosmic fireball with particles that are now part of a distant star, and particles that form the body of some living creature on some distant, undiscovered planet.

Indeed, the particles that make up my body once jostled in close proximity and interacted with the particles that now make up your body. We are as much parts of a system as the two photons flying out of the heart of the Aspect experiment.

Theorists such as d'Espagnat and Bohm argue that we must accept that, literally, *everything is connected to everything else*, and only a *"holistic"* approach to the Universe is likely to explain phenomena such as human consciousness. It is too early yet for the physicists and philosophers groping toward such a new picture of consciousness and the Universe to have produced a satisfactory outline of its likely shape, and speculative discussion of the many possibilities touted would be out of place here. But I can provide an example from my own background, rooted in the solid traditions of physics and astronomy. One of the great puzzles of physics is the property of inertia, the resistance of an object, not to motion but to changes in its motion.

In free space, any object keeps moving in a straight line at constant speed until it is pushed by some outside force – this was one of Newton's great discovers. The amount of push needed to move the object depends on how much material it contains. But how does the object "know" that it is moving at constant speed in a straight line – what does it measure its velocity against? Since the time of Newton, philosophers have been well aware that the standard against which inertia seems to be measured is the frame of reference of what used to be called the "fixed stars," although we would now speak in terms of distant galaxies. The Earth spinning in space, a long foucault pendulum like the ones seen in so many science museums, an astronaut, or an atom, they all "know" what the average distribution of matter in the Universe is.

Nobody knows why or how the effects works, and it has led to some intriguing if fruitless speculations. If there were just one particle in an empty Universe, it couldn't have any inertia, because there would be nothing against which to measure its movement or resistance to movement. But if there were just two particles, in an otherwise empty Universe, would they each have the same inertia as if they were in our Universe? If we could magically remove half of the matter in our Universe, would the rest still have the same inertia, or half as much?

The puzzle is as great today as it was three hundred years ago, but perhaps the death of *local realistic* views of the world gives us a clue. If everything that everything that ever interacted in the Big Bang maintains its connection with everything it interacted with, then every particle in every star and galaxy that we can see "knows" about the existence of every other particle.

Inertia becomes a puzzle not for cosmologists and relativists to debate, but one firmly in the arena of quantum mechanics. Does it seem paradoxical? Richard Feynman summed up the situation succinctly in his Lectures: 'The paradox is only a conflict between reality and your feelings of what reality ought to be.' Does it seem pointless, like the debate about the number of "angles" that can dance on a pinhead? Already, early in 1983, just a few weeks after the publication of the Aspect team's results,

Courtesy of California Institute of Technology and the Hebrew University of Jerusalem scientists at the University of Sussex in England were announcing the results of experiments that not only provide independent confirmation of the "connectedness" of things at the quantum level, but offer scope for practical applications including a new generation of computers – as much improved over present solid-state technology as the transistor radio itself is an improvement over the semaphore flag as a signaling device.

7.22 Confirmation and applications

The Sussex team, headed by Terry Clark, has tackled the problem of making measurements of quantum reality the other way around. Instead of trying to construct experiments that operate on the scale of normal quantum particles – the scale of atoms or smaller – they have attempted to construct "*quantum particles*" that are more nearly the size of conventional measuring devices. Their technique depends upon the property of superconductivity, and used a ring of superconducting material, about half a centimeter across, in which there is a constriction at one point, a narrowing of the ring to just one ten-millionth of a square centimeter in cross section. This "*weak link*," invented by Brian Josephson who developed the Josephson junction, makes the ring of superconducting material act like an open-ended cylinder such as an organ pipe or a tin can with both ends removed.

The Schrodinger waves describing the behavior of superconducting electrons in the ring act rather like the standing sound waves in an organ pipe, and they can be "tuned" by applying a varying electromagnetic field at radio frequencies. In effect, the electron wave around the whole of the ring replicates a single quantum particle, and by using a sensitive radio-frequency detector the team is able to observe the effects of a quantum transition of the electron wave in the ring. It is, for all practical purposes, as if they had a single quantum particle half a centimeter across with which to work – a similar, but even more dramatic, example to the little bucket of superfluid helium mentioned earlier."

The experiment provides direct measurements of single quantum transitions, and it also provides further clear evidence of non-locality. Because the electrons in the superconductor act like one boson, the Schrodinger wave that makes a quantum transition is spread out around the whole ring. The whole of this "pseudo-boson" makes the transition at the same time. It is not observed that one side of the ring makes the transition first, and that the other side only catches up when a signal moving at the speed of light has had time to travel around the ring and influence the rest of the "particle." In some ways, this experiment is even more powerful than the Aspect the of Bell's inequality. That test depends upon arguments which, although mathematically unambiguous, are not easy for the layman to follow. It is much easier to grasp the concept of a single "particle" that is half a centimeter across and yet behaves like a single quantum particle, and that responds, in its entirety, instantaneously to any prodding it receives from the outside.

Already, Clark and his colleagues are working on the next logical development. They hope to construct a large "macro-atom," perhaps in the form of a straight cylinder 6 meters long. If this device responds to outside stimulation in the way expected, then there may indeed be a crack opened in the door that leads to faster-than-light communication. A detector at one end of the cylinder, measuring its quantum state, will respond instantly to a change in the quantum state triggered by a prod at the other end of the cylinder. This is still not much use for conventional signaling – we couldn't build a macro-atom reaching from here to the Moon, say, and use it to eliminate that annoying lag in communications between lunar explorers and ground control here on Earth. But it would have direct, practical applications.

In the most advanced modern computers, one of the key limiting factors on performance is the speed with which electrons can get around the circuitry from one component to another. The time delays involved are small, down to the nanosecond range, but very significant.

The prospect of communicating instantaneously across great distances is in no way made more likely by the Sussex experiments, but the prospect of building computers in which all the components respond instantaneously to a in the state of one component is brought within the realm of possibility. It is this prospect that has encouraged Terry Clark to make the claim that "when its rules are translated into circuit hardware it will make the already amazing electronics of the 20th century look like semaphore by comparison."

So not only is the Copenhagen interpretation fully vindicated for all practical purposes by the experiments, it looks as if there are developments in store as far beyond those developments are beyond classical devices. But still the Copenhagen interpretation is intellectually unsatisfying. What happens to all those "ghostly" quantum worlds that collapse with their wave function when we make a measurement of a subatomic system? How can an overlapping reality, no more and no less real than the one we eventually measure, simply disappear when the measurement is made? The best answer is that the alternative realities do not disappear, and that Schrodinger's cat really is both alive and dead at the same time, but in two or more different worlds. The Copenhagen interpretation and its practical implications are fully contained within a more complete view of reality, the many worlds interpretation.

7.23 Many Worlds Interpretation

Most of the physicists who bother to think about such things at all are happy with the collapsing wave functions of the Copenhagen interpretation. But it is a respectable minority view, and it has the merit of including the Copenhagen interpretation within itself. The uncomfortable feature that has prevented this improved interpretation from taking the world of physics by storm is that it implies the existence of many other worlds – possibly an infinite number of them – existing in some way sideways across time from our reality, parallel to our own Universe but forever cut off from it.

7.24 Who observes the observers

This many-worlds interpretation of quantum mechanics originated in the work of Hugh Everett a graduate student at Princeton University in the 1950s. Puzzling over the peculiar way in which the Copenhagen interpretation requires wave functions to collapse magically when observed, he discussed alternatives with many people, including John Wheeler who encouraged Everett to develop his alternative approach as his PhD thesis. This alternative view starts from a very simple question that is the logical culmination of considering the successive collapses of the wave function implied when I carry out an experiment in a closed room, then come out and tell you the result, which you tell a friend in New York, who reports it to someone else, and so on. At each step, the wave function becomes more complex, and embraces more of the "real world." But at each stage the alternatives remain equally valid, overlapping realities, until the news of the outcome of the experiment arrives. We can imagine the news spreading across the whole Universe in this way, until the whole Universe is in a state of overlapping wave functions, alternative realities that only collapse into one world when observed. But who observes the Universe?

By definition, the Universe is closed or "*self-contained*." It includes everything, so there is no outside observer who notices the existence of the Universe and thereby collapses its complex web of interacting alternative realities into one wave function. Wheeler's idea of consciousness – ourselves – as the crucial observer operating through reverse causality back to the Big Bang is one way out of this dilemma, but it involves a circular argument as puzzling as the puzzle it is supposed to eliminate. I would prefer even the solipsist argument, that there is only one observer in the Universe, myself, and that my observations are the all-important factor that crystallizes reality out of the web of quantum possibilities – but extreme solipsism is a deeply unsatisfying philosophy for someone whose own contribution to the world is writing books to be read by other people.

Everett's many worlds interpretation is another, more satisfying and more complete possibility. Everett's interpretation is that the overlapping wave functions of the whole Universe, the alternative realities that interact to produce measureable interference at the quantum level, do not collapse. All of them are equally real, and exist in their own parts of "super-space" (and super-time).

What happens when we make a measurement at the quantum level is that we are forced by the process of observation to select one of these alternatives, which becomes part of what we see as the "real" world; the act of observation cuts the ties that bind alternative realities together, and allows them to go on their own separate ways through super-space, each alternative reality containing its own observer who has made the same observation but got a different quantum "answer" and thinks that he has "collapsed the wave function" into one single quantum alternative.

7.25 More on Schrodinger's Cats

It is hard to grasp what this means when we talk about the collapse of the *wave function* of the whole Universe, but much easier to see why Everett's approach represents a step forward if we look at a more homely example. Our search for the real cat hidden inside Schrodinger's paradoxical box has, at last, come to an end, for that box provides just the example I need to demonstrate the power of the many-worlds interpretation of quantum mechanics. The surprise is that the trail leads not to one real cat, but two.

The equations of quantum mechanics tell us that inside the box of Schrodinger's famous thought experiment there are versions of a "live cat" and "dead cat" wave function that are equally real. The conventional, Copenhagen interpretation looks at these possibilities from a different perspective, and says, in effect, that both wave functions are equally unreal, and that only one of them crystallizes as into reality when we look inside the box.

Everett's interpretation accepts the quantum equations entirely at face value and says that both cats are real. There is a live cat, and there is a dead cat; but they are located in different worlds. It is not that the radioactive atom inside the box either did or didn't decay, but that it did both. Faced with a decision, the whole world – the Universe – split into two versions of itself, identical in all respects except that in one version the atom decayed and the cat died, while in the other the atom did not decay and the cat lived.

It sounds like science fiction, but it goes far deeper than any science fiction, and it is based on impeccable mathematical equations, a consistent and logical consequence of taking quantum mechanics literally.

7.26 Everett's work

The importance of Everett's work, published in 1957, is that he took this seemingly outrageous idea and put it on a secure mathematical foundation using the established rules of quantum theory. It is one thing to speculate about the nature of the Universe, but quite another to develop those speculations into a complete, self-consistent theory of reality. Indeed, Everett was not the first person to speculate in this way, although he seems to have produced his ideas totally independently of any earlier suggestions about multiple realities and parallel worlds.

Most of those earlier speculations – and many more since 1957 – have indeed appeared in the pages of science fiction. The earliest version I have been able to trace is in Jack Williamson's *The Legion of Time*, first published as a magazine serial in 1938.

Many SF stories are set in "parallel" realities, where the South won the American Civil War, or the Spanish Armada succeeded in conquering England, and so on. Some describe the adventures of a hero who travels sideways through time from one alternative reality to another; a few describe, with suitable gobbledygook language, how such an alternative world may split off from our won. Williamson's original story achieve concrete reality until some key action is taken at a crucial time in the past where the courses of the two worlds diverge (there is also "conventional" time travel in this story, and the action is as circular as the argument). The idea has echoes of the collapse of a wave function as described by the conventional Copenhagen interpretation, and Williamson's familiarity with the new ideas of the 1930s is clear from the passage in which a character explains what is going on: With the substitution of waves of probability for concrete particles, the world lines of objects are no longer the fixed and simple paths they once were. Geodesics have an infinite proliferation of possible branches, at the whim of subatomic indeterminism. Williamson's world is a world of 'ghost realities,' where the heroic action takes place, with one of them collapsing and disappearing when the crucial decision is made and another of the ghosts is selected to become concrete reality. Everett's worlds are equally real, and where, alas, not even heroes can move from one reality to its neighbor. But Everett's version is science fact, not science fiction.

Let's get back to the fundamental experiment in quantum physics, the two-hole experiment. Even within the framework of the conventional Copenhagen interpretation, although few quantum cooks realize it, the interference pattern produced on the screen of that experiment when just one particle travels through the apparatus is explained as interference from two alternative realities, in one of which the particle goes through hole *A*, in the other of which it goes through hole *B*. When we look at the holes, we find the particle one of them, and there is only goes through? On the Copenhagen interpretation, it choose at random in accordance with the quantum probabilities – God does play dice with the Universe. On the many-worlds interpretation, it doesn't choose. Faced with a choice at the quantum level not only the particle itself but the entire Universe splits into two versions. In one Universe, the particle goes through hole *A*, in the other it goes through hole *B*. In each Universe there is an observer who sees the particle go through just one hole. And forever afterward the two Universes are completely separate and non-interacting – which is why there is no interference on the screen of the experiment. The phrase "parallel worlds" suggests alternative realities lying side by side in "*superspace-time.*" This is a false picture. A better picture sees the Universe constantly splitting, like a branching tree. But this is still a false image. Multiply this picture by the number of quantum events going on all the time in every region of the Universe and you get some idea of why conventional physicists balk at the notion. And yet, as Everett established twenty-five years ago, it is a logical, self-consistent description of quantum reality that conflicts with experimental or observational evidence.

In spite of its impeccable mathematics, Everett's new interpretation of quantum mechanics made scarcely a ripple as it fell into the pool of scientific knowledge in 1957. A version of the work appeared in "*Reviews of Modern physics,*" and alongside it there appeared a paper from Wheeler drawing attention to the importance of Everett's work. But the ideas remained largely ignored until they were taken up by Bryce DeWitt, of the University of North Carolina, more than ten years later.

It isn't clear why the idea should have taken so long to catch on, even in the minor way that it achieved success in the 1970s. Apart from the heavy math, Everett carefully explained in his "*Reviews of Modern Physics*" paper that the argument that the splitting of the Universe into many worlds cannot be real because we have no experience of it doesn't hold water. All the separate elements of a superposition of states obey the wave equation with complete indifference as to the actuality of the other elements, and the total lack of any effect of one branch on another implies that no observer can ever be aware of the splitting process. Arguing otherwise is like arguing otherwise is like arguing that the Earth cannot possibly be in orbit around the Sun, because if it were we would feel the motion. "In both cases," says Everett, "the theory itself predicts that our experience will be what in fact it is."

7.27 Beyond Einstein

In the case of the many-worlds interpretation, the theory is conceptually simple, causal, and gives predictions in accordance with experience. Wheeler did his best to make sure people noticed the new idea. It is difficult to make clear how decisively the relative state formulation drops classical concepts. One's initial unhappiness at this step can be matched but a few times in history: when Newton described gravity by anything so preposterous as action at a distance; when Maxwell described anything as natural as action at a distance in terms as unnatural as field theory; when Einstein denied a privileged character to any coordinate system . . . nothing quite comparable can be cited from the rest of physics except the principle in general relativity that all regular coordinate systems are equally justified.

Apart from Everett's concept," Wheeler concluded, "no self-consistent system of ideas is at hand to explain what one shall mean by quantizing a closed system like the Universe of general relativity." Strong words, indeed; but the Everett interpretation suffers from one major defect in trying to oust the Copenhagen interpretation from its established place in physics. The many-worlds version of quantum mechanics makes exactly the same predictions as the Copenhagen view when assessing the likely outcome of any experiment or observation. That is both a strength and a weakness. Since the Copenhagen interpretation has never yet been found wanting in these practical terms, any new interpretation must give the same "answers" as the Copenhagen interpretation wherever it can be tested; so the Everett interpretation wherever it can be tested; so the Everett interpretation passes its first test.

But it only improves upon the Copenhagen view by removing the seemingly paradoxical features from double-slit experiments or from tests of the kind invented by Einstein, Podolsky and Rosen. From the viewpoint of all the quantum experiments, it is hard to see the difference between the two interpretations, and the natural inclination is to stick with the familiar. For anyone who has studied the *EPR* thought experiments, and now the various tests of Bell's inequality, however, the attraction of the Everett interpretation, it is not that our choice of which spin component to measure forces the spin component of another particle, far away across the Universe to magically take up a complementary state, but rather that by choosing which spin component to measure we are choosing which branch of reality we are living in. In that branch of *superspace*, the spin of the other particle always is complementary to the one we measure. It is choice that decides which of the quantum worlds we measure in our experiments, and therefore which one we inhabit, not chance. Where all possible outcomes of an experiment actually do occur, and each possible outcome is observed by its own set of observers, it is no surprise to find that what we observe is one of the possible outcomes of the experiment.

7.28 A deeper look at the many worlds interpretation

The many-worlds interpretation of quantum mechanics was almost studiously ignored by the physics community until DeWitt took the idea up in the late 1960s, writing about the concept himself and encouraging his student Neill Graham to develop an extension of Everett's work as his own *PhD* thesis. As DeWitt explained in an article in *Physics Today* in 1970, the Everett interpretation has an immediate appeal when applied to the paradox of Schrodinger's cat. We no longer have to worry about the puzzle of a cat that is both dead and alive, neither alive nor dead. Instead, we know that in our world the box contains a cat that is either alive or dead, and that in the world next door there is another observer who has an identical box that contains a cat that is either dead or alive. But if the Universe is "constantly splitting into a stupendous number of branches," then "every quantum transition taking place on every star, in every galaxy, in every remote corner of the Universe is splitting our local world on Earth into myriads of copies of itself."

Dewitt recalls the shock he experienced on first encountering this concept, the 'idea of 10^{100} slightly imperfect copies of oneself all constantly splitting into further copies.' But he was persuaded by his own work, by Everett's thesis and by Graham's renewed study of the phenomenon. He even considers how far the splitting can actually continue to take place. In a finite universe – and there are good reasons for believing that if general relativity is a good description of reality then the Universe is finite – there must only be a finite number of "branches" on the quantum tree, and super-space simply may not have enough room to accommodate the more bizarre possibilities, the fine structure of what DeWitt call "maverick worlds," realities with strangely distorted patterns of behavior. In any case, although the strict Everett interpretation says that anything that is possible does occur in some version of reality, somewhere in super-space, that is not the same thing as saying anything imaginable can occur. We can imagine impossible things, and the real worlds could not accommodate them. In a world otherwise identical to our own, even if pigs (otherwise identical to our own pigs) had wings they would not be able to fly; heroes, no matter how super, cannot slip sideways through cracks in time to visit alternative realities, even though *SciFi* writers speculate about the consequences of such actions; and so on.

DeWitt's conclusion is as dramatic as the earlier conclusion of Wheeler: The view from where Everett, Wheeler and Graham sit is truly impressive. Yet it is a completely casual view, which even Einstein might have accepted . . . it has a better claim than most to be the natural end product of the interpretation program begun by Heisenberg in 1925. Perhaps it is only fair, at this point, to mention that Wheeler himself has recently expressed doubts about the whole business. In response to a questioner at a symposium held to mark the century of Einstein's birth, he said of the many-worlds theory, 'I confess that I have reluctantly had to give up my support of that point of view in the end – much as I advocated it in the beginning - because I am afraid it carries too great a load of *metaphysical* baggage.'

General relativity is a theory that describes "closed systems," and Einstein originally envisaged the Universe as a closed, finite system. Although people talk about open, infinite universes, strictly speaking such descriptions are not properly covered by relativity theory. The way for our Universe to be closed is if it contains enough matter for gravity to bend space-time around on itself, like the bending of space-time around a black-hole. That needs more matter than we can see in the visible galaxies but most observations of the dynamics of the Universe suggest that it is in fact in a state very close to being closed – either "just closed" or "just open."

In that case, there is no observational justification for rejecting the fundamental relativistic implications that the Universe is closed and finite, and there is every reason to seek the dark matter that holds it together gravitationally.

This shouldn't be read as pulling the rug from under the Everett interpretation; the fact that Einstein changed his mind about the statistical basis of quantum mechanics didn't pull the rug from under that interpretation. Nor does it mean that what Wheeler said in 1957 is no longer true. It is still true, in 2011, that apart from Everett's theory no self-consistent system of ideas is at hand to explain what is meant by *quantizing* the Universe. But Wheeler's change of heart does show how hard many people find it to accept the "many-worlds" theory.

The load of *meta-physical* baggage required can be seen as being far less disturbing then that of the Copenhagen interpretation or Schrodinger's experiment with the cat, or the requirement of three times as many dimensions of "phase space" as there are particles in the Universe. The concepts are no stranger than other concepts that seem familiar just because they are discussed so widely, and the many-worlds interpretation offers new insights into why the Universe we live in should be the way it is. The theory is far from being played out and still deserves serious attention.

7.29 Beyond Everett

Cosmologists today talk quite happily about events that occurred just after the Universe was born in a Big Bang, and they calculate the reactions that occurred when the age of the Universe was 10^{-35} second or less. The reactions involve a maelstrom of particles and radiation, pair production and annihilation. The assumptions about how these reactions take place come from a mixture of theory and the observations of the way particles interact in giant accelerators, like the one run by CERN in Geneva. According to these calculations, the laws of physics determined from our puny experiments here on Earth can explain in a logical and self-consistent fashion how the Universe got from a state almost infinite density into the state of almost infinite density into the state we see it in today. The theories even make a stab at predicting the balance between matter and antimatter in the Universe, and between matter and radiation. Everyone interested in science, however mild and passing their interest has heard of the Big Bang theory of the origin of the Universe.

Theorists happily play with numbers describing events that allegedly occurred during split seconds some 15 thousand million years ago. But who today stops to think what these ideas really mean? It is absolutely mind-blowing to attempt to understand the implications of these ideas.

Who can appreciate what a number like 10^{-35} of a second really means, let alone comprehends the nature of the Universe when it was 10^{-35} second old? Scientists who deal with such bizarre extremes of nature really should not find it too difficult to stretch their minds to accommodate the concept of parallel worlds.

In fact, that felicitous-sounding expression, borrowed from science fiction, is quite inappropriate. The natural image of alternative realities is as alternative branches fanning out from a main stem and running alongside one another through super space, like the branching lines of a complex railway junction. Like some super-superhighway, with millions of parallel lanes, the SciFi writers imagine all the worlds proceeding side by side through time, our near neighbors almost identical to our won world, but with the differences becoming clearer and more distinct the further we move "sideways in time." This is the image that leads naturally to speculation about the possibility of changing lanes on the superhighway, slipping across into the world next door. Unfortunately, the math isn't quite like this neat picture.

Mathematicians have no trouble handling more *dimensions* than the familiar three space dimensions so important in our everyday lives. The whole of our world, one branch of "Everett's many-worlds" reality, is described mathematically in four dimensions, three of space and one of time, all at right angles to one another, and the math to describe more dimensions all at right angles to each other and to our own four is routine number juggling. This is where the alternative realities actually lie, not parallel to our own world but at right angles to it, perpendicular worlds branching off "sideways" through super-space. The picture is hard to visualize, but it does make it easier to see why slipping sideways into an alternative reality is impossible. If you set off at right angles to our world – sideways – you would be creating a new world of your won. Indeed, on the many-worlds theory this is what happens every time the Universe is faced with a quantum choice. The only way you could get in to one of the alternative realities created by such a splitting of the Universe as a result of a *cat-in-the-box* experiment, or a *two-holes* experiment, would be to go back in time in our own four-dimensional reality to the time of the experiment, and then to go forward in time along the alternative branch, at right angles to our own four-dimensional world. This might be impossible.

Conventional wisdom has it that true time travel must be impossible, because of the paradoxes involved, like the one where you go back in time and kill your grandfather before your own father has been conceived. On the other hand, at the quantum level particles seem to be involved in "time travel" at the time, and Frank Tipler has shown that the equations of general relativity permit time travel. It is possible to conceive of a kind of genuine travel forward and backward in time that does not permit paradoxes, and such a form of time travel depends on the reality of alternative universes.

And if you have trouble believing this, you may be beginning to feel that the good old Schrodinger equation is more cozy and familiar. Far from it. The wave interpretation of quantum mechanics does start out from a simple wave equation familiar from other areas of physics, and for a single particle the correct quantum-mechanical description does involve a save in three dimensions, though not in our everyday space but in something called "configuration space." Unfortunately, you need three different dimensions for the wave for each particle involved in the description. To describe two particles interacting, you need six dimensions; to describe a system of three particles you need nine dimensions; and so on. The wave function of the entire Universe, whatever that means, is a wave function involving three times as many dimensions as there are particles in the Universe. Physicists who dismiss the Everett interpretation of reality as carrying too much excess baggage conveniently forget that the wave equations they use every day can only be accepted as a good description of the Universe by invoking an equally mind-boggling load of extra-dimensional baggage.

David Gerrold explored these possibilities in an entertaining Sci-Fi book "*The Man Who Folded Himself*," well worth reading as a guide to the complexities and subtleties of a many-worlds reality. The point is that, taking the classic example, if you go back in time and kill your grandfather you are creating, or entering (depending on your point of view) an alternative world branching off at "right angles" to the world in which you started. In that "new" reality, your father, and yourself, are never born, but there is no paradox because you are still born in the "original" reality, and make the journey back through time and into an alternative branch. Go back again to undo the mischief you have done, and all you do is reenter the original branch of reality, or at least one rather like it.

But even Gerrold does not "explain" the bizarre events that happen to his main character in terms of perpendicular realities, and as far as we know, this physical explanation of the mathematics if the Everett interpretation is original – certainly a new twist to the time travel saga that the SciFi writers have not, as yet, embraced.

The point that is worth stressing is that the alternative realities do not, on this picture, lie "alongside" ours, where they can be slipped into and out of with very little other branches. There may be a world in which Bonaparte's given name was Pierre, not Napoleon, but where history otherwise flowed essentially as in our own branch of reality; there may be a world in which that particular Bonaparte never existed. Both are equally remote and inaccessible from our own world. Neither can be reached except by traveling backward in time in our own world to the appropriate branching point, and then by striking out again forward in time at right angles (one of the many right angles.) to our own reality.

The concept can be extended to remove the paradoxical nature of any of the time-travel paradoxes beloved of science fiction writers and reads, and argued over by philosophers. All possible thinks do happen, in some branch of reality. The key to entering those possible realities is not travel sideways in time, but backward and then forward into another branch. Possibly, the best science-fiction novel ever written makes use of the many-worlds interpretation as author Gregory Benford did so consciously. In his book "Timescape the fate of a world" is fundamentally altered as a result of messages that are sent back to the 1960s from the 1900s. The story is beautifully crafted, gripping and stands up in its own right even without the *SciFi* theme. But the one point we want to pick up here is that because the world is changed as a result of actions taken by the people who receive the messages from the future, the future from which those messages came does not exist for them. So where did the messages come from? You could, perhaps, make a case on the old Copenhagen interpretation for a ghost world sending back ghost messages that affect the way the wave function collapses, but you'd be hard pressed to make that argument stand up.

On the other hand, in the many-worlds interpretation it is straightforward to visualize messages from one reality going back in time to a branching point where they are received by people who then move forward in time into their own, different branch of reality. Both alternative worlds exist, and communication between them is broken once the critical decisions that affect the future have been made. Timescape, as well as being a good read, actually contains a "thought experiment" every bit as intriguing and relevant to the quantum mechanics debate as the *EPR* experiment or Schrodinger' cat. Everett himself may not have appreciated it, but a many-worlds reality is exactly the kind of reality that permits time travel. It is also the kind of reality that explains why we should be here debating such issues.

There is another element here that is worth emphasizing. Even if time travel is theoretically possible, there may be insuperable practical difficulties to prevent us from sending material objects through time. But sending messages through time could be a relatively simple matter if we can find a way to make use of the particles that travel backward in time in Feynman's interpretation of reality.

7.30 Our place in the multiverse

According to the interpretation of the "many-worlds theory," the future is not determined, as far as our conscious perception of the world is concerned, but the past is. By the act of observation we have selected a "real" history out of the many realities, and once someone has seen a tree in our world it stays there even when nobody is looking at it. This applies all the way back to the Big Bang. At every junction in the quantum highway there may have been many new realities created, but the path that leads to us is clear and unambiguous. There are many routes into the future, however, and some version of "us" will follow each of them. Each version of ourselves will think is following a unique path, and will look back on a unique past, but it is impossible to know the future since there are so many of them. We may even receive messages from the future, either by mechanical means as in *Timescape*, or, if you wish to imagine the possibility, through dreams and extrasensory perception. But those messages are unlikely to do us much good. Because there is a multiplicity of future worlds, any such messages must be expected to be confused and contradictory. If we act on them, we are more likely than not going to deflect ourselves into a branch of reality different from the one the "messages" came from, so that it is highly unlikely that they can ever "come true." The people who suggest that quantum theory offers a key to practical *ESP*, telepathy, and all the rest are only deluding themselves.

The picture of the Universe as an unrolling Feynman diagram across which the instant "now" moves at a steady rate is an oversimplification. The real picture is of a multidimensional Feynman diagram, all the possible worlds, with "now" unrolling across all of them, up every branch and detour. The greatest question left to answer within this framework is why our perception of reality should be what it is – why should the choice of paths through the quantum maze that started out in the Big Bang and leads to us have been just the right kind of path for the appearance of intelligence in the Universe?

The answer lies in an idea often referred to as the "**anthropic principle**." This says that the conditions that exits in our Universe are the only conditions, apart from small variations, that could have allowed life like us to evolve, and so it is inevitable that any intelligent species like us should look out upon a Universe like the one we see about us. If the Universe wasn't the way it is, we wouldn't be here to observe it. We can imagine the Universe taking many different quantum paths forward from the Big Bang. In some of those worlds, because of differences in the quantum choices made near the beginning of the universal expansion, stars and planets never form, and life as we know it does not exit. Taking a specific example, in our Universe there seems to be a preponderance of matter particles and little or no antimatter. There may be no fundamental reason for this – it may be just an accident of the way the reactions worked out during the fireball phase of the Big Bang.

It is just as likely that the Universe should be empty, or that it should consist chiefly of what we call antimatter, with little or no matter present. In the empty Universe, there would be no life as we know it; in the antimatter Universe there could be life just like us, a kind of looking-glass world made real. The puzzle is why a world ideal for life should have appeared out of the Big Bang. The anthropic principle says that many possible worlds may exist, and that we are an inevitable product of our kind of Universe. But where are the other worlds? Are they ghosts, like the interacting worlds of the Copenhagen interpretation? Do they correspond to different life cycles of the whole Universe, before the Big Bang that began time and space as we know them? Or could they be Everett's many worlds, all existing at right angles to our own?

It seems that this is by far the best explanation available today, and that the resolution of the fundamental puzzle of why we see the Universe the way it is amply compensates for the load of baggage carried by the Everett interpretation. Most of the alternative quantum realities are unsuitable for life and are empty. The conditions that are just right for life are special, so when living beings look back down the quantum path that has produced themselves they see special events, branches in the quantum road that may not even be the most likely on a statistical basis, but are the ones that lead to intelligent life. The multiplicity of worlds like our own but with different histories – in which Britain still rules all of its North American colonies; or in which the North American natives colonized Europe – together make up just one small corner of a much vaster reality. It is not chance that has selected the special conditions suitable for life out of the array of quantum possibilities, but choice. All worlds are equally real, but only suitable worlds contain observers.

The success of the Aspect team's experiments to test the Bell inequality has eliminated all but two of the possible interpretations of quantum mechanics ever put forward. Either we have to accept the Copenhagen interpretation with its *ghost* realities and half-dead cats, or we have to accept the Everett interpretation with its many worlds. It is, of course, conceivable that neither of the two "*best buys*" in the science supermarket is correct, and that both these alternatives are wrong. There may yet be another interpretation of quantum mechanical reality which resolves all of the puzzles that the Copenhagen interpretation and the Everett interpretation resolve, including the Bell Test, and which goes beyond our present understanding – in the same way, perhaps, that general relativity transcends and incorporates special relativity. But if you think this is the soft option, an easy route out of the dilemma, remember that any such "new" interpretation must explain everything that we have learned since

Planck's leap in the dark, and that it must explain everything as well as, or better than, the two current explanations. That is a very tall order indeed, and it is not the way of science to sit idly back and hope that someone will come up with a "better" answer to our problems. In the absence of a better answer, we have to face up to the implications of the best answer, we've got.

After more than half a century of intensive effort devoted to the puzzle of quantum reality by the best brains of the 20th century, we have to accept that science can at present only offer these two alternatives explanations of the way the world is constructed. Neither of them seems very palatable at first sight. In simple language, either nothing is real or everything is real.

The issue may never be resolved, because it may be impossible to devise an experiment to distinguish between the two interpretations, short of time travel. But it is quite clear that Max Jammer, one of the ablest of quantum philosophers, was not exaggerating when he said that 'the multi-Universe theory is undoubtedly one of the most daring and most ambitious theories ever constructed in the history of science.' Quite literally, it explains everything, including the life and death of cats. As an incurable optimist, it is the interpretation of quantum mechanics that appeals most to me. All things are possible, and by our actions we choose our own paths through the many worlds of the quantum. In the world in which we live, what you see is what you get; there are no hidden variables; *God doesn't play dice; and everything is real*. One of the anecdotes told and retold about Niels Bohr is that when someone came to him with a wild idea purporting to resolve one of the puzzles of quantum theory in the 1920s, he replied, 'Your theory is crazy but it's not crazy enough to be true.' In this view Everett's theory is crazy enough to be true.

7.31 Anti-worlds

Schwarzschild black holes is not a physical barrier, but merely a gateway to a weird space-time region beyond. Just how weird became clear when the algebra describing this region was scrutinized. According to the idealized mathematical description, the "other universe" is a mirror image of itself, stretching away to infinity. There is, however, a crucial difference. The direction of time in the other universe is reversed relative to ours. The idea that there may exist a sort of parallel universe with time running the other way - an anti-world, if you like - has a certain appeal. We have encountered such a speculation before, in the context of the kaon.

All quantum systems are subject to inherent uncertainty. As a typical system evolves, there are many possible outcomes, many contending realities, on offer. For instance, in the various laser experiments described earlier, a photon had a choice of which path to follow through a piece of apparatus. In the case of a lab experiment, the observer will always see a single, specific, concrete reality selected from among the phantom contenders. Thus a measurement of the photon's path will always yield a result or either one path or the other, but never both. When it comes to the Universe as a whole, there is no outside observer, because the Universe is all that there is, so quantum cosmology runs up against a major problem of interpretation. The favored way out is to assume that all the contending quantum realities enjoy equal status.

They are not mere "phantom worlds," or "potential realities," but "really real" - all of them. Each reality corresponds to an entire Universe, complete with its own space and time. These many Universes are not connected through space and time, but are somehow "parallel," existing alongside each other. In general, there will be an infinity of them.

The philosophy underlying this way of thinking was largely developed by Niels Bohr and others working in Denmark in the 1920s and 1930s and for this reason has become known as the "*Copenhagen interpretation*" of quantum physics. It was heavily criticized by Albert Einstein, among others, and is not without its critics today, as we shall see later in this chapter. However, it is still the orthodox interpretation accepted by most working physicists and we shall spend some more time developing it further before explaining what some perceive to be its weaknesses and discussing some alternative approaches.

Underlying Bohr's philosophy is a form of what is known as "positivism," which can be summed up in a phrase from the philosopher Wittgenstein: 'whereof we cannot speak, thereof we should remain silent.' In the present context this can be interpreted as saying that if something is unobservable (e.g. simultaneous knowledge of the *HV* and +/- 45-degree polarization of a photon), we should not assume that it has any reality. In some philosophical contexts this is matter of choice – we can imagine that there are angels dancing on a pinhead if we want to – but, from the Copenhagen point of view at least, in quantum physics it is a matter of necessity. This kind of thinking is certainly counter-intuitive to anyone familiar with classical physics; for example, we find it difficult not to believe that an object must always be 'somewhere'.

When we have learned about wave-particle duality, we may well be prepared to accept this and say something like 'When the particle is not being observed it is actually a wave.'

But this statement also attributes reality to something that is unobservable. We showed in an earlier chapter that the wave propertied emerge when we perform experiments on a large number of particles. This is the case, for example, when an electromagnetic wave is detected by a television set or when we observe an interference experiment after many photons have arrived at the screen (even though they may have passed through the apparatus one at a time). However, if we say the wave is 'real' when we are considering an individual object, we are again suggesting that something unobservable is 'really there.' The wave function should not be interpreted as a physical wave; it is a mathematical construction, which we use to predict the probabilities of possible experimental outcomes.

Philosophers often refer to the properties of an object as its 'attributes': those of a typical classical object include permanent quantities such as its mass, charge and volume, as well as others, such as position and speed, that may change during the particle's motion. In the Copenhagen interpretation, the attributes an object possesses depend on the context in which it is being observed. Thus a photon that is observed emerging from the *H* channel of an *HV* polarizer possesses the attribute of horizontal polarization, but if it is then passed through a +/- 45-degree polarizer it loses this attribute and acquires the attribute of +45-degree or -45-degree polarization.

It may be harder for us to accept that an attribute such as position has a similarly limited application, but quantum physics forces us to adopt this "counter-intuitive" way of thinking.

We can develop these ideas further by giving some consideration to the meaning and purpose of a scientific theory. A useful analogy is a map that might be used to navigate around a strange city.

Though the map is normally much smaller than the physical area it represents, it aims to be a faithful representation of the terrain it is modeling: depictions of streets and buildings on the map are related to each other in the same way as they are in reality. Clearly, the map is not the same thing as the terrain it models and indeed is different from it in important respects: for example, it is usually of a different size from the area it represents and is typically composed of paper and ink rather than Earth and stones.

A scientific theory also attempts to model reality. Leaving quantum physics aside for the moment, a classical theory attempts to construct a "map" of physical events. Consider the simple case of an object, such as an apple, being released and allowed to fall under gravity: the apple is at rest, is released, accelerates, and stops when it reached the ground. At every stage of the motion all the relevant attributes of the object are represented in our map (e.g. the time, height and speed) are denoted by algebraic variables and to construct a map of the object's motion, we carry out some mathematical calculations. However, the real apple falls to the floor in the predicted time even though it is quite incapable of performing the simplest mathematical calculation! This can require extensive use of mathematics, which is used to construct a map of reality, but the map is not reality itself. We also have to take care to choose a map that is appropriate to the physical situation we are addressing. Thus, a map based on Maxwell's theory of electromagnetism will be of little use to us if we are trying to understand the fall of an apple under gravity. Even if we do choose a map based on Newton's laws, it still has to use appropriate parameters: for example, it should include a representation of the effect of air resistance unless our apple is falling in a vacuum. When we come to quantum physics, we have to accept that there is no single map of the quantum world. Rather, quantum theory provides a number of maps; which we should use in a particular situation depends on the experimental context or even on the experimental outcome, and may change as the system evolves in time. Pursuing our analogy a little further, we might say that quantum physics enables us to construct a 'map book' and that we must look up the page that is appropriate to the particular situation we are considering. Let's return to our example of a 45-degree photon passing through an *HV* polarizer. Before the photon reaches the polarizer the appropriate map is one in which a +45-degree polarized photon moves from left to right, whereas once it emerges from the polarizer the appropriate map is one representing the two possibilities of the photon being horizontally or vertically polarized, respectively; and when it is finally detected the appropriate map is one that describes only the actual outcome.

7.32 Alternative Interpretations (Subjectivism)

One reaction to the *quantum measurement problem* is to retreat into 'subjective idealism'. In doing this, wed simply accept that quantum physics implies that it is impossible to give an objective account of physical reality. The only thing we know that must be real is our personal subjective experience: the counter may both fire and not fire, the cat may be both alive and dead, but when the information reaches our mind through my brain we certainly know which has really occurred. Quantum physics may apply to photons, counters and cats, but it does not apply to you or me! Of course, we can't be sure that the states of our mind are real either, so we are in danger of relapsing into 'solipsism', wherein only we and our minds have any realty. Philosophers have long argued about whether they could prove the existence of an external physical world, but the aim of science is not to answer this question but rather to provide a consistent account of any objective world that does exist. It would be ironic if quantum physics were to finally destroy this mission. Most of us would much rather search for an alternative way forward.

7.33 The Unfinished Business of Physics

The story of the quantum as have been told here seems neatly cut and dried, except for the semi-philosophical question of whether you prefer the Copenhagen interpretation or the many-worlds version. That is the best way to present the story in a book, but it isn't the whole truth. The story of the quantum is not yet finished, and theorists today are grappling with problems that may lead to a step forward as fundamental as the step Bohr took when he quantized the atom. Trying to write about this unfinished business is messy and unsatisfying; the accepted views of what is important and what can safely be ignored may change completely by the time the report gets into print. But to give you a flavor of how things may be progressing, included in this epilogue of this book is an account of the unfinished aspects of the quantum story and some hints about what to watch out for in the future.

Ch 8 Quantum Entanglement (Supplemental)

The quantum phenomena called entanglement with its implications are what happens when two quantum particles interact with each other. What results from that is that those particles, once interacted, become are entangled. This means that their quantum states are interdependent. There is a correlation between them, so that if one is in a particular state, then the other one has to have some other particular state depending on the kind of entanglement that they have.

8.1 So a classic example would involve, say, two electrons and the electrons have a quantum property called spin. The spin of an electron can have two values. Its like a quantum heads or tails. So one value could be spin Up or spin Down. And if the electrons become entangled, then this may create a situation where those two spins are correlated such that if one of the electrons has a spin up then the other one must have a spin down. This is a prediction of quantum mechanics that was pointed out by Albert Einstein in 1935 that the phenomena of entanglement can happen. He realized that there was something wrong about the phenomena.

One can make the analogy of quantum phenomena as having two gloves. A left-handed glove and a right handed glove. So imagine that you have those two gloves and you send them out to two people from different sides of the world, for instance. We will call these two people Alica and Bob. And they know these gloves are a pair. So as soon as Alice receives her glove and opens the package and finds she has the left-handed glove, then she knows right away that Bob must have the right-handed glove. However, because there is this correlation between them, there is nothing magical about how she gets that information, its just simple logic. However, this is the complication in quantum mechanics regarding this phenomenon. Neils Bohr, the Danish physicist, who pioneered early quantum mechanics suggested that in the case of quantum particles, its not the fact that they have this property of spin all the time, they only have that property (the fixed value) when we observe it. So for these two entangled photons, if we think of sending those out to Alica and Bob, Bohr said, while they are going outwards they don't have a fixed orientation of their spin. All we can say is that those spins are correlated. So if Alice then measures her electron and finds it has spin UP then Bob's electron will have spin Down. But that spin wasn't determined until Alice measured it. And this is where Einstein felt that there was a problem with entanglement. Because it seemed to indicate that the act of Alice measuring her electron spin somehow affected Bob's spin. So once Alice found that her electron had spin UP, somehow magically or what Einstein called "Spookily", that electron with spin UP sort of somehow transmitted some kind of influence to Bob's spin to make sure that it was spin DOWN.

Alice could have equally have measured her electron and found that it had spin DOWN, in which case Bob's would be spin UP. So there seemed to be what Einstein called "a spooky action at a distance", implied by Bohr's interpretation of quantum mechanics.

8.2 Einstein and two colleagues, Poldosky and Rosen suggested in 1935 that actually there has to be some alternative to the phenomena, because "Spooky action at a distance" should not be allowed in physics. Einstein had showed that it isn't possible for any signal or any information to be transmitted faster than light. And so, you can't have this instantaneous action at a distance. There has to be some time for a signal to span the distance. So Einstein suggested that what must be going on is that all along these two electrons had some property that somehow fixed their spins already. It was just that it was a property that we just could not measure. He called them hidden variables. So he thought that you could not find out in any experiment which of the two possible spins Alice's electron had as it was going towards her, but never the less it was fixed. That then reduces the situation to being like the left an right handed glove analogy. Which were left-handed and right-handed all along.

Bohr's response to EPR - When he received the EPR paper, almost a decade after the Copenhagen Interpretation was developed. Bohr had not yet realized the implications of quantum theory to which EPR objected. He did not realize that the theory claimed that observation, in and of itself, *without any physical disturbance* can instantaneously affect a remote physical system - .

So there were these two possibilities for what entanglement was about. Bohr's view verses Einstein's view. The problem was that there was no obvious way of distinguishing between them because both views predicted the same outcome. Which is that we would measure, rather Alice and Bob would measure all of these correlations that exist between the spins of these two entangled electrons. How do know that it was do to hidden variables or due to something else that Bohr was suggesting?

All of that changed in 1964 when the Irish physicist John Bell suggested an experiment. However, at that stage, it was just a thought experiment that he said would allow us to distinguish between these two possibilities. And it was John Bell's experiment that two proverbial "quantum Boxes" can mimic in experiment. This experiment requires combinations to get a set of outputs that satisfy the three required rule of Bell's through experiment.

With classical or mechanical type combinations, the best you can do is 75% of success rate or the square-root of 2 (3 times out of 4 rules being satisfied).

However with quantum type combinations, the best you can do is an 85% success rate with quantum entanglement correlation. So quantum rules give a better outcome then do classical rules. This is because we used the Bohr instantaneous communication property.

So the classical limit or the hidden variables limit of 75 percent correlation between the particles is 10 percent less reliable than the quantum approach. This suggests that in Einstein's idea that these things called hidden variables, which fix the properties of quantum particles before they are measured does not apply. So with this thought experiment of Bell and the subsequent physical experiment done in 1984 bore fruit for Neils Bohr. So it seems that Bohr was correct and Einstein was wrong. Quantum entanglement is real.

However, this does not imply that there is some "spooky action at a distance". Because this so called "spooky action at a distance" is what Einstein implied with respect to entanglement. But actually, there is a better way to think about entanglement by saying something a little different. One way to think about it is to say that once the two boxes became entangled they are no longer separate objects. So what happens over at box A is completely independent to what happens at box B. They are in some quantum since the same object and that remains the case no matter how far apart they are. Even if the tow particles we measure are on opposites sides of the galaxy they remain in some since a single quantum entity.

Another way of thinking of it is to say there is some kind of sharing information between them. Physicists call this quantum non-locality. And its distinct form the notion that some how making a measurement on a particle is transmitting information or a signal to the other particle to fix what the value of what its property is. However, that does not happen as was alluded to earlier. If it did happen, it would violate special relativity as Einstein suggested, meaning nothing including signals can travel faster then light. So quantum mechanics tells you something else. It tells you that there is this property called "quantum non-locality" which is a real property in the universe.

Ch 9 Delayed Choice Quantum Eraser Experiment

From earlier 20th century double-hole experiments, the shocking conclusions don't end there. In recent years technology has allowed scientists to perform a fascinating variation of the test. Its results call into question our perception of time itself. This is considered to be a high-tech version of the *double slit experiment.* Electrons are being fired toward a barrier with two holes in it. But the scientists can ***delay*** their decision about whether to observe the electrons until after they have passed through the holes, but before they hit the screen. So imagine a electron particles being shot towards a barrier with two holes in it. Initially, my eyes are closed, so it goes through both holes and it behaves like a wave. But then, at the last moment before it hits the back screen, I open my eyes and decide to observe it. At that moment the electrons in essence become particles and seemingly always were particles from the time they left the electron gun. So its as though they went *back in time* to before they went through the holes, then decided to go through one or the other, not through both had they been behaving like waves.

That is the enigma, that our choice of experiment to perform determines the prior state of the electron. Some how or other, we have an *influence* on it to which it makes it seem as to appear to *travel backwards in time.* However, by further analysis, the electron has not gone backwards in time, although on the face of it, the observation does seems to appear that way.

9.1 The Diagram of the experiment:

The experiment begins in the top left corner with the laser firing two entangled photons either through slit A or slit B. Now it is truly random to which slit it will fire the pair of photons through. And because we have no measuring device at either slit we cannot know which slit the laser fires through until they pass through and are detected on either side. So at each firing, it will either go through A or B. As the twined particle is fired, the first prism splits the entangled photons.

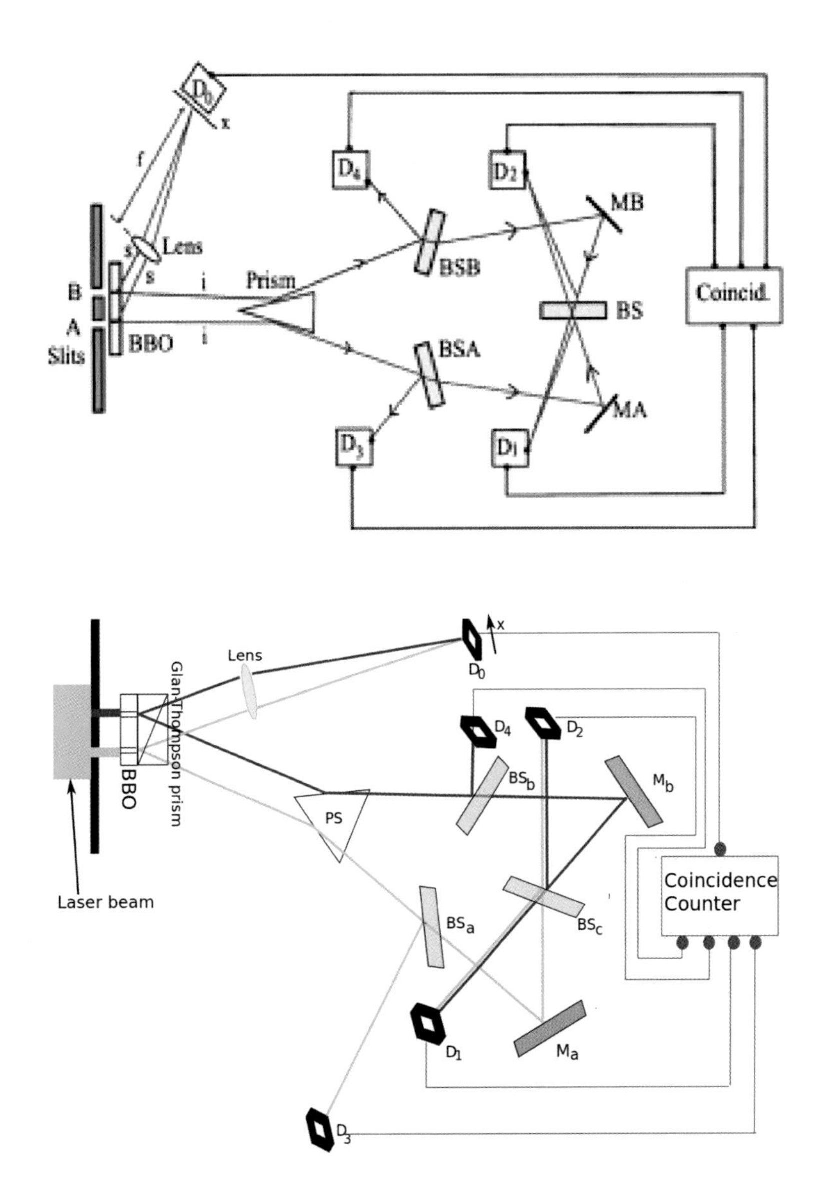

Fig. 9-1 The Delayed Choice Quantum Eraser

This is how it works:

When two subatomic particles interact they can become entangled. That means their spins, positions, and other properties become linked through a process unknown to modern science. You then make a measurement of one of the particles instantaneously, which determines what the behavior of the other entangled particle should be. And when the experiment is done, its found that indeed the other particle's quantum state is exactly determined once you made a measurement of the partner particle's quantum state.

9.2 Back to the experiment:

Now the first prism splits the entangled photons and sends them in different directions. The top one goes to detector-D0. Now if the photons only hit D0 you don't know the path information so when the photon arrives it could have either come from path A or Path B. So because we don't know the path information it should produce an interference pattern if it only came from there. If we palace a measuring device in front of the slit we would know the path information. But with just the result from D0 we don't know the path information. So because of the lack of knowledge that we would have about the system, the particle would act in a way as if it goes through both becoming the wave of possibilities that it could have been instead of one of these possibilities if we knew the definite path information. But the other entangled photon goes the other way and because it is entangled, it will affect the result of its twin that went to D0. Now the other photon from either A or B goes through the prism and hits either BSa or BSb. At both of these, there is a 50/50 chance that it will either pass through or bounce off. And it either goes to D3 or D4. Now if the photon hits either one of these detectors, we obtain which path information, because the only way the photon can hit D3 is if it came out of B. And the only way it could hit D4 is if it came out of A. There is no way a photon that came out of B could hit D4 and visa-versa. So if a photon hits D3 or D4 we will know the path information that it took and we will get a **club pattern** (meaning the, *we know the which path information – that the particle went through*).

Now if the photon only randomly passes through BSa or BSd it will either bounce off of PS or between the lens and it has a 50/50 chance of passing through BSc or bouncing off of it. So if the photon passes through BSa or BSb we lose the path information. Because if it hits D1 or D2 it could either have come from A or B. We could never trace the path information back to A or B. So when they hit D1 or D2 we should get an interference pattern demonstrating that the photon went through both slits since we don't have definite path information. Now here's one of the important implications of this experiment about what is causing collapse. Some argue physical interaction from the detector is what is causing the collapse. But if that was the case D1 and D2 should cause collapse every time. But that is not what happens. If a photon makes it to D1 or D2 they always display an interference pattern. And every time a photon hits D3 or D4 a club pattern is formed.

But the only difference is what we observers know about these two stations. Because of the experiment setup, we know that if a photon hits D3, it will always be a club pattern showing the photon only went through one slit. If it hits D1 we know it will always be an interference pattern showing that the photon acted in a way as if it went through slits. But the only difference between these two is what we the conscious observers know about the system. *Our knowledge of the system causes different results in how the photon will act.* If it was all random and not caused by our knowledge we should get some club patterns at D1 and some interference patterns in D3. But this is not what the experiment shows. We always get a club pattern at D3 and D4 because whenever a photon hits here at BSa we always know the pattern information. And we will always get an interference pattern at D1 and D2 because we can never know the actual path information so the photon acts in a way as if it goes through both. Where as at D3 and D4 the photon only goes through one slit because we know the path information. The particles acted in a way that correlates to our knowledge. **What causes collapse is knowledge**. And knowledge requires a knower.

Sir Rudolf Peierls said "the moment at which you can throw away one possibility and keep only the other is when you finally become conscious of the fact that the experiment has given one result. You see, the quantum mechanical description is in terms of knowledge, and knowledge requires somebody who knows."

Book - The Ghost in the Atom page 73-74.

Now the other implication is even more mind-boggling. Because a photon knew beforehand where it would end up. How do we know this? Well it's because of how its twin acts at D0. The twin photon registers at D0 before its entangled twin ends up at a detector. And whatever registers at D0 always correlates to wherever its twin ends up. So if the twin hits D3 or D4, D0 always registers a club pattern to correlate. And an interference pattern of its twin always lands at D1 or D2.

Note: A **club-pattern** = meaning the *we know "the which path information."*

9.3 A closer look at the *delayed choice quantum eraser*

The question of weather the detectors in double-slit experiments physically caused the wave function to collapse was settled by experiments like the 1999 delayed-choice quantum eraser experiment as mentioned above. The experiment showed that the wave properties of a photon could not possibly be collapsed into a particle by some physical effect of the detectors. That's because there were no detectors between the slits and the screen so that the "which-path" information was detected after the photons were already registered on the screen.

* *Quantum Entanglement* – is a physical phenomenon that occurs when pairs or groups of particles are generated or interact in ways such that the quantum state of each particle cannot be described independently – instead, a quantum state must be described for the system as a whole. Measurements of physical properties such as position, momentum, spin, polarization etc ... performed on entangled particles are found to be appropriately correlated.

* *Fringe patterns* – the interference pattern detected on the back screen.

* *Non-locality* – acting at any *distance*, be it large or small. (does not violate causality)

Coherence is what allows the interference pattern to occur in the first place.

So when you place an object in front of the *coherence wave* you disrupt the coherence of the wave-particle and change the pattern that reaches the screen.

This de-coherence destroys the coherence fringe or wavelike patterns which would eventually be found on the back screen.

The double-slit experiment in which you try to determine which slit it traversed the is called a *"which-way experiment."* And if the test is done on the far side of the slits, it's called a *delayed choice experiment.*

Physicists have come up with clever ways in which to measure which way the particle travelled while still preserving coherence. This is the fore-mentioned "Delayed Quantum Eraser" experiment performed in 1999. The apparatus makes use of a very special type of *crystal* that absorbs the incoming photon and creates two new twin photons (photons that are now entangled, an entangled pair), each with half the energy of the original. They are now coherent entangled pairs of photons passing through the slits.

So by placing the crystal in front of the double slit, to make coherent entangled pairs of any photons passing through. Send one of each pair off to the screen to produce an interference pattern and use the other to figure out which slit the original photon passed through.

The part of the apparatus containing devices A and B are the beam splitters (half silver mirrors). They are the quantum eraser part. Its job is to destroy any information about the path of the photons. They work by allowing about 50% of the photons to pass through while reflecting the other 50%, due to the *half-silvered mirrors*. Now you have a new possible outcome. So instead of being reflected to detectors A or B half of the photons end up in detectors C or D.

This clever arrangement ensures that if C or D light up, then we have no idea which slit that photon came from. So if we only look at the photons who twins end up either detector C or D we do see an interference pattern. It seems like the simple act of scrambling the which-way information, retroactively sends message "the observer lost the information on which slit we went through so it's safe to have gone through both again." So perhaps "observation" is just entanglement between the observer and the experiment.

Note: the apparatus is not *erasing* the "which-path" information, but rather it's officiating. It's not allowing you to know what the which-path was.

So in conclusion, since we know that the experiment showed that the wave property of a photon could not possibly be collapsed into a particle by some physical effect of the detectors. This leads us to ponder the question "does *consciousness* lead to the collapse of the wave function through *knowledge* that we have obtained with respect to the apparatus? The up coming chapters in this book will help to shed some light on this most strange phenomena of nature.

Ch 10 Extra Dimensions

10.1 The Kaluza-Klein Theory - A look at Extra Dimensions

Extra dimensions have changed the way physicists think about the Universe. And because the connections of extra dimensions to the world could tie into many more well-established physics ideas, extra dimensions are a way to approach older, already verified facts about the Universe.

One of the earliest attempts to "unify" gravity and electromagnetic forces came in the form of Kaluza-Klein theory, a short-lived theory that attempted to unify the forces of nature by introducing an *extra space dimension*. In this theory, the extra space dimension was curled up to a "microscopic" size. Though the theory failed, many of the same concepts were eventually applied in the study of string theory.

Einstein's theory of gravity (General Relativity) had proved so elegant in explaining gravity that physicists wanted to apply it to the other force known at the time — the electromagnetic force. Was it possible that this other force was also a manifestation of the geometry of space-time? In 1915, even before Einstein completed his general relativity field equations, the British mathematician David Hilbert said that research by Nordström and others indicated 'that gravitation and electrodynamics are not really different.' Einstein responded, 'I have often tortured my mind in order to bridge the gap between gravitation and electromagnetism.' One theory in this regard was developed and presented to Einstein in 1919 by German mathematician Theodor Kaluza. In 1914, Nordström had written Maxwell's equations in five dimensions and had obtained the gravity equations (see the section "Pulled in another direction: Einstein's competition for a theory of gravity"). Kaluza took the gravitational field equations of general relativity and wrote them in five dimensions, obtaining results that included Maxwell's equations of electromagnetism! When Kaluza wrote to Einstein to present the idea, the founder of relativity replied by saying that increasing the dimensions "never dawned on me" (which means he must have been unaware of Nordström's attempt to unify electromagnetism and gravity, even though he was clearly aware of Nordström's theory of gravity). In Kaluza's view, the universe was a 5-dimensional cylinder and our 4-dimensional world was a projection on its surface. Einstein wasn't quite ready to take that leap without any evidence for the extra dimension.

Still, he incorporated some of Kaluza's concepts into his own unified field theory that he published and almost immediately recanted in 1925. A year later, in 1926, Swedish physicist Oskar Klein dusted off Kaluza's theory and reworked it into the form that has come to be known as the Kaluza-Klein theory. Klein introduced the idea that the fourth space dimension was rolled up into a tiny circle, so small that there was essentially no way for us to detect it directly.

In Kaluza-Klein theory, the geometry of this extra, hidden space dimension dictated the properties of the electromagnetic force — the size of the circle, and a particle's motion in that extra dimension, related to the electrical charge of a particle. The physics fell apart on this level because the predictions of an electron's charge and mass never worked out to match the true value. Also, many physicists initially intrigued with the Kaluza-Klein theory became far more intrigued with the growing field of quantum mechanics, which had actual experimental evidence (as you see in Chapter 7). Another problem with the theory is that it predicted a particle with zero mass, zero spin, and zero charge. Not only was this particle never observed (despite the fact that it should have been, because it's a low-energy particle), but the particle corresponded to the radius of the extra dimensions. It didn't make sense to add a theory with extra dimensions and then have a result be that the extra dimensions effectively didn't exist.

There is another (though less conventional) way to describe the failure of Kaluza-Klein theory, viewing it as a fundamental theoretical limitation: For electromagnetism to work, the extra dimension's geometry had to be completely fixed. In this view, tacking an extra dimension onto a theory of dynamic space should result in a theory that is still dynamic. Having a fifth dimension that's fixed (while the other four dimensions are flexible) just doesn't make sense from this point of view. This concept, called background dependence, returns as a serious criticism of string theory in Chapter 17. Whatever the ultimate reason for its failure, Kaluza-Klein theory lasted for only a short time, although there are indications that Einstein continued to tinker with it off and on until the early 1940s, incorporating elements into his various failed unified field theory attempts. In the 1970s, as physicists began to realize that string theory contained extra dimensions, the original Kaluza-Klein theory served as an example from the past. Physicists once again curled up the extra dimensions, as Klein had done, so they were essentially undetectable. Such theories are called Kaluza-Klein theories.

10.2 Defining Dimensionality

No physical theory we know off dictates that there should be only three dimensions of space. Dismissing the possibility of extra dimensions before even considering their existence might be very premature. Just as "up-down" is a different direction from "left-right" or "forward-backward," other completely new dimensions could exist in our Cosmos. Although we can't see or feel them, additional dimensions of space are a logical possibility.

Such hypothetical unseen dimensions didn't yet have a name. But should they exist, they would be new directions along which something might travel.

These passages could be flat, like the dimensions we are accustomed to. Or they could be warped, like reflections in a fun-house mirror. They might be tiny – far smaller than an atom – until recently, that's what anyone who believed in extra dimensions assumed.

But new research has shown that extra dimensions might also be big, or even infinite in size, yet still be hard to see. Our senses register only three large dimensions, so an infinite extra dimension might sound incredible. But and infinite unseen dimension is one of many possibilities for what might exist in the Cosmos.

Research into extra dimensions has also led to other remarkable concepts – ones that might fulfill a science fiction fantasy – such as parallel universes, warped geometry, and three-dimensional sinkholes.

10.3 How Unseen dimensions can help our understanding

Resent advances suggest that extra dimensions, not yet experienced and not yet entirely understood, might nonetheless resolve some of the most basic mysteries of our Universe. Extra dimensions could have implications for the world se see, and ideas about them might ultimately reveal connections that we miss in three-dimensional space.

In addition, the connections that occur with additional dimensions of space might illuminate perplexing aspects of particle physics, shedding light on decades-old mysteries. Relationships between particle properties and forces that seem inexplicable when space is shackled in three dimensions seen to fit together elegantly in a world with more dimensions of space.

For many years, physicists have relied on a theory called the Standard Model of particle physics, which tells us about the fundamental nature of mater and the forces through which elementary constituents interact. Physicists have tested the Standard Model by creating particles that have not been present in the world since the earliest seconds of the Universe, and they've found that the Standard Model describes many of their properties extremely well. However the theory leaves some fundamental questions left unanswered. The study and investigation of extra dimensions could solve some of those puzzling questions.

10.4 Why is gravity so weak

One of the chief mysteries is why gravity is so much weaker than the other known forces. Gravity might not feel weak when you're hiking up a mountain, but that's because the entire Earth is pulling on you. A tiny magnet can lift a paper clip, even though all the mass of the Earth is pulling it in the opposite direction. Why is gravity so defenseless against the small tug of a tiny magnt? In standard three-dimensional particle physics, the wakness of gravity is a huge puzzle. But extra dimensions might provide an answer.

In 1998 Physicists Lisa Randel and Raman Sundrum developed a model of gravity. The model is based on "warped" geometry," a notion that Randel says arises in Einstein's General theory of Relativity. According to the theory, space and time are integrated into a single spacetime fabric that gets distorted, or warped, by matter and energy.

They found a configuration in which spacetime warps so severely that even if gravity is strong in one region of space, it is feeble everywhere else. They also found something even more remarkable. Although physicists have assumed for over eighty years that extra dimensions must be tiny in order to explain why we haven't seen them, in 199 Randel and Sundrum discovered that not only can warped space explain gravity's feebleness, but also that an invisible extra dimension can stretch out to infinity, provided it is suitably distorted in a curved spacetime. An extra dimension cn e infinite in size – but nevertheless be hidden.

The Randel-Sundrum model suggests that we could be living in a three-dimensional pocket of space, even though the rest of the universe behaves as if it is higher-dimensional. This result opens a host of new possibilities for the fabric of spacetime, which could consist of distinct regions, each appearing to contain a different number of dimensions.

Not only are we not in the center of the Universe, a Copernicus shocked the world by suggesting five hundred years ago, but we just might be living in an isolated neighborhood with three spatial dimensions that's part of a higher-dimensional Cosmos.

The newly studied membrane-like objects called branes are important components of the rich higher-dimensional landscapes. If extra dimensions are a physicist's playground, then braneworlds – hypothesized universes in which we live on a brane – are the tatalizing, multi-layereed, multi-faceted jungle gym. This chapter will take you to braneworlds and universes with curled-up, warped, large, and infinite dimensions, some of which contain a single brane and others of which have multiple branes housing unseen worlds. All of these are within the realm of possibility.

In 2007, the Large Hadron Collider (LHC) was turned on and since then it has produced tremendously energetic particles that could be new types of matter. If any of these extra-dimensional theories is right, it could leave visible signs at the LHC. The evidence would include particles called Kaluza-Klein modes, which travel in the extra dimensions yet leave traces of their existence here in the familiar three dimensions. Kaluza-Klein modes would be fingerprints of extra dimensions in our three-dimensional world. And if we're very lucky, experiments will register other clues as well, perhaps even higher-dimensional black holes.

10.5 What are dimensions

Working with spaces that have many dimensions is actually something everyone does every day, although admittedly most of us don't think of it that way.

The number of dimensions is the number of quantities you need to know to completely pin down a point in a space. The multidimensional space might be an abstract one, such as the space of features you are looking for in a house, or it might be concrete, like the real physical space we will soon consider. But when buying a house, you can think of the number of dimensions as the number of quantities you would record in each entry in a database – the number of quantities you find worth investigating.

A space if a particular dimension is a space requiring a particular number of quantities to specify a point. In one dimension, that would be a point on a plot with a single x axis; in two dimensions, a point on a plot with an x and a y axis; in three dimensions, it would be a point on a plot with and x, a y, and a z axis.

In three-dimensional space, three numbers are all you ever need to know your precise location. The numbers you specify might be latitude, longitude, and altitude; or length, width, and height; or you might have a different way to choose your three numbers. The critical thing is that three dimensions means you need precisely three numbers. In two-dimensional space you need two numbers, and in higher-dimensional space you need more.

More dimensions means freedom (degree of freedom) to move in a greater number of completely different directions. A point in a four-dimensional space simply requires one additional axis – difficult to draw.

String theory suggests even more dimensions: it postulates six or seven extra spatial dimensions, meaning that six or seven additional coordinates are needed to plot a point. And very recent work in string theory has shown that there could be even more dimensions than that. Many of the concepts about extra dimensions that I will describe apply to any number of extra dimensions.

A *metric* gives a measuring rod that reveals your choice of units in order to set the scale, just like on a map, where a half-inch might represent one mile, r as in the metric system, which gives us a meter stick we all agree on.

But that is not all a metric specifies. It also tells us whether space bends or curls around, like the surface of a balloon when it is blown up into a sphere. The metric contains all the information about the shape of space. A metric for curved space tells us about both distances and angles. Just as an inch can represent different distances, an angle can correspond to different shapes.

As physics has evolved, so has the amount of information stored in the metric. When Einstein developed relativity, he recognized that a fourth dimension – time – is inseparable from the three dimensions of space. Time, too, needs a scale, so Einstein formulated gravity, by using a *metric* for four-dimensional *space-time*, adding the dimension of time to the three dimensions of space.

And more recent developments have shown that additional spatial dimensions might also exist. In that case, the true space-time metric will involve more than three dimensions of space. The number of dimensions and the metric for those dimensions is how ne describes such a multidimensional space.

For example, when you view a hose spread over a football field from above, it looks it has one dimension. But when you view it up close, you see that the surface has two dimensions and the volume it encloses has three.

Imagine again a garden hose, which can be thought of as a long sheet of rubber rolled up into a tube with a small circular cross-section. This time, we'll think of the hose as the entire universe (not an object inside the universe). If the universe were shaped like this garden hose, we would have one very long dimension and one very small, rolled-up dimension – exactly what we want.

For a little creature – a flat bug, say – that lived in the garden-hose universe, the universe would look two-dimensional. (In this scenario, our bug has to stick to the surface of the hose – the two-dimensional universe doesn't include the interior for which is three-dimensional. The bug could crawl in two directions: along the length of the hose or around it. The bug that started somewhere along the hose could crawl around and eventually return to where it started. Because the second dimension is small, the bug wouldn't travel very far before it returned.

If a population of bugs living on the hose experienced forces, such as the electric force or gravity, the forces would be able to attract or repel bugs in any direction on the surface of the hose. Bugs could be separated from one another either along the length of the hose or around the hose's circumference, and would experience any force that was present on the hose. Once there is sufficient was present on the hose. Once there is sufficient resolution to distinguish distances s small as the diameter of the hose, forces, and objects exhibit both of the dimensions they actually have.

However, if our bug could observe its surroundings, it would notice that the two dimensions were very different. The one along the length of the hose would be very big. It could even be infinity long. The other dimension, on the other hand, would be very small. Two bugs could never get very far from each other in the direction around the hose. And a bug that tried to take a long trip in that direction would quickly end up back where it started. A thoughtful bug that liked to stretch its legs would know that its universe was two-dimensional, and that one dimension extended a long way while the other was very small and rolled up into a circle.

But the bug's perspective is nothing like the one that creatures like us would have in Klein's universe, in which the extra dimension is rolled up to an extremely small size, 10^{-33} cm. Unlike the bug, we are not small enough to detect – never mind travel in – a dimension of such a tiny size.

Moreover, physical effects wouldn't betray the extra dimension's existence. Big beings in the garden-hose universe would fill out the entire second, small dimension and would never know that this dimension was there. Without the ability to detect structure or variations along the extra dimension, such as wiggles or undulations of matter or energy, they could never register its existence

Just as with our two-dimensional garden-hose universe, a four-dimensional Kaluza-Klein universe with a single tiny, rolled-up dimension would appear to us to have one dimension fewer than the four it actually has. Because we wouldn't know about the additional spatial dimension unless we could detect evidence of structure on its minute sale, the Kaluza-Klein universe would look, three-dimensional. Rolled-up, or compactified, extra dimensions will never be detected if they are sufficiently tiny. In the chapters that follow, we'll investigate just how tiny, but for now, rest assured that the Planck length is well below the threshold of detectability.

10.6 Extra dimensions

Another mathematical result of string theory is that the theory only makes sense in a world with more than three space dimensions. (Our Universe has three dimensions of space – left/right, up/down, and front/back.) Two possible explanations currently exist for the location of the extra dimensions:

1. The extra space dimensions (generally six of them) are curled up (*compactified*, in string theory terminology) to incredibly small sizes, so we never perceive them.
2. We are "stuck" on a 3-dimensional brane, and the extra dimensions extend off of it and are inaccessible to us.

A major area of research among string theorists is on mathematical models of how these extra dimensions could be related to our own. Some of these recent results have predicted that scientists may soon be able to detect these extra dimensions (if they exist) in upcoming experiments, because they may be larger than previously expected.

10.7 Higher Dimensions

We live in a world with three *"spatial"* dimensions. In a nutshell this means that there are three distinct directions through which movement is possible: up-down, left-right and forward-backward. In addition, we have the flow of time (forward only as far as we know). Mathematically, this gives us the relativistic description of coordinates (x, y, z, t).

It is possible to imagine a world where one of the spatial directions or dimensions have been removed (say up-down). Such a two dimensional world was described by Edward Abbott in his classic Flatland. What if instead, we added dimensions? This idea is actually pretty useful in physics, because it provides a pathway toward unifying different physical theories.

This kind of thinking was originally put forward by two physicists named Kaluza and Klein in the 1920s. Their idea was to bring gravity and electromagnetism into a single theoretical framework by imagining that these two theories were four-dimensional limits of a five-dimensional super-theory.

This idea did not work out, because back then people did not know about quantum field theory and so did not have a complete picture of particle interactions, and did not know that the fully correct description of electromagnetic interactions is provided by quantum electrodynamics. But this idea has a lot of appeal and reemerged in string theory.

Kaluza and Klein had to explain why we don't see the higher dimension, and hit upon the idea of "compactification" – a procedure where we make the higher dimensions so small they are not detectable at lower energy (i.e., on the kind of energy scales that we live in). If they are small enough, the extra dimensions can't be noticed or detected scientifically without the existence of the appropriate technology. If they are so small that they are on the Planck scale, we might not be able to see them at all. Compactification explains why we may not be aware of extra spatial dimensions even if they exist. If, for example, the radius of a three-dimensional cylinder is very small, from a distance it looks like a line (a two-dimensional object).

String theory requires the existence of extra spatial dimensions for technical reasons. An interesting side effect of these extra dimensions is that another mystery of particle physics is done away with. Experimentalists have worked out that there are three families of particles. For example, when considering leptons, there is the electron and its corresponding neutrino. But there are also the "heavy electrons" known as the *muon* and the *tau*, together with their corresponding neutrinos, that are really just duplicates of the electron. The same situation exists for the quarks. Why are there three particle families? And why are there the types of particle interactions that we see? It turns out that higher spatial dimensions together with string theory may provide an answer. The way that you *compactify* the extra dimensions (the topology) determines the numbers and types of particles seen in the Universe.

In string theory this results from the way that the strings can wrap around the compactified dimensions determining what vibrational modes are possible in the string and hence what types of particles are possible.

One important compactified manifold that is called *Calabi-Yau manifold.* A Calabi-Yau manifold that compactifies six spatial dimensions and leaves three spatial dimensions "macroscopic" plus time gives a ten-dimensional universe as required by most of the string theories. A key aspect of *Calabi-Yau Manifolds* is that they *"break symmetries."* Thus another mystery of particle physics is explained, so-called spontaneous symmetry breaking in quantum field theory.

10.10 Quantum Units of Nature – Planck Units

Physicists occasionally use a system of natural units, called *Planck units*, which are calculated based on fundamental constants of nature like Planck's constant, the gravitational constant, an the speed of light.

Planck's constant comes up often in discussing quantum physics. In fact, if you were to perform the mathematics of quantum physics, you'd find that little *h* variable all over the place. Physicists have even found that you can define a set of quantities in terms of Planck's constant and other fundamental constants, such as the speed of light, the gravitational constant, and the charge of an electron.

These Planck units come in a variety of forms. There is a Planck charge and a Planck temperature, and you can use various Planck units to derive other units such as the Planck momentum, Planck pressure, and Planck force.

For the purposes of the discussion of string theory, only a few Planck units are relevant. They are created by combining the gravitational constant with the speed of light and Planck's constant, which makes them the natural units to use when talking about phenomena that involve those three constants, such as quantum gravity. The exact values aren't important, but here are the general scales of the relevant Planck units:

1. Planck length: 10^{-35} meters (if a hydrogen atom were as big as our galaxy, the Planck length would be the size of a human hair)
2. Plank time: 10^{-43} seconds (the time light takes to travel the Planck length – a very, very short period of time)
3. Planck mass: 10^{-8} kilograms (about the same as a large bacteria or vary small insect)
4. Planck energy: 10^{28} electron-volts (roughly equivalent to a ton of TNT explosive)

Keep in mind that the exponents represent the number of zeroes, so the Plank energy is a 1 followed by 28 zeroes, in electron-volts. The most powerful particle accelerator on Earth, the Large Hadron Collider that came online in 2008 can produce energy only in the realm of TeV – that is, a 1 followed by 12 zeros, in electro-volts. The negative exponents, in turn represent the number of decimal places in very small numbers, so the Planck time has 42 zeroes between the decimal point and the first non-zero digit. It's a very small amount of time. Some of these units were first proposed in 1899 by Max Planck himself before either relativity or quantum physics. Such proposals for natural units – units based on fundamental constants of nature – had been made at least as far back as 1881.

Planck's constant makes its first appearance in the physicist's 1899 paper. The constant would later show up in his paper on the quantum solution to the ultraviolet catastrophe. Planck units can be calculated in relation to each other. For example, it takes exactly the Planck time for light to travel the Planck length. The Planck energy is calculated by taking the Plank mass an applying Einstein's $E = mc^2$ (meaning that the Planck mass and Planck energy are basically two ways of writing the same value).

In quantum physics and cosmology, these Planck units sneak up all the time. Planck mass represents the amount of mass needed to be crammed into the Planck length to create a black hole. A field in quantum gravity theory would be expected to have a vacuum energy with a density roughly equal to one Planck energy per cubic density roughly equal to one Planck energy per cubic Planck length – in other words, it's 1 Planck unit of energy density.

Why are these quantities so important to string theory?

The *Planck length* represents the distance where the smoothness of relativity's space-time and the quantum nature of reality begin to rub up against each other. This is the "quantum foam." It's the distance where the two theories each, in their own way, fall apart. Gravity *explodes* to become incredibly powerful, while "quantum fluctuations" and vacuum energy run rampant. This is the realm where a theory of quantum gravity, such as sting theory, is needed to explain what's going on.

10.11 Planck units and Zeno's paradox

If the Planck length represents the shortest distance allowed in nature, it could be used to solve the ancient Greek puzzle called *Zeno's paradox*. Here is the paradox:

You want to cross a river, so you get in your boat. To reach the other side, you must cross half the river. Then you must cross half of what's left. Now cross half of what's left. No matter how close you get to the other side of the river, you will always have to cover half that distance, so it will take you forever to get across the river, because you have to cross an infinite number of halves.

The traditional way to solve this problem is with calculus, where you can show that even though there are an infinite number of halves, it's possible to cross them all in a finite amount of time. (Unfortunately for generations of stymied philosophers, calculus was invented by Newton and Leibnitz 2,000 years after Zeno posed his problem.)

As it turns out, ***the Planck length is really the shortest distance allowed by nature, the quantum of distance***, it offered a physical resolution to the paradox.

When your distance from the opposite shore reaches the Planck length, you can't go half anymore. Your only options are to go the whole Planck length or go nowhere. In essence, you can't cut the distance of the Planck scale in halves. So you in principle finally reach you destination.

In some sense, these units are sometimes considered to be quantum quantities of time and space, and perhaps some of the other quantities as well. Mass and energy clearly come in smaller scales, but time and distance don't seem to get much smaller that the Planck time and Planck length. Quantum fluctuations, due to the uncertainty principle, become so great that it becomes meaningless to even talk about something smaller.

In most string theories, the length of the strings or length of compactified extra space dimensions are calculated to be roughly the size of the Planck length. The problem with this is that the *Planck length* and the *Planck energy* are connected through the uncertainty principle, which means that to explore the Planck length – the possible length of a "string" in string theory – with precision, you'd introduce an "uncertainty" in energy equal to the Planck energy.

This is an energy 16 orders of magnitude (add 16 zeros) more powerful than the newest, most powerful particle accelerator on Earth can reach. Exploring such small distances required a vast amount of energy, far more energy than we can produce with present technology.

Ch 11 The Fabric of Space-Time

11.1 Quantum Reality

The second anomaly to which Lord Kelvin referred led to the quantum revolution, one of the greatest upheavals to which modern human understanding has been subjected. By the time the fires subsided and the smoke cleared, the veneer of classical physics had been singed off the newly emerging framework of quantum reality.

A core feature of classical physics is that if you know the positions and velocities of all objects at a particular moment, Newton's equations, together with their *Maxwellian* updating, can tell you their positions and velocities at any other moment, past or future. Without equivocation, classical physics declares that the past and future are etched into the present. This is also shared by both special and general relativity. Although the relativistic concepts of past and future are subtler than their familiar classical counterparts, the equations of relativity, together with a complete assessment of the present, determine them just as completely.

By the 1930s, however, physicists were forced to introduce a whole new conceptual schema called quantum mechanics. Quite unexpectedly, they found that only quantum laws were capable of resolving a host of puzzles and explaining a variety of data newly acquired from the atomic and subatomic realm. But according to the quantum laws, even if you make the most perfect measurements possible of how things are today, the best you can ever hop to do is predict the *probability* that things will be one way or another at some chosen time in the future, or that things were one way or another at some chosen time in the past. The Universe, according to quantum mechanics, is not etched into the present; the Universe, according to quantum mechanics, participates in a game of chance.

Although there is still controversy over precisely how these developments should be interpreted, most physicists agree that probability is deeply woven into the fabric of quantum reality. Whereas human intuition, and its embodiment in classical physics, envision a reality in which things are always definitely one way *or* another, quantum mechanics describes a reality in which things sometimes hover in a haze of being partly one way *and* partly another. Things become definite only when a suitable observation forces them to relinquish quantum possibilities and settle on a specific outcome.

The outcome that's realized, though, cannot be predicted – we can predict only the odds that things will turn out one way or another. This, plainly speaking, is weird. We are unused to a reality that remains ambiguous until perceived. But the oddity of quantum mechanics does not stop here. At least as astounding is a feature that goes back to a paper Einstein wrote in 1935 with two younger colleagues, Nathan Rosen and Boris Podolsky, that was intended as an attack on quantum theory.

11.2 Entangling Space – What does it mean to be separate in a quantum universe

To accept the special and general relativity is to abandon Newtonian absolute space and absolute time. While it's not easy, you can train you mind to do this. Whenever you move around, imagine your now shifting away from the "nows" experienced by all others not moving with you. While you are driving along a highway, imagine your watch ticking away at a different rate compared with timepieces in the homes you are speeding past. While you are gazing out from a mountaintop, imagine that because of the warping of space-time, time passes more quickly for you than for those subject to stronger gravity on the ground far below. I say "imagine" because in ordinary circumstances such as these, the effects of relativity are so tiny that they go completely unnoticed. Everyday experience thus fails to reveal how the Universe really works, and that's why a hundred years after Einstein, almost no one, not even professional physicists, feels relativity in their bones. This isn't surprising; one is hard pressed to find the survival advantage offered by a solid grasp of relativity. Newton's flawed conceptions of absolute space and absolute time work wonderfully well at the slow speeds (non-superluminal) and moderate gravity we encounter in daily life, so our senses are under no evolutionary pressure to develop relativistic acumen. Deep awareness and true understanding therefore require that we diligently use our intellect to fill in the gaps left by our senses.

11.4 Probability and the Laws of Physics

If an individual electron is also a wave, what is it that is waving? Erwin Schrodinger weighted in with the first guess: maybe the stuff of which electrons are made can be smeared out in space and it's this smeared electron essence that does the waving. An electron particle, from this point of view, would be a sharp spike in an electron mist. It was quickly realized, though, that this suggestion couldn't be correct because even a sharply spiked wave shape – such as a giant tidal wave – ultimately spreads out. And if the spiked electron wave were to spread we would expect to find part of a single electron's electric charge over here or part of its mass over there. But we never do.

When we locate an electron, we always find all of its mass and all of its charge concentrated in one tiny, point-like region. In 1927, Max Born put forward a different suggestion, one that turned out to be the decisive step that forced physics to enter a radically new realm. The wave, he claimed, is not a smeared-out electron, nor is it anything ever previously encountered in science. The wave, Born proposed, is a *probability* wave.

To understand what this means, picture a snapshot of water wave that shows regions of high intensity (near the peaks and troughs) and regions of low intensity (near the flatter transition regions between peaks and troughs). The higher the intensity, the greater the potential the water wave has for exerting force on nearby ships or on coastline structures. The probability waves envisioned by Born also have regions of high and low intensity, but the meaning he ascribed to these wave shapes was unexpected: the size of a wave at a given point in space is proportional to the probability that the electron is located at that point in space. Places where the probability wave is large are locations where the electron is most likely to be found. Places where the probability wave is small are locations where the electron is unlikely to be found. And places where the probability wave is zero are locations where the electron will not be found.

Figure 11-1 gives a "snapshot" of a probabilistic interpretation. Unlike a photograph of water waves, though, this image could not actually have been made with a camera. No one has ever directly seen a probability wave, and conventional quantum mechanical reasoning say that no one ever will. Instead, we use mathematical equations (developed by Schrodinger, Niels Bohr, Werner Heisenberg, Paul Dirac, and others) to figure out what the probability wave should look like in a given situation.

We then test such theoretical calculations by comparing them with experimental results in the following way. After calculating the purported probability wave for the electron in a given experimental setup, we carry out identical versions of the experiment over and over again from scratch, each time recording the measured position of the electron. In contrast to what Newton would have expected, identical experiments and starting conditions do not necessarily lead to identical measurements. Instead, our measurements yield a variety of measured locations. Sometimes we find the electron here, sometimes there, and every so often we find it *way* over there. If quantum mechanics is right, the number of times we find the electron at a given point should be proportional to the size (actually, the square of the size), at that point, of the probability wave that we calculated.

Eight decades of experiments have shown that the predictions of quantum mechanics are confirmed to spectacular precision. Only a small portion of an electron's probability wave is shown in Figure 9-1: according to quantum mechanics, every probability wave extends throughout all of space, throughout the entire Universe. In many circumstances, though, a particle's probability wave quickly drops very close to zero outside some small region, indicating the overwhelming likelihood that the particle is in that region. In such cases, the part of the probability wave left out (the part extending throughout the rest of the Universe) looks very much like the part near the edges of the figure: quite flat and near the value zero. Nevertheless, so long as the probability wave somewhere in the Andromeda galaxy has a nonzero value, no matter how small, there is a tiny but genuine – nonzero – chance that the electron could be found there.

Thus, the success of quantum mechanics forces us to accept that the electron, a constituent of matter that we normally envision as occupying a tiny, point-like region of space, also has a description involving a wave that, to the contrary, is spread through the entire Universe. Moreover, according to quantum mechanics, this particle-wave fusion holds for all of nature's constituents, not just electrons: protons are both particle-like and wavelike; neutrons are both particle-like and wavelike, and experiments in the early 1900s even established that light – which demonstrably behaves like a wave

11.5 Unification in Higher Dimensions

In 1910, Einstein received a paper that could easily have ben dismissed as the ravings of a crank. It was written by a little-known German mathematician named Theodor Kaluza, and in a few brief pages it laid out an approach foe unifying the two forces known at the time, gravity and electromagnetism. To achieve this goal, Kaluza proposed a radical departure from something so basic, so completely taken for granted, that it seemed beyond questioning. He proposed that the Universe does not have three space dimensions. Instead, Kaluza asked Einstein and the rest of the physics community to entrain the possibility that the Universe has *four* space dimensions so that, together with time, it has a total of five space-time dimensions.

First off, what in the world does that mean? Well, when we say that there are there are three space dimensions we mean that there are three "independent" directions or axes along which you can move. From your current position you can delineate these as left/right, back/forth, and up/down; in a universe with three space dimensions, any motion you undertake is some combination of motion along these three directions. Equivalently, in a universe with three space dimensions you need precisely three pieces of information to specify a location. In a city, for example, you need a building's street, its cross street, and a floor number to specify the whereabouts of a dinner party. And if you want people to show up while the food is still hot, you also need to specify a fourth piece of data: a time. That's what we mean by space-time's being four-dimensional.

Kaluza proposed that in addition to left/right, back/forth. And up/down, *the universe actually has one more spatial dimension that, for some reason, no one has ever seen.* If correct, this would mean that there is another independent direction in which things can move, and therefore that we need to give pieces of information to specify a precise location in space, and a total of five pieces of information if we also specify a time.

Kaluza realized that the equations of Einstein's general theory of relativity could fairly easily be extended mathematically to a universe that had one more space dimension. Kaluza undertook this extension and found, naturally enough, that the higher-dimensional version of general relativity not only included Einstein's original gravity equations but, because of the extra space dimension, also had extra equations. When Kaluza studied these extra equations, he discovered something extraordinary: the extra equations were none other than the equations Maxwell had discovered in the nineteenth century for describing the electromagnetic field. By imaging a universe with one new space dimension, Kaluza had proposed a solution to what Einstein viewed as one of the most important problems in all of physics. Kaluza had found a framework that combined Einstein's original equations of general relativity with those of Maxwell's equations of electromagnetism. That's why Einstein didn't throw Kaluza's paper away.

Intuitively, you can think of Kaluza's proposal like this. In General Relativity, Einstein awakened space and time. As they flexed and stretched, Einstein realized that he'd found the geometrical embodiment of the gravitational force. Kaluza's paper suggested that the geometrical reach of space and time was greater still. Whereas Einstein realized that gravitational fields can be described as warps and ripples in the usual three space and one time dimensions, Kaluza realized that in a universe with an additional space dimension there would be additional space dimension warps and ripples. And those warps and ripples, his analysis showed, would be just right to describe, electromagnetic fields. In Kaluza's hands, Einstein's own geometrical approach to the universe proved powerful enough to unite gravity and electromagnetism.

Of course, there was still a problem. Although the mathematics worked, there was – and still is – no evidence of a spatial dimension beyond the three we all know about. So was Kaluza's discovery a mere curiosity, or was it somehow relevant to our universe? Kaluza had a powerful trust in theory – he had, for example, learned to swim by studying a treastise on swimming and then diving into the sea – abut the idea of an invisible space dimension, no matter how compelling the theory, still sounded outrageous. Then, in 1926, the Swedish physicist Oskar Klein injected a new twist into Kaluza's idea, on that suggested where the extra dimension might be hiding.

Klein's contribution was to suggest that what's true for an object within the Universe might be true for the fabric of the Universe itself. Namely, just as a tightrope's surface has both large and small dimensions (having a large or long dimension as well as a small curled up dimension), so does the fabric of space. Maybe the three dimensions we all know about – left/right, back/forth, and up/down – are like the horizontal extent of the tightrope, dimensions of the big, easy-to-see variety. But just as the surface of the tightrope has an additional, small, curled-up, circular dimension, maybe the fabric of space also has a small, curled-up, circular dimension, one so small that no one has powerful enough magnifying equipment to reveal its existence. Because of its tiny size, Klein argued, the dimension would be hidden.

How small would the dimension be? Well, by incorporating certain features of quantum mechanics into Kaluza's original proposal, Klein's mathematical analysis revealed that the radius of an extra circular spatial dimension would likely be roughly the Planck length, certainly much too small for experimental accessibility (current state-of-the-art equipment cannot resolve anything smaller than about a thousandth the size of an atomic nucleus, falling short of the Planck length by more than a factor of a million billion. Yet, to a imaginary, Planck-sized creature, this tiny, curled-up circular dimension would provide a new direction on which it could roam just as freely as an ordinary size creature negotiates the circular dimensions of the tightrope.

Or course, just as an ordinary creature finds that there isn't much room to explore in the clockwise direction before it finds itself back at its starting point, a *Planck-sized* creature walking along a curled-up dimension of space would also repeat the path and circle back to its starting point. But aside from the length of the travel it permitted, a curled-up dimension would provide a direction in which the tiny creature could move just as easily as it does in the three familiar unfurled dimensions.

With this modification to Kaluza's original idea, Klein provided an answer to how the universe might have more than the three space dimensions of common experience that could remain hidden, a framework that has since become known as *Kaluza-Klein theory.* And since an extra dimension of space was all Kaluza needed to merge general relativity and electromagnetism, Kaluza-Klein theory would seem to be just what Einstein was looking for. Indeed, Einstein and many other became quite excited about unification through a new, hidden space dimension, and a vigorous effort was launched to see whether this approach would work in complete detail. But it was not long before Kaluza-Klein theory encountered its own problems. Perhaps most glaring of all, attempts to incorporate the electron into the extra-dimensional picture prived unworkable. Einstein continued to dabble in the Kaluza-Klein framework until at least the early 1940s, but the initial promise of the approach failed to materialize, and interest gradually died out.

Within a few decades, though, Kaluza-Klein theory would make a spectacular comeback – and that would be with String Theory.

11.6 The Shape of Hidden Dimensions (*Calabi-Yau shape)*

The equations of string theory actually determine more than just the number of spatial dimensions. They also determine the kinds of shapes the extra dimensions can assume.

In the figures above, we focused on the simplest of shapes – circles, hollow spheres, solid balls – but the equations of string theory pick out significantly more complicated class of six-dimensional shapes known as *Calabi-Yau shapes* or *Calabi-Yau spaces*. These shapes are named after two mathematicians, Eugenio Calabi and Shing-Tung Yau, who discovered them mathematically long before their relevance to string theory was realized.

Bear in mind that in this figure a two-dimensional graphic illustrates a six-dimensional object, and this results in a variety of significant distortions. Even so, the pictures gives a rough sense of what these shapes look like.

If the particular Calabi-Yau shape in Figure 9-2 constituted the extra six dimensions in string theory, on ultramicroscopic scales space would have the form illustrated in Figure 12.b. As the Calabi-Yau shape would be tacked on to every point in the usual three dimensions, you and I and everyone else would right now be surrounded by and filled with these little shapes.

Figure 9-2: (a One example of a Calabi-Yau shape. (b) A highly magnified portion of space with additional dimensions in the form of a tiny Calabi-Yau shape.

Literally, as you walk from one place to another, your body would move through all nine dimensions, rapidly and repeatedly circumnavigating the entire shape, on average making it seem as if you weren't moving through the extra six dimensions at all. If these ideas are right, the ultramicroscopic fabric of the Cosmos is embroidered with the richest of textures.

11.7 String Physics and Extra Dimensions

The beauty of general relativity is that the physics of gravity is controlled by the geometry of space. With the extra spatial dimensions proposed by string theory, you'd naturally guess that the power of geometry to determine physics would substantially increase. And it does.

Why does string theory require ten space-time dimensions? This is a tough question to answer non-mathematically, but le me explain enough to illustrate how it comes down to an interplay of geometry and physics.

Imagine a string that's constrained to vibrate only on the two-dimensional surface of a flat tabletop.

The string will be able to execute a variety of vibrational patterns, but only those involving motion in the left/right and back/forth directions of the table's surface. If the string is then released to vibrate in the third dimension, motion in the up/down dimension that leaves the table's surface, additional vibrational patterns become accessible. Now, although it is hard to picture in more than three dimensions, this conclusion – more dimensions means more vibrational patterns – is general. If a string can vibrate in a fourth spatial dimension, it can execute more vibrational patterns than it could in only three; if a string can vibrate in a fifth spatial dimension, it can execute more vibrational patterns than it could in only four; and so on.

This is an important realization, because there is an equation in string theory that demands that the number of independent vibrational patterns meet a very precise constraint. If the constraint is violated, the mathematics of string theory falls apart and its equations are rendered meaningless.

In a universe with three space dimensions, the number of vibrational patterns is *too* small and the constraint is not met; with four space dimensions, the number of vibrational patterns is still too small; with five, six, seven, or eight dimensions it is still too small; but with nine space dimensions, the constraint on the number of vibrational patterns is satisfied perfectly. And that's how string theory determines the number of space dimensions.

While this illustrates well the interplay of geometry and physics, their association within string theory goes further and, in fact, provides a way to address a critical problem encountered earlier. Recall that, in trying to make detailed contact between string vibrational patterns and the known particle species, physicists ran into trouble. They found that there were far too many massless string vibrational patterns and, moreover, the detailed properties of the vibrational patterns did not match those of the known matter an force particles.

But what I didn't mention earlier, because we hadn't yet discussed the idea of extra dimensions, is that although those calculations took account of the *number* of extra dimensions (explaining, in part, why so many string vibrational patterns were found), they did not take account of the small size and complex *shape* of the extra dimensions – they assumed that all space dimensions were flat and fully unfurled – and that makes a substantial difference.

Strings are so small that even when the extra six dimensions are crumpled up into a *Calabi-Yau* shape, the strings still vibrate into those directions. For two reasons, that's extremely important. First, it ensures that the strings always vibrate in all nine space dimensions, and hence the constraint on the number of vibrational patterns continues to be satisfied, even when the extra dimensions are tightly curled up.

Second, just as the vibrational patterns of air streams blown through a tuba are affected by the twists and turns of the instrument, the vibrational patterns of strings are influenced by the twist and turns in the geometry of the extra six dimensions. If you were to change the shape of a tuba by making a passageway narrower or by making a chamber longer, the air's vibrational patterns and hence the sound of the instrument would change.

Similarly, if the shape and size of the extra dimensions were modified, the precise properties of each possible vibrational pattern of a string would also be significantly affected. And since a string's vibrational pattern determines its mass and charge, this means that the extra dimensions play a pivotal role in determining particles properties. This is a key realization. The precise size and shape of the extra dimensions has a profound impact on string vibrational patterns and hence on particle properties. As the basic structure of the universe – from the formation of galaxies and stars to the existence of life as we know it – depends sensitively on the particle properties, the code of the Cosmos may well be written in the geometry of a Calabi-Yau shape.

We saw one example of a Calabi-Yau shape in figure 9-2. But are at least hundreds of thousands of other possibilities. The question, then, is which Calabi-Yau shape, if any, constitutes the extra-dimensional part of the spacetime fabric. This is one of the most important questions string theory faces since only with a definite choice of Calabi-Yau shape are the detailed features of string vibrational patterns determined. To date, the question remains unanswered. The reason is that the current understanding of string theory's equations provides no insight into how to pick one shape from the many; from the point of view of the known equations, each Calabi-Yau shape is as valid as any other. The equations don't even determine the size of the extra dimensions. Since we don't see the extra dimensions, they must be small, but precisely how small remains an open question.

Is this a fatal flaw of string theory? Possibly. But I don't think so. The exact equations of string theory have eluded theorists for many years and so much work has used *approximate* equations. These have afforded insight into a great many features of string theory, but for certain question – including the exact size and shape of the extra dimensions – the approximate equations, determining the form of the extra dimensions is a prime – and in my goal opinion attainable – objective. So far, this goal remains beyond reach.

Nevertheless, we can still ask whether *any* choice of Calabi-Yau shape yields string vibrational patterns that closely approximate the known particles. And here the answer is quite gratifying.

Although we are far from having investigated every possibility, examples of Calabi-Yau shapes have been found that give rise to string vibrational patterns in rough agreement.

For instance, in the mid-1980s Philip Candelas, Gary Horowitz, Andrew Strominger, and Edward Witten (the team of physicists who realized the relevance of Calabi-Yau shapes for string theory) discovered that each hole the term is used in a precisely defined mathematical sense – contained within a Calabi-Yau shape gives rise to a family of lowest-energy string vibrational patterns.

A Calabi-Yau shape with three holes would therefore provide an explanation for the repetitive structure of three families of elementary particles in figure 9-2. Indeed, a number of such three-holed Calabi-Yau shapes have been found. Moreover, among these preferred Calabi-Yau shapes are ones that also yield just the right number of messenger particles as well as just the right electric charges and figure 9-2.

This is an extremely encouraging result; by no means was it ensured. In merging general relativity and quantum mechanics, string theory might have achieved one goal only to find it impossible to come anywhere near the equally important goal of explaining the properties of the known matter and force particles. Researchers take heart in the theory's having blazed past that disappointing possibility. Going further and calculating the precise masses of the particles is significantly more challenging. As we discussed, some particles have masses that are deviated from the lowest-energy string vibrations – zero times the Planck mass – by less than one part in a million billion. Calculating such *infinitesimal deviations* requires a level of precision way beyond what we can muster our current understanding of string theory's equations.

As a matter of fact, many string theorists believe that the tiny masses rise in string theory much as they do in the standard model. Recall that in the standard model, a *Higgs field* takes on a nonzero value throughout all space, and the mass of a particle depends on how much drag force it experiences as it wades through the *Higgs ocean.* A similar scenario likely plays out in string theory. If a huge collection of strings all vibrate in just the right coordinated way throughout all of space, they can provide a uniform background that for all intents and purposes would be indistinguishable from a Higgs ocean. String vibrations that initially yielded zero mass would then acquire tiny nonzero masses through the drag force they experience as they move through the drag force they experience as they move and vibrate though the string theory version of the Higgs ocean.

Notice, though, that in the standard model, the drag force experienced by a given particle – and hence the mass it requires – is determined by experimental measurement and specified as an input to the theory. In the string theory version, the drag force – and hence the masses of the vibrational patterns – would be traced back to interactions between strings (since the Higgs ocean would be made of strings) and should be calculable.

String theory, at least in principle, allows all particle properties to be determined by the theory itself.

No one has accomplished this, but as emphasized, string theory is still very much a work in progress. In time, researchers hope to realize fully the vast potential of this approach to unification. The motivation is strong because the potential payoff is big. With hard work and substantial luck, sting theory may one day explain the fundamental particle properties and, in turn, explain why the Universe is the way it is.

11.8 The Fabric of the Cosmos according to String Theory

Even though much about string theory still lies beyond the bounds of our comprehension, it has already exposed dramatic new vistas. Most strikingly, in mending the rift between general relativity and quantum mechanics, string theory has revealed that the fabric of the Cosmos may have many more dimensions than we perceive directly – dimensions that may be the key to resolving some of the Universe's deepest mysteries. Moreover, the theory intimates that the familiar notions of space and time do not extend into the sub-*Planckian* realm, which suggests that space and time as we currently understand them may be mere approximations to more fundamental concepts that still await our discovery.

In the Universe's initial moments, these features of the space-time fabric that, today, can be accessed only mathematically, would have been manifest. Early on, when the tree familiar spatial dimensions were also small, there would likely have been little or no distinction between what we now call the big and the curled-up dimensions of string theory.

Their current size disparity would be due to cosmological evolution which, in a way that we don't yet understand, would have had to pick three of the spatial dimensions as special, and subject only them to the 14 billion years of expansion discussed in earlier observable Universe would have shrunk into the sub-Planckian domain, so that what we've been referring to as the fuzzy patch.

we can now identify as the realm where familiar space and time have yet to emerge from the more fundamental entities – whatever they may be – that current research is struggling to comprehend.

Further progress in understanding the primordial Universe, and hence in assessing the origin of space, time, and time's arrow, requires a significant honing of the theoretical tools we use to understand string theory – a goal that, not too long ago, seemed noble yet distant.

As we'll now see, with the development of M-theory, progress has exceeded many of even the optimists' most optimistic predictions.

11.10 The Second Superstring Revolution

Over the last three decades, not one but five distinct versions (approximations) of string theory have been developed. While their names are not of the essence, they are called Type I, Type IIA, Type IIB, Heterotic-O and Heterotic-E. All share the essential features – the basic ingredients are strands of vibrating energy – and, as calculations in the 1970 and 1980s revealed, each theory requires six extra space dimensions; but when they are analyzed in detail, significant differences appear. For example, the Type I theory includes the vibrating string loops. So-called *closed strings*, but unlike the other string theories, it also contains *open strings*, vibrating string snippets that have two loose ends. Furthermore, calculations show that the list of string vibrational patterns and the way each pattern interacts and influences others differ from one formulation to another.

The most optimistic of string theorists envisioned that these differences would serve to eliminate four of the five versions when detailed comparisons to experimental data could one day be carried out. But, frankly, the mere existence of five different formulations of string theory was a source of quit discomfort. The dream of unification is one in which scientists are led to a unique theory of the Universe. If research established that only one theoretical framework could embrace both quantum mechanics and general relativity, theorists would reach unification *nirvana*. They would have a strong case for the framework's validity even in the absence of direct experimental verification.

After all, a wealth of experimental support for both quantum mechanics and general relativity already exists, and it seems plain as day that the laws governing the Universe should be mutually compatible. If a particular theory were the unique *mathematically consistent* arch spanning the two experimentally confirmed pillars of twentieth-century physics, that would provide powerful, albeit indirect, evidence for the theory's inevitability.

But the fact that there are five versions of string theory, superficially similar yet distinct in detail would seem to mean that string theory fails the uniqueness test. Even if the optimists are someday vindicated and only of the five string theories is confirmed experimentally, we would still be vexed be the nagging question of why there are four simply be mathematical curiosities? Would they have any significance for the physical world? Might their existence b the tip of a theoretical iceberg in which clever scientists would subsequently show that there are actually five other versions, or six, or seven, or perhaps even an endless number of distinct mathematical variations on a theme of strings?

During the late 1980s and early 190s, with many physicists hotly pursuing and understanding of one or another of the string theories, the enigma of the five versions was not a problem researchers typically dealt with on a day-to-day basis. Instead, it was one of those quiet questions that everyone assumed would be addressed in the distance future, when the understanding of each individual string theory had become significantly more refined.

But in the spring of 1995, with little warning, these modest hopes were wildly exceeded. Drawing on h work of a number of string theorists including Chris Hull, Paul Townsend, John Schwarz, and others), Edward Witten – who for two decades has been the world's most renowned string theorist – uncovered a hidden unity that tied all five string theories together. Witten showed that rather than being distinct, the five theories are actually just five different ways of mathematically analyzing a *single* theory. Much as the translations of a book into five different languages might seem, to a *monolingual* reader, to be five distinct texts, the five string formulations appeared distinct only because Witten had yet to write the dictionary for translating among them. But once revealed, the dictionary provided a convincing demonstration that – like a single master theory links all five string formations.

The unifying master theory has tentatively been called *M-theory*, M being a tantalizing placeholder whose meaning – Master? Majestic? Mother? Magic? Mystery? Matrix? – awaits the outcome of a vigorous worldwide research effort now seeking to complete the new vision illuminated by Witten's powerful insight.

This revolutionary discovery was a gratifying leap forward. String theory, Witten demonstrated in one of the field's most prized papers (and in important follow-up work with Petr Horava), is a single theory. No longer did string theorists have to qualify their candidate for the unified theory Einstein sought by adding, with a tinge of embarrassment, that the proposed unified framework lacked unity because it came in five different versions. How fitting, by contrast, for the farthest-reaching proposal for a unified theory to be, itself the subject of a meta-unification. Through Witten's work, the unity embodied by each individual string theory was extended to the whole string framework.

Chapter 10 sketches the status of the five string theories before and after Witten's discovery, and is a good summary image to keep in mind. It illustrates that M-theory is not a new approach, per se, but that, by clearing the clouds, it promises a more refined and complete formulation of physical law than is provided by any one of the individual string theories. M-theory links together and embraces equally all five string theories by showing that each is part of a grander theoretical synthesis.

11.11 Eleven Dimensions

So, with our newfound power to analyze string theory, what insights have emerged? There have been many. I will focus on those that have had the greatest impact on the story of space an time.

Of primary importance, Witten's work revealed that the approximate string theory equations used in the 1970s an 1980s to conclude that the Universe must have nine space dimensions missed the true number by one. The exact answer, his analysis showed, is that the Universe according to M-theory has ten space dimensions, that is, eleven space-time dimensions.

Much as Kaluza found that a universe with five space-time dimensions provided a framework for unifying electromagnetism and gravity, and much as string theorists found that a universe with ten space-time dimensions provided a framework for unifying quantum mechanics and general relativity, Witten found that a universe with eleven space-time dimensions provided a framework for unifying all string theories. Like five villages that appear, viewed from ground level, to be completely separate but, when viewed from a mountaintop – are seen to be connected by a web of paths and roadways, the additional space dimension emerging from Witten's analysis was crucial to finding connections between all five string theories.

While Witten's discovery surely fit the historical pattern of achieving unity through more dimensions, when he announced the result at the annual international string theory conference in 1995, it shook the foundations of the field. Researchers had thought long and hard about the approximate equations being used, and everyone was confident that the analyses had given the final word on the number of dimensions. But Witten revealed something startling.

Witten showed that all of the previous analyses had made a mathematical simplification tantamount to assuming that a hitherto unrecognized tenth spatial dimension would be extremely small, much smaller than all others. So small, in fact, that the approximate string theory equations that all researchers were using lacked the resolving power to reveal even a mathematical hint of the dimension's existence. And that led everyone to conclude that string theory had only nine space dimensions. But with the new insights of the unified M-theoretic framework, Witten was able to go beyond the approximate equations, probe more finely, and demonstrate that one space dimension had been overlooked all along. Thus, Witten showed that the five ten-dimensional frameworks that string theorists had developed for more than a decade were actually five approximate descriptions of a single, underlying eleven-dimensional theory.

You might wonder whether this unexpected realization invalidated previous work in string theory. By and large, it didn't. The newfound tenth spatial dimension added and unanticipated feature to the theory, but if string/M-theory is correct, and should the tenth spatial dimension turn out to be much smaller than all others – as, for a long time, had been unwittingly assumed – previous wok would remain valid. However, because the known equations are still unable to nail down the sizes or shapes of extra dimensions, string theorists have expanded much effort over the last few years investigating the new possibility of a not-so-small tenth spatial dimension. Among other things, the wide-ranging results of these studies have put the unifying power of M-theory on a firm mathematical foundation.

11.12 Testing String theory

Testing string theory is a challenge because strings are extremely small. But remember the physics that determined the string's size. The messenger particle of gravity – the graviton – is among the lowest-energy string vibrational patterns, and the strength of the gravitational force it communicates is proportional to the length of the string. Since gravity is such a weak force, the string's length must b tiny; calculations show that it must be within a factor of a hundred or so of h Planck length for the string's graviton vibrational pattern to communicate a gravitational force with the observed strength.

Given this explanation, we see that a highly energetic sting is not constrained to be tiny, since it no longer has any direct connection to the graviton particle (the graviton is a low-energy, zero-mass vibrational pattern). In fact, as more and more energy is pumped into a string, at first it will vibrate more and more frantically. But after a certain point, additional energy will have a different effect: it will cause the string's length to increase, and there's no limit to how long it can grow. By pumping enough energy into a string, you could even make it grow to macroscopic size. With today's technology we couldn't come anywhere near achieving this, but it's possible that in the *searingly* hot, extremely energetic aftermath of the Big Bang, long strings were produced. If some have managed to survive until today, they could very well stretch clear across the sky. Although a long shot, it's even possible that such long strings could leave tiny but detectable imprints on the data we receive from space, perhaps allowing string theory to be confirmed one day through astronomical observations.

Higher –dimensional *p*-branes need not be tiny, either, and because they have more dimensions than strings o, a qualitatively new possibility opens up. When we picture a long – perhaps infinitely long – string, we envision a long on-dimensional object that exists within the three large space dimensions of everyday life. A power line stretched as far as the eye can see provides a reasonable image. Similarly, if we picture a large – perhaps infinitely large – two-brane, we envision a large two-dimensional surface that exists within the three large space dimensions of common experience.

I don't know of a realistic analogy, but a ridiculously huge drive-in movie screen, extremely thin but as high and as wide as the eye can see, offers a visual image to latch on to. When it comes to large three-brane, though, we find ourselves in a qualitatively new situation.

A three-brane has three dimensions, so if it were large – perhaps infinitely large – it would fill all three big spatial dimensions. Whereas a one-brane and a two-bran, like the power line and movie screen, are objects that exist within our three large space dimensions, a large three-brane would occupy all the space of which we've aware.

This raises an intriguing possibility. Might we, right now, be living within a three-brane? Like Snow White, whose exists within a two-dimensional movie screen – a two-brane – that itself resides within a higher-dimensional universe (the three space dimensions of the movie theater), might everything we know exist within a three-dimensional screen – a three-brane – that itself resides within the higher-dimensional universe of string/M-theory?

Could it b that what Newton, Leibniz, Mach, and Einstein called three-dimensional space is actually a particular three-dimensional entity in string/M-theory? Or, in more relativistic language, could it be that the four-dimensional spacetime developed by Minkowski and Einstein is actually the wake of a three-brane as it evolves through time? In short, might the Universe as we know it be a brane?

The possibility that we are living within a three-brane – the so-called braneworld scenario – is the latest twist in string/M-theory's story. It provides a qualitatively new way of thinking about string/M-Theory, with numerous and far-reaching ramifications. The essential physics is that branes are rather like cosmic "Velcro;" in a particular way, they are very sticky.

11.13 String Theory Confronts Experiment

The possibility that we are living within a large three-brane is, of course, just that: a possibility. And, within the brane-world scenario, the possibility that the extra dimensions could be much larger than once thought – and the related possibility that strings could also be much larger than once thought – are also just that: possibilities. But they are tremendously exciting possibilities. True, even if the brane-world scenario I right, the extra dimensions and the string size could still be *Planckian*. But the possibility within string/M-theory for strings and the extra dimensions to be much larger – to be just beyond the reach of today's technology – is fantastic. It means that there is at least a chance that in the next few years, string/M-theory will make contact with observable physics and become an experimental science.

How big a chance? I don't know, and nor does anyone else. My intuition tells me it's unlikely, but my intuition is informed by a decade and a half of working within the conventional framework of Planck-sized strings and Planck-sized extra dimensions. Perhaps my instincts are old-fashioned.

Thankfully, the question will be settled without the slightest concern for anyone's intuition. If the strings are big, or if some of the extra dimensions are big, the implications for upcoming experiments are spectacular.

There currently are a variety of experiments that will test, among other things, the possibilities of comparatively large strings and large extra dimensions. If strings are as large as a billionth of a billionth (10^{-18}) of a meter, the particles corresponding to the higher harmonic vibrations will not have enormous masses, in excess of the Planck mass, as in the standard scenario. Instead, their masses will be only a thousand to a few thousand times that of a proton, and that's low enough to be within reach of the Large Hadron Collider now being built at CERN. If these string vibrations were to be excited through high-energy collisions, the accelerator's detectors would light up with immense energy. A whole host of never-before-seen particles would be produced, and their masses would be related o one another's much as the various harmonics are related on a cello. String theory's signature would be etched across the data with a flourish that would have impressed John Hancock. Researchers wouldn't be able to miss it, even without their glasses.

Moreover, in the brane-world scenario, high-energy collisions might even produce – get this – miniature black holes. Although we normally think of black holes as gargantuan structures out in deep space, it's been known since the early days of general relativity that if you crammed enough matter together in the palm of your hand, you'd create a tiny black hole. This doesn't happen because no one's grip – and no mechanical device – is even remotely strong enough to exert a sufficient compression force. Instead, the only accepted mechanism for black hole production involves the gravitational pull of an enormously massive star's overcoming the outward pressure normally exerted by the star's nuclear fusion processes, causing the star to collapse in on itself. But it gravity's intrinsic strength on small scales is far greater than previously thought, tiny black holes could be produced with significantly less compression force than previously believed. Calculations show that the Large Hadron Collider may have just enough squeezing power to create a *cornucopia* of microscopic black holes through high-energy collisions between protons. Think about how amazing that would be. The Large Hadron Collider might turn out to be a factor for producing microscopic black holes. These black holes would be so small and would last for such a short time that they wouldn't pose us the slightest threat (years ago, Steven Hawking showed that all black holes disintegrate via quantum processes – big ones very slowly, tiny ones very quickly), but their production would provide confirmation of some of the most exotic ideas ever contemplated.

11.14 Braneworld Cosmology

A primary goal of current research, one that is being hotly pursued by scientists worldwide is to formulate an understanding of cosmology that incorporates the new insights of string/M-theory. The reason is clear: not only does cosmology grapple with big questions, and not only have we come to realize that aspects of familiar experience – such as the arrow of tine – are bound up with conditions at the Universe's birth, but cosmology also provides a theorist with a proving ground. If a theory can make it in the extreme conditions characteristic of the Universe's earliest moments, it can make it anywhere.

As of today, cosmology according to string/M-theory is a work in progress, with researchers heading down two main pathways. The first and more conventional approach imagines that just as inflation provided a brief but profound front end to the standard Big Bang theory, string/M-theory provides a yet earlier and perhaps ye more profound front end to inflation. The vision is that string/M-theory will *unfuzz* the fuzzy patch we've used to denote our ignorance of the Universe's earliest moments, and after that, the cosmological drama will unfold according to inflationary theory's remarkable successful script, recounted in earlier chapters.

11.16 A brief Assessment – Criticisms of the models

The brane-world scenario and the cyclic cosmology model is spawned along with the inflationary model are all highly speculative.

At their present levels of development, both the inflationary and the cyclic models provide insightful cosmological frameworks, but neither offers a complete theory. Ignorance of the prevailing conditions during the Universe's earliest moments forces proponents of inflationary cosmology to simply assume, without theoretical justification, that the conditions required for initiating inflation arose. If he did, the theory resolves numerous cosmological conundrums and launches time's arrow. But such successes hinge on inflation's happening in the first place.

What's more, inflationary cosmology has not been seamlessly embedded within string theory and so is not yet part of a consistent merger of quantum mechanics and general relativity.

The cyclic model has its own share of short-comings. As with Tolman's model, consideration of entropy buildup (and also of quantum mechanics) ensures that the cyclic model's cycles could not have gone on forever. Instead, the cycles begin at some definite time in the past, and so, as with inflation, we need an explanation of hoe the first cycle got started. If it did, then the theory, also like inflation, resolves the key cosmological problems and sets time's arrow pointing from each low-entropy splat forward through the ensuing stages of Figure 13.8. But, as it's currently conceived, the cyclic model offers no explanation of how or why the Universe finds itself in the necessary configuration.

Why, for instance, do *six* dimensions curl themselves up into a particular Calabi-Yau shape while one of the extra dimensions dutifully takes the shape of a spatial segment separating two three-branes? How is it that the two end-of-the-world three-branes line up so perfectly and attract each other with just the right force so that the stages proceed as we've described? And, of critical importance, what actually happens when the two three-brane collide in the cyclic model's version of a bang?

On this last question, there is hope that the cyclic model's *splat* is less problematic than the *singularity* encountered at time zero in inflationary cosmology. Instead of all of space being infinitely compresses, in the cyclic approach only the single dimension between the branes gets squeezed down; the branes themselves experience overall expansion, not contraction, during each cycle. And this, Steinhardt, Turok, and their collaborators have argued, implies finite temperature and finite densities on the branes themselves. But this is a highly tentative conclusion because, so far, no one has been able to get the better of the equations and figure out what would happen should branes slam together. In fact, the analyses so far completed point toward the splat being subject to the same problem that afflicts the inflationary theory at time zero: the mathematics breaks down.

Thus, cosmology is still in need of a rigorous resolution of its singular start – be it the true start of the Universe, or the start of our current cycle. The most compelling feature of the cyclic model is the way it incorporates dark energy and the observed accelerated expansion. In 1998, when it was discovered that the Universe is undergoing accelerated expansion, it was quite a surprise to most physicists and astronomers.

While it can be incorporated into the inflationary cosmological picture by assuming that the Universe contains precisely the right amount of dark energy, accelerated expansion seems like a clumsy add-on. In the cyclic model, by contrast, dark energy's role is natural and pivotal. The trillion-year period of slow but steadily accelerated expansion is crucial for wiping the slate clean, for diluting the observable Universe to near nothingness, and for resetting conditions in preparation for the next cycle. From this point of view, both the inflationary model and he cyclic model rely on accelerated expansion – the inflationary model rely on accelerated expansion – the inflationary model near its beginning and the cyclic model at the end of each of its cycles – but only the latter has direct observational support. (Remember, the cyclic approach is designed so that we are just entering the trillion-year phase of accelerated expansion, and such expansion has been recently observed.) That's a tick in the cyclic model's column, but it also means that should accelerated expansion fail to be confirmed by future observations, the inflationary model could survive (although the puzzle of the missing 70 percent of the Universe's energy budget would emerge anew) but the cyclic model could not.

Ch 12 Parallel Worlds

12.1 The Cosmos composed of Parallel Worlds

Cosmology is the study of the Universe as a whole, including its birth and perhaps its ultimate fate. Not surprisingly, it has undergone many transformations in it slow, painful evolution, an evolution often overshadowed by religious dogma and superstition.

The first revolution in cosmology was ushered in by the introduction of the telescope in the 1600s. With the aid of the telescope, Galileo Galilei, building on the work of the great astronomers Nicolaus Copernicus and Johannes Kepler, was able to open up the splendor of the heavens for the first time to serious scientific investigation. The advancement of this first stage of cosmology culminated in the work of Isaac Newton, who finally laid down the fundamental laws governing the motion of the celestial bodies. Instead of magic and mysticism, the laws of heavenly bodies were now seen to be subject to forces that were computable and reproducible.

A second revolution in cosmology was initiated by the introduction of the great telescopes of the twentieth century, such as the one at Mount Wilson with its huge 100-inch reflecting mirror. In the 1920s, astronomer Edwin Hubble used this giant telescope to overturn centuries of dogma, which stated that the universe was static and eternal, by demonstrating that the galaxies in the heavens are moving away from the Earth at tremendous velocities – that is, the universe is expanding. This confirmed the results of Einstein's theory of general relativity, in which the architecture of space-time, instead of being flat and linear, is dynamic and curved. This gave the first plausible explanation of the origin of the universe, hat the universe began with a cataclysmic explosion called the "big bang," which sent the stars and galaxies hurtling outward in space. With the pioneering work of George Gamow and his colleagues on the big bang theory and Fred Hyle on the origin of the elements, a scaffolding was emerging giving the broad outlines of the evolution of the universe.

A third revolution is now under way. It is only about ten years old and it has been ushered in by a battery of new high-tech instruments, such as space satellites, lasers, gravity wave detectors, X-ray telescopes, and high-speed supercomputers.

We now have the most authoritative data yet on the nature of the universe, including its age, its composition, and perhaps even its future and eventual death.

Astronomers now realize that the universe is expanding in a runaway mode, accelerating without limit, becoming colder and colder with time. If this continue, we face the prospect of the "big freeze," when the universe is plunged into darkness an cold, and all intelligent life dies out.

12.2 Trapped on Branes

It is very unlikely that you will explore all the space available to you. There are probably places that you wish to go to, like outer-space, but can't, although, in principle you can. There is no physical law that makes it impossible.

However, if you lived in a black hole (until it decayed away) you would literally be trapped there. The black hole would keep you trapped in its interior, and you would never be able to escape.

There are many more familiar examples of things with restricted freedom of movement fro which there are regions of space that are truly inaccessible. A charge on a wire is an object that lives in a three-dimensional world, but travel in only one of its dimensions. There are also commonplace things that are confined to two-dimensional surfaces. Water droplets on a shower curtain travel only along the curtain's two-dimensional surface, for example.

Branes like shower curtains trap things on lower-dimensional surfaces. They introduce the possibility that in a world with additional dimensions, not all matter is free to travel everywhere. Just as the water droplets on the curtain are bound to a two-dimensional brane sitting inside a higher-dimensional world. But unlike the droplets on the curtain, they are truly trapped.

Particles confined to branes are truly trapped on those branes by physical laws. Brane-bound objects never venture into the extra dimensions that extend of the brane. Not all particles will be trapped on branes; some particles might be free to travel through the bulk. But what distinguishes theories with branes from multidimensional theories without them are the particles on the branes – the ones that don't travel through all the dimensions.

In principle, branes and the bulk could have any number of dimensions, so long as a brane never has more dimensions than the bulk.

The *dimensionality* of a brane is the number of dimensions in which brane-confined particles are permitted to travel. Although there are many possibilities, the branes that will be most interesting to us later on will be the three-dimensional ones.

We don' know why three dimensions should appear to be so special. But branes with three spatial dimensions could be relevant to our world because they could extend along the three spatial dimensions we know. Such branes could appear in a bulk space with any number of dimensions that is more than three – four, five, or more dimensions.

Even if the universe does have many dimensions, if the particles and forces with which we are familiar are trapped on a brane that extends in three dimensions, they would still behave as if they lived in only three. Particles confined to branes would travel only along the brane. And if light were also stuck to the brane, light rays would spread out only along the brane. In a three-dimensional brane, light would behave exactly as it would in a truly three-dimensional universe.

Furthermore, forces trapped on a brane influence only particles confined to this brane. The material of which we are composed, such as nuclei and electrons, and the forces through which these building blocks interact, such as the electric force, might be confined on a three-dimensional brane. Brane-bound forces would spread out only along their brane, and bran-bound particles would be exchanged and would travel solely along the dimensions of the brane.

So if you lived in such a three-dimensional brane, you would be able to travel freely along its dimensions much as you do in three dimensions now. Anything confined within a three dimensional brane would look just the same as it would if the world were truly three-dimensional. The other dimensions would exist adjacent to the brane, but things struck to a three-dimensional brane would never penetrate the higher-dimensional bulk.

But although forces and matter can be stuck on a brane, brane-worlds are interesting precisely because we know that not everything is confined to a single brane. Gravity, for example, is never confined to a brane. According to general relativity, gravity is woven into the framework of space and time. That means that gravity must be exerted throughout space and in every dimension. If it could be confined to a single brane, we would have to abandon general relativity.

Fortunately this is not the case. Even if branes exist, gravity will be felt everywhere, on and off branes. This is important because it means that brane-worlds have to interact with the bulk, even if only via gravity. Because gravity extends into the bulk, and everything interacts via gravity, brane-worlds will always be connected t the extra dimensions. Brane-worlds do not exist in isolation: they are part of a larger whole with which they interact. In addition to gravity, there could conceivably exist other particles and forces in the bulk. If there are, such particles could also interact with particles confined to a brane and connect brane-bound particles to the higher-dimensional bulk. The string theory branes that we will briefly consider later on have specific properties aside from the ones I have mentioned: they can carry particular charges, and they will respond in particular ways when something pushes on them. Branes are lower-dimensional surfaces that can house forces and particles, and they can be the boundaries of higher-dimensional space.

Because branes could trap most particles and forces, the universe we live in could conceivably be house on a three-dimensional brane, floating in an extra-dimensional sea. Gravity would extend into the extra dimensions, but stars, planets, people, and everything else around us that we sense could be confined to a three-dimensional brane. We would then be living on a brane. A brane might be our habitat.

12.3 Branes Worlds

If there can be one brane suspended in a higher-dimensional space-time, there is no denying the possibility of many more. Brane-world scenarios often involve more than a single brane. We don't yet know the number of types of branes that could be present in the Cosmos. *Multiverse* is a name that is sometimes attracted to theories with more than one brane. People often use the word to describe a cosmos with non-interacting or only weakly interacting pieces. So, the Universe can contain multiple branse that interact only via gravity or don't interact at all.

It is possible to have different branes that are too far apart ever to communicate with one another, or that can communicate with one another only wealky, through mediating particles that travel between them. Particles on distant branes, then, would experience entirely different forces, and brane-bound particles would never have direct contact with particles bound on another brane.

So when there is more than one brane with no force in common aside from gravity, we can refer to the Universe housing them both as a multiverse. Other branes migh be parallel to ours and might house parallel worlds. But many other types of brane-world might exist too. Branes could intersect and particles could be trapped at the intersections. Branes could have different *dimensionality.* They could curve. They could move. They could wrap around unseen invisible dimensions. It is not impossible that such geometries exist in the Cosmos.

In a world in which branes are embedded into a higher-dimensional bulk there could be some paricles that explore the higher dimensions and others that stay trapped on branes. If the bulk separates one brane from another, some particles can be on the first brane, some on the other, and some in the middle.

Theories tell us about many ways in which particles and forces might be distributed among different branes and the bulk. Even for branes derived from string theory, we don't yet know why string theory should single out any particular allocation of particles and forces. Brane-worlds introduce new physical scenarios that might describe both the world we think we know and other worlds we don't know on other branes we don't know, separated from our world un unseen dimensions.

New forces confined to distant branes might exist. New particles with which we will never directly interact might propagate on such other branes. Additional stuff accounting for dark matter and dark energy – the matter and energy that we surmise from their gravitational effects but whose identity is a mystery – might be distributed among different branes, or even in both the bulk and on other branes. Gravity might even influence particles differently as you go from one brane to the next.

If there is life on another brane, those beings, imprisoned in an entirely different environment, most likely experience entirely different forces that are detected by different senses. Our senses are attuned to the chemistry, light, and sound surrounding us. Because fundamental forces and particles are likely to be different, the creatures of other branes, should they exist, are unlikely to bear much resemblance to the life of our brane. The other branes will probably be nothing like our own. The only necessarily shared force is gravity, and even gravity's influence can vary.

The consequences of a brane-world will depend on the number and types of branes, and where they are located. Unfortunately for the curious, particles and forces confined to distant branes are not required to influence us very strongly.

They might merely determine what travels in the bulk, and emit weak signals which might never even reach us. Therefore many conceivable brane-worlds will be very difficult to detect, even if they do exist. After all, gravity is the only interaction that we know for sure is shared between the stuff on our brane and the stuff on any other brane, and gravity is an extremely weak force. Without direct evidence, other branes will remain cloistered in the realm of theory and conjecture.

Ch 13 In Search of the Multiverse

13.1 The Mulitverse

In some experiments, the behavior of light seems to be like the behavior of ripples on a pond; in other experiments, it seems to be like a stream of tiny billiard balls. But this does not mean that light 'is' a wave or 'is' a particle, nor even that it is a mixture of wave and particle. It is something we cannot envisage, which if asked one question will respond like a wave, which if asked another question will respond like a particle. The same is true of electrons and all other quantum entities. Perhaps, limited by our human experience, we are asking the wrong questions. But we are struck with the questions and answers we've got.

As long ago as 1929, the physicist Arthur Eddington summed the situation up, in his book *The Nature of the Physical World.* 'No familiar concepts can be woven around the electron,' he wrote; 'something unknown is doing we don't know what.'

In this regard, nothing has changed in the past eighty years. We still don't know what electrons (or other quantum entities) are, nor how they do the things they do.

Indeed, by breaking our mental link with things like waves and particles, it might be more helpful to translate all of quantum physics into a new language; it would certainty make just as much sense. Which makes it all the more remarkable that without knowing what quantum entities are or how they do the things they do, by knowing that they do certain things when prodded in certain ways physicists are able to use quantum entities. This is a bit like the way you can learn to drive a car by learning how to manipulate the controls, without having the faintest idea what is going on under the bonnet.

To take just two examples, quantum physics is essential for the design of computer chips, which are in everything from your mobile phone to supercomputers used in weather forecasting, and quantum physics explains how large molecules like DNA and RNA, the molecules of life, work. Studying quantum physics is not just an esoteric hobby for unworldly elements that has direct practical benefits. There are also esoteric and philosophical implications of quantum theory. And nothing could be stranger than the story os Schrodinger's cat and her successors.

It's a sign of how inadequate our everyday experiences are as a guide to the quantum world that, having cautioned you that quantum entities are neither waves, nor particles, nor a mixture of wave and particle, the best way to begin to get some insight into what goes on at the sub-atomic level is to consider the ways in which such entities behave *like* waves or particles. This provides at least some insight into one of the most important but also non-commonsensical, features of the quantum world – uncertainty.

13.2 A Quantum of Uncertainty

In quantum physics, uncertainty is a precise thing. For a quantum entity, there are pairs of parameters, known as *conjugate variables*, for which it is impossible to have a precisely determined value of each member of the pair at the same time. The more accurately you know property A, the less accurately you know property B, and vice versa. This is not the fault of our inadequate measuring equipment. It is a law of nature, discovered by the physicist Werner Heisenberg in 1927, and known as Heisenberg's Uncertainty Principle. The most important of these pairs of conjugate variables are position/momentum, an energy/time.

The position/momentum relationship is the archetypal example described by Heisenberg. Momentum, in this context, is equivalent to velocity, and velocity describes both the speed and direction that something is moving in. Heisenberg found that the uncertainty in the position of an entity such as an electron, multiplied by the uncertainty in its momentum, is always bigger than a certain (tiny!) number, Planck's constant, divided by 2π. In principle, you can get as near to this limit as you like. But the more precisely the position of, say, an electron is pinned down, the more uncertainty there is about where the electron is going.

The more accurately its momentum (or velocity) is determined, the less accurately is its position defined. This uncertainty is a property of the electron (or other quantum entity) itself. An electron itself does not 'know' both where it is and where it is going at the same time.

This is where the wave and particle analogies are useful. But remember that they are only analogies. A wave is a spread out thing. It might well be travelling in a definite direction at a definite speed, but it cannot be located at a point. A particle, if it is small enough, can very nearly be located at a point, provided that it is not moving with a well-defined momentum. But if it moves – if it has a certain momentum – it is no longer located at a point. The more a quantum entity is constrained by circumstances to act like a *wave*, the less certain it is where the entity is located. The more it is constrained to act like a *particle*, the less certainty there is about where it is going.

The standard way to describe this is in terms of probabilities. If an electron is fired off from an electron gun in the direction of phosphorescent screen, as in an old-fashioned TV cathode ray tube, the moment the electron leaves that gin the wave representing it begins to spread out through space, because its position is uncertain. In principle, the laws of quantum physics tell us, the electron could end up anywhere in the Universe; but there is a very high probability that it will strike the screen and make a spot of light there. The instant it does so, the uncertainty in its position shrinks dramatically to the size of the spot on the TV screen. This is called the collapse of the wave function. Then, the wave begins to spread out again from the new location.

Unless the electron has got tied up in an atom, r trapped in some other way, its position becomes more and more uncertain as time passes. If it has got tied up in an atom, it is still constrained by quantum uncertainty; but that is not directly relevant to the search for the Multiverse.

13.3 The only mystery

All this is hard to get your head round. But the essence of the quantum world can be summed up in terms of one simple experiment, involving sending light, or a beam of electrons, through two holes in a blank obstruction. Richard Feynman, who won a Nobel Prize for his work on quantum theory, said that this experiment 'has in it the heart of quantum mechanics. In reality, it contains the only mystery.'

The experiment with two holes is also called the double-slit experiment, because when it is carried out using light the holes can simply be two parallel slits made in a piece of card or paper with a razor. Light is shone through the two slits in a darkened room, and spreads out on the other side before arriving at a second sheet of card, where it makes a pattern. The pattern is one of alternating light and dark stripes. In the nineteenth century this was explained, or interpreted, as the result of waves spreading out from the two slits and interfering with one another, like overlapping ripples spreading out from two pebbles dropped into a still pond simultaneously. Where the waves are moving *in step*, they combine their strength to make a bright stripe; where the waves are moving *out of step*, they cancel each other and leave a dark stripe. This seemed to be definitive proof that light is a wave.

But at the beginning of the twentieth century Albert Einstein proved that light behaves like a stream of particles. In a process known as the photoelectric effect, light falling onto a metal surface knocks electrons out of the surface. The energy of the ejected electrons only has certain values, and Einstein interpreted this as the result of light arriving at the surface in the form of little particles, now called photons, each with a certain energy. It was, incidentally, for this work, not either of his theories of relativity, that Einstein received the Nobel Prize.

So there are two sorts of experiment you can do with light, one which shows light behaving as a wave and one which shows light behaving as a stream of particles. Exactly the same thing happened with the investigation of electrons, but the other way round.

Today, variations on the experiment with two holes are so subtle that they can be carried out by shooting single entities, photons or electrons in different experiments, through the holes one at a time. I'll describe the results for electrons, but exactly the same kind of experiments have been carried out for photons as well. Instead of a sheet of card on the other side of the experiment, there is a detector screen like a computer monitor which records a spot of light every time an electron arrives, and allows these spots to stay and build up into a pattern as more and more electrons arrive. When researchers do this, each electron arrives as a particle and makes a single spot of light on the screen. But as hundreds and thousands of electrons are fired through the experiment one after another, the pattern that builds up on the screen is an interference pattern, the typical pattern for waves moving through the experiment.

Each electron seems not only to go through both holes at once and interfere with itself, but then to find its place in the interference pattern alongside all the electrons that have gone before an all the electrons that are still to come. Entities in the quantum world seem to know about the whole experiment, both in terms of space the two holes) and in terms of time.

There's more. The experimenters can set up detectors to look at the two holes, and monitor which one each electron goes through. When they do this, they never see the electron going through both holes at once. They see it go through one hole or the other. And when they do this, there is no interference pattern. The spots on the screen form two blobs, one behind each hole, just as you would expect if they were made by particles. The electrons also seem to know if they are being watched or not – and the same is true for photons and all other quantum entities.

This is why Feynman said that the experiment with two holes has in it the heart of quantum physics, and that nobody knows how it can be like that. Even though we cannot *understand* what is going on in the quantum world, the equations of quantum mechanics make it possible to *describe* what is going on, with great precision. By knowing, for example, the circumstances in which electrons seem to move like waves and the circumstances in which they seem to behave like particles, we can design computer chips. It may be crazy, but it works.

13.4 Interpreting the unimaginable

So people try to come up with images of how it works in terms that human beings can comprehend. These are called interpretations of quantum physics. The first of these aids to the imagination to be developed is called the Copenhagen Interpretation, because it was largely developed by scientists working in that city. It was the standard way of thinking about the quantum world from the 1930s to the 1980s, and is still widely taught. But it raises at least as many questions as it answers.

According to the Copenhagen Interpretation, it is meaningless to ask what atoms, electrons and other quantum entities are doing when we are not looking at them. And we can never be certain what the precise outcome of a quantum experiment will be. All we can do is calculate the probability that a particular experiment will come up with a particular result. This is exactly like the way that if you roll a pair of true dice there will be a certain probability of getting a score of 12, another probability of getting a total of 5, and so on. You also know you will never get a total of 17, or 4.3. The same sort of thing happens in quantum experiments. Some outcomes are more likely, some are less likely, and some are impossible.

With dice, you may not know what total you will get in advance, but at least you know the dice are there even if you are not looking at them. When quantum entities are not being observed, the Copenhagen Interpretation says, they dissolve into a mixture of waves (sometimes called a wave function) representing the various probabilities.

This mixture is called a *superposition of states*. When a measurement is made, the act of measuring forces the quantum entity to choose one of these states, in line with the various probabilities, and the wave function collapses. But as soon as the measurement has been made, the quantum entity once again begins to dissolve into a mixture, a new superposition of states.

Looking specifically at the experiment with two holes, the Copenhagen Interpretation says that as soon as the electron leaves the gun on one side of the experiment it dissolves into a superposition of states, waves that pass through both holes. Once the waves have gone through the holes, they interfere with one another to produce a new superposition of states. Then, when the superposition reaches the target screen the wave function collapses into a single point and the electron becomes a real particle, at least temporarily.

But if we set the experiment up to see which hole the electron goes through, the act of observation forces the wave function to collapse at one of the holes, and it then spreads out on the other side from a single site, with no interference, creating a different kind of pattern.

If there was anything better than the Copenhagen Interpretation, it would have been discarded long ago. But there isn't anything better. There are only alternative interpretations, which are precisely as good as the Copenhagen Interpretation, in the sense that they are just as good at predicting the outcomes of quantum experiments, but no better, because they do not predict anything the Copenhagen Interpretation does not predict.

And they each involve inevitable components of quantum weirdness, such as signals that travel backwards in time or instantaneous communication between quantum entities across great distances. So which quantum interpretation you choose to work with is simply a preference based on which aspect of quantum weirdness you feel most comfortable (or least uncomfortable) with. The one that is relevant to the search for the Multiverse, and which is exactly as good as all the other interpretations, including the Copenhagen Interpretation, as far as any experimental test goes, is the Many Worlds Interpretation, developed by Hugh Everett in the 1950s.

13.5 The Branching Tree of History

Communication between the different branches of Everett's Multiverse, sometimes known as 'parallel worlds', would be impossible, according to the same equations that describe the existence of such multiple realities. Except for one intriguing possibility, which is strictly outside the scope of this book, but too enticing to resist mentioning briefly.

That possibility is time travel. The idea of "parallel worlds" is one of many which appeared in science fiction long before it became respectable science; another is the idea of time travel. In many parallel-universe stories, either the protagonists are somehow shifted 'sideways in time' into another universe, or the entire story is set in a parallel reality, an alternative history that has branched off from our own timeline at some critical point in the past – for example, where the Axis powers won the Second World War.

The resulting image we have is of history as a tree with many branches, representing different universes that exist because they branch off at different times according to different such outcomes; the analogy is actually far from perfect, since if Everett is right there is no 'main trunk' for the tree, and the branching is more complex, but it will do. Many time-travel stories involve travelers who go back in time and, either deliberately or accidentally, change history so that they return to a 'present' very different from the one they set out from.

Combining these two ideas, a traveller might go back in time down one branch of history (in one universe), then forward in time up another branch (another universe). It isn't that he or she has changed history; both versions always existed.

In the classic 'granny paradox', for example, a traveler goes back in time and accidentally causes the death of her own grandmother, before the traveller's mother has been born.

In that case, if there is only one timeline, the traveler is never born, so she never goes back in time, so granny survives – and so on. In the many worlds version of the story, the traveler goes back an granny is killed, but this is the branching point for another universe. The traveler might go forward in time up the alternative branch to find a present where she had never existed, or she might go back up her original branch to find, to her surprise, that granny wasn't dead after all. Either way, there is no paradox.

It all makes for entertaining fiction. The surprise for us is that there is nothing in the known laws of physics to prevent time travel, although it would be extremely difficult to build a time machine. The equations of the *general theory of relativity* (the best theory of space and time we have, which has passed every test yet devised) allow for the possibility of time travel; but they don't allow the possibility of travelling back in time before the moment that the time machine is built. This is why we are not overrun with visitors from the future. The time machine hasn't been built – yet.

But getting back to the Multiverse, the granny paradox reminds us of the parable of Schrodinger's cat. This is just the kind of puzzle that Everett's Many Worlds Interpretation resolves.

Because of the way he original cat puzzle is set up, there are only two possible outcomes – two 'eigenvalues' in the language of quantum physics. The cat is either dead or alive. According to Everett's interpretation, this means that there are two equally real worlds, superimposed on one another, but never able to influence each other – a universe with a dead cat and a universe with a live cat. It is easy to imagine a more complicated scenario, with the outcome determined by the equivalent of rolling dice, in which there might be many more possible results. Extending Schrodinger's own idea, there might be a couple of dozen cats housed in their own compartments, and which one gets killed would depend on the outcome of the roll of the dice. There would be a corresponding number of parallel universes after the experiment was carried out, with a whole variety of quantum cats to consider. But not, as is often assumed, an infinite number of parallel universes. Although the 'many' in the Many Worlds Interpretation would be an incomprehensibly large number, there is no reason to think that it would actually be infinity.

In this vast stack of parallel universes, the universe next door would be almost indistinguishable from our more, universes slightly farther away would be more different, having branched off from ours at earlier times, and universes far away across the Multiverse would be utterly different from our own.

The best reason for taking the 'Many Worlds Interpretation' seriously is that nobody has ever found any other way to describe the entire Universe in quantum terms. Wheeler realized this from the very beginning; in his 1957 Reviews of Modern Physics paper, his final sentence read:

Apart from Everett's concept of relative states, no self-consistent system of ideas is at hand to explain what one shall mean by quantizing a closed system like the universe of general relativity.

The more sure cosmologists are that they understand the Universe we see around us, the more obvious it is that the only way to reconcile this view with quantum physics is to take on board the Many Worlds Interpretation – which is why Everett's ideas are now regarded as more respectable than ever.

13.6 Three Dimensions work well

Most people would never think of questioning the fact that the Universe exists in the three dimensions of space plus one of time. That's just the way things are. But some of the most important discoveries in science come from asking why things we take for granted just happen to be that way – to take a famous example, why does an apple fall from a tree, and why does the Moon go round the Earth?

This example, of the kind of question that led Isaac Newton to his insights about the nature of gravity, is apposite because the fact that space has three dimensions is intimately related to one of the most important things allowing planets like the Earth to exist in stable orbits around stars like the Sun – the *inverse-square* law of gravity, described by Newton in the 1680s.

According to this law, the force of gravity attracting two objects to one another is proportional to 1 divided by the square of the distance between the two objects. Einstein's general theory of relativity explains this relationship in terms of curved space, but this does not affect the basic observational fact that the law operates in this way.

The general theory does not overthrow Newton's description of gravity, but includes Newton's description within itself – apples didn't start falling any differently, and the Moon didn't change its orbit, when Einstein came along.

Intriguingly, an inverse-square law is the only kind of law that allows stable orbits to exist. In our Universe, if the Earth's orbit were to shift a tiny bit either way, speeding up or slowing down as it moves around the Sun, the operation of the inverse-square law would shift it back to its present orbit, in an example of negative feedback. This is because of a trade-off between the speed of the Earth in its orbit and the force of attraction it feels from the Sun – in everyday language, a balance between centrifugal force and gravity. But in a universe where the law of gravity were an inverse cube, for example, planetary orbits would be unstable.

A planet that slowed slightly and moved a little closer to its sum would feel a stronger force that pulled it inwards in a spiral of doom (gravity wins), while a planet that speeded up slightly and moved a little bit farther out from its sun would experience a weakening force that allowed it to drift away into space (centrifugal force wins). Even tiny changes, like those caused by the impact of a meteorite, would be disastrous, as a result of positive feedback. Similar things happen with other kinds of laws of gravity. The trade-off allowing stable planetary orbits only works for an inverse-square law.

Among other things, the general theory of relativity explains the fact of the inverse-square law by showing that the dimensionality of the law of gravity is always one less than the dimensionality of space. In a space of two dimensions, the force of gravity between two objects is proportional to 1 over the distance between them; in a space of four dimensions, the force of gravity between two objects is proportional to 1 over the cube of the distance between them; and so on. So planetary orbits are only stable in a space with three dimensions.

Around the time that researchers were making this discovery, in the first quarter of the twentieth century, they also discovered that the equations of electromagnetism, discovered by the Scot James Clerk Maxwell in the nineteenth century, also only work in a universe which has three dimensions of space an one of time. In our Universe, gravity is what keeps planets in their orbits, and electromagnetism is what holds atoms and molecules together to make people. In 1955, just before Hugh Everett came up with his many world idea, the British cosmologist Gerald Whitrow proposed that the reason we observe the Universe we live in to have three spatial dimensions is that observers can only exist in universes that have three dimensions of space (plus one of time). If life can only exist in three-dimensional space, and we are alive, it is no surprise to find ourselves in a spatially three-dimensional Universe.

3.8 The Quantum Cat

Quantum physics is the set of laws that govern the behavior of things on small scales – essentially, the size of atoms and smaller. To put that scale in perspective, it would take roughly ten million atoms lined up side by side to stretch across the gap between two of the points in the serrated edge of a postage stamp.

Newtonian physics describes the behavior of things like billiard balls rolling across a table and colliding with one another, waves rippling across the surface of a pond, or the launch of a rocket ship on its way to Mars. But at another level, it is utterly astonishing hat quantum physics turns out to be dramatically different from Newtonian physics – different in its very nature, not just in minor ways. Because, after all, things like billiard balls, water in a pond, and rocket ships are all made of atoms. How can the whole behave so differently from the sum of its parts?

There is no single satisfactory answer to that question. There are several possible answers, all equally valid, which is an unsatisfactory situation in itself. And none of those answers 'make sense' in terms of our everyday experience of the world.

This is the single most important thing to take on board concerning quantum physics. It is totally outside our everyday experience. There is no way that the human mind can understand what quantum entities such as light or electrons 'really are'. All we can do is carry out experiments, and interpret the results of those experiments by making analogies with things we think we understanding in the everyday world.

13.9 Neither wave nor particle

In some experiments, the behavior of light seems to be like the behavior of ripples on a pond; in other experiments, it seems to be like a steam of tiny billiard balls. But this does not mean that light 'is' a wave or 'is' a particle, nor even that it is a mixture of wave and particle. It is something we cannot envisage, which if asked one question will respond like a wave, which if asked another question will respond like a particle. The same is true of electrons and all other quantum entities. Perhaps, limited by our human experience, we are asking the wrong questions. But we are stuck with the questions and answers we've got.

13.10 Worlds Collide

Branes float about in the higher dimensions of space, just as a two-dimensional membrane, or sheet, can move about in three dimensions. Rather than being stacked up like sheets in a pile of paper, branes (whole universes) behave in the higher dimensions like material objects in three dimensions – they can move about and collide with one another, or orbit around each other like the Moon orbiting the Earth, or planets orbiting the Sun. A better analogy than a stack of paper sitting quietly on my desk would be the turmoil of the same sheets of paper blown about in a strong wind.

Although we cannot detect the extra dimensions of space directly, there are fields which ought to be affected by the geometry of higher-dimensional space and by the proximity of the nearest brane. These might one day be detectable; meanwhile, physicists have tried to find ways in which these fields might drive inflation.

Pushing two brains together takes energy, just as it takes energy to push two positively charged atomic nuclei together. But if branes did smash together at high speed, their energy of motion might be converted into other forms, just as nuclear energy is released when a heavy atomic nucleus is struck by a fast-moving particle and fissions. At the end of the 1990s, Georgi Dvali, of New York University, and Henry Tye, of Cornell University, suggested that in a head-on collision between branes, part of the of kinetic energy of the collision might be converted into the kind of energy needed to trigger inflation. It turned out that the energy involved is far too small to do the trick; but that wasn't to the end of the story.

A breakthrough came in 2001, when a large team of researchers started thinking about another way to get energy out of brane collisions. Just as atomic energy pales into insignificance compared with the energy that is released when matter particles and antimatter particles annihilate – in such collisions, all of the mass is converted into energy, in line with Einstein's famous equation – so, they reasoned, vastly more energy than that in an ordinary brane-brane collision would be released if a brane met up with anti-brane and annihilated.

The equations allow for existence of both branes and antibranes, and just as an electron is attracted to its antimatter counterpart, the positron, so a brane is attracted to an anti-brane, ensuring that they will indeed collide and annihilate if they get close to one another.

This might not seem to be much use, since there would be no matter left over to inflate. But so much energy is released in brane-anti-brane annihilation that some of it spills over into any nearby branes, providing more than enough spare energy to trigger inflation. There's a bonus. This process of annihilation naturally tends to produce a variety of universes with relatively few dimensions, like our own Universe.

Ch 14 The Philosophy of Neuroscience

14.1 The relationship between mind and matter

The problem at issue here is the nature of connection between the features of the experiments described in psychological/mentalistic terms and the features described in *spacio-temporally-based* physical terms. This question is an aspect of the long-standing problem of the relationship between mind and matter, which has a history dating back to the time of the ancient Greeks. The issue was rekindled by the rise of Newtonian physics during the seventeenth century, and it generated a huge body of speculation and argumentation during the second half of the twentieth century.

It is neither appropriate nor feasible try to review or explain here the complexities of contemporary philosophical opinions on this question, except to say that the reigning view is "materialism," and that: (1), there is no agreement among its proponents as to how to make rational good sense of this doctrine (Horgan, 1994); and (2), the doctrine, and its supporters, seem, nevertheless, to have strongly influenced the thinking of many neuroscientists.

The central thesis of materialism is that:

"The human body is a causally complete physico-chemical system: although the body is highly susceptible to external causal influences, all physical events in the body, and all bodily movements are fully explainable in physico-chemical terms." (Horgan, 1994:472)

This thesis is, from a contemporary physics point of view, very obscure: What does "causally complete physico-chemical system" mean? It seems to be referring back to the seventeenth/nineteenth century classical-physics conception of nature, which is now known to fundamentally false. If the doctrine does indeed assume a classical-physics-type of conception of nature, then every bodily movement would indeed be in principle fully explainable in physico-chemical terms. But no reference to mental states or events occurs in classical physics. Hence one is faced with a question that the proponents of materialism have debated at length: how does one inject causally efficacious mind into this causally complete physical description of nature that make no mention of mind.

The ingenuity of philosophers has provided many possible answers, but every proposal seems flawed to other philosophers, and debate continues.

In addition to the problem of trying to attach causally efficacious mind onto a causally complete mindless theory, there is a still more basic problem: classical physical theory is known to be fundamentally false, and in principle inapplicable to systems, such as brains, that depend sensitively upon, for example, the motions of ions. And there is also the problem that the physico-classical aspect of contemporary physical theory, by itself, does not yield any predictions about empirically accessible data. To obtain even statistical predictions in quantum theory one needs to bring in "The Observer," who is described, fundamentally, in psychological/mentalistic terms. And this "Observer" plays a participatory role in the quantum formulation: he is not the purely passive observer of classical physics. He enters into the dynamics in a causally efficacious way.

In view of these profound, and apparently profoundly relevant, deviations of contemporary physics from what is essentially the seventeenth century materialist creation of Isaac Newton, it is strange that twenty-first century neuroscientists should adhere so unwaveringly to that seemingly inappropriate and ill-defined doctrine based on a now-known-to-be-fundamentally-false physical theory.

14.2 Orthodox Quantum

What orthodox quantum theory affirms, and *materialism* appears to deny, is that psychologically described realities enter fundamentally into the scientific description of phenomena, and have, per se, causal effects that are not explained solely in terms of the physical laws that are the generalization of the physical laws of classical physical theory. These causal effects are, however, explained by using other laws of quantum physics, which have no analog in classical physics.

The big problem with materialism as a basis for neuroscience is that it elevates to primary status, by fiat, the (actually nonexistent) physical entities of classical physics, and relegates to secondary status the experiential realities that are the primary variables of pragmatic empirical science. Classical physics does not explain how functional properties come to be connected to experiential realities, and hence fails to explain the causal efficacy of our thoughts. The power of our thought to influence bodily action seems, therefore, from the standpoint of classical physics, to be some sort of illusion, or at least unexplained mystery. *Quantum theory,* on the other hand, displays the mechanism of conscious control, and hence dispels the mystery. The details are given in the following chapters

Ch 15 Altered States of Consciousness

15.1 Sleep and dreams

Everyone dreams, though some people claim never to. The proof is easy to come by. If a self-professed non-dreamer is woken up when their brain is showing the characteristic signs of *REM* (rapid eye movement) sleep, then they will almost certainly report a dream. So the non-dreamer is really just forgetful; it is dream recall that varies widely, rather than dreaming itself. Another way to demonstrate this is to give them a pencil and paper and ask them, every morning, to write down any recollections they have on waking. Anyone can do this, and the usual effect is a dramatic increase in dream recall. Within a few days most people find themselves swamped by dreams and quite happy to go back to a little less reliable recall. In a typical night's sleep the brain cycles through four stages of non-REM sleep; first going down through stages 1–4, then back up to stage 1, and then into a REM period, repeating this pattern four or five times a night. If people are woken at the different stages they describe different experiences. In REM sleep they will usually, though not always, say they are dreaming, while in non-REM sleep they may describe thinking, mulling something over, watching rather static images, or nothing at all. Children, and even babies, show the same physiological stages, but the capacity for complex and *"vivid" dreams* develops only gradually, as cognitive skills and imagination develop.

A great deal is known about the physiology of sleep and some of the ways in which it can go wrong. But this knowledge hides a much less certain picture when it comes to considering sleep as a state of consciousness. Like much to do with consciousness, the notion of states of consciousness and altered states of consciousness (ASCs) seems superficially obvious. For example, we all know that it feels different from normal to be drunk or delirious with fever, and we may guess, even without experiencing it, that it feels different again to be high on drugs, or to be in a mystical state. So we can call all these ASCs. Yet any attempt to define ASCs immediately runs into trouble. There are two obvious ways to try.

First, there are objective measures, such as how much alcohol a person has drunk, or which method of hypnosis was used on them. This is not ideal because two people may drink the same amount and one become completely inebriated while the other is hardly affected.

Similarly, induction techniques affect different people differently, and some not at all. Few states of consciousness are associated with unique physiological patterns, and measuring brain states gives confusing results. Measures of behavior can be unhelpful because people can claim to have been in profoundly altered states without their behavior apparently changing at all. In any case, all these objective measures really seem to miss the point that an altered state is how you feel it is, and is something private to the person having it. For this reason, subjective definitions are usually preferred. For example, psychologist Charles Tart defines an ASC as 'a qualitative alteration in the overall pattern of mental functioning, such that the experiencer feels his consciousness is radically different from the way it functions ordinarily'. This certainly captures the idea of ASCs but also creates problems, such as knowing what a 'normal' state is, and dealing with cases in which people are obviously (to anyone else) in a strange state but claim to feel completely normal. Also, curiously, this definition hits a problem when we look at the most obvious state of all – dreaming. One of the most characteristic features of an ordinary dream is that we do not feel that our 'consciousness is radically different' – at least not at the time. It is only afterwards that we wake up and say 'I must have been dreaming'. For this reason, some people even doubt that dreams can be counted as experiences. After all, we do not seem to be experiencing them at the time – only remembering them afterwards. So did they really happen as they seemed to or might they have been concocted at the moment of waking up? And can we know? Interestingly, there are ways of finding out. For example, it is possible to incorporate features into people's dreams by, for example, playing sounds to them or dripping water on their skin. Sometimes they will later report having dreamed of church bells or waterfalls. By asking them to estimate the timing of these events, it has been shown that dreams do take about the time they seem to. An even better method is to use those rare people who can have "*lucid dreams*" at will. A lucid dream is when you know, during the dream, that it is a dream. In surveys, about 50% of people claim to have had a lucid dream, and 20% have them fairly frequently. For those who have never had one they sound rather strange. A typical lucid dream starts when something peculiar happens and the dreamer starts to have doubts – how did I get on top of this building, and why is my grandmother here when I thought she was dead? Instead of accepting the peculiarity, as we usually do in dreams, the dreamer realizes it cannot be real.

With that realization everything changes. The dream scenery seems more vivid, the dreamer feels more like their normal waking self, and may even take control of the dream. At this point many people start to fly and have fun, but lucid dreams rarely last long and most people lapse back into the ignorance of ordinary dreaming very quickly. A few, very rare, expert lucid dreamers have taken part in laboratory experiments and learned to signal from their dreams. In *REM* sleep almost all of the body's muscles are paralyzed, otherwise you would act out your dreams, but the eyes still move and breathing carries on, so lucid dreamers can sometimes signal by moving their eyes. This allows experimenters to time their dreams and to observe brain activity during the dreams. Generally speaking, this confirms the realistic timing of dreams and also shows that the brain is behaving very much as it would if the person were really running down the street, playing tennis, singing a song – or whatever they are dreaming about. The difference is that they are not physically doing it. In *sleep paralysis* all the body muscles are paralyzed and only the eyes can move.

Most cultures have sleep paralysis myths such as the "*Old Hag of Newfoundland*;" a demonic-looking old woman who appears in the night and may crush the sleeper's chest so as to prevent them from moving. Are dreams experiences? There is no generally accepted theory of dreams, and some very odd facts to be explained. For example, on waking up, we remember having had dreams of which we were not conscious at the time. While experiments suggest that dreams go along in real time, many anecdotes describe dreams that were concocted at the moment of waking up. The most famous is that of the French physician Alfred Maury (1817–1892), who dreamed of being dragged through the French Revolution to the guillotine only to wake up with the bed head falling on his neck. One theory allows both these to be true.

During REM sleep numerous brain processes go on in parallel, and none is either 'in' or 'out' of consciousness. On waking up, any number of stories can be concocted backwards by selecting one of many possible threads through the multiple scraps of memory that remain. The chosen story is only one of many such stories that might have been selected. There is no actual dream; no story that really happened 'in consciousness'. On this 'retro-selection' theory, dreams are not streams of experiences passing through the sleeping mind.

This REM paralysis has another consequence. Sometimes people wake up before the paralysis has worn off and find they cannot move. This is known as sleep paralysis and can be a very frightening experience if you don't know what it is. Often it includes rumbling or grinding noises, eerie lights, and the powerful sense that there is someone close by. Most cultures have their sleep paralysis myths, such as the Old Hag of Newfoundland who comes and sits on people's chests in the night, or the incubus and succubus of medieval lore. Alien abduction experiences may be the modern equivalent – *a vivid experience concocted in that unpleasant paralyzed state between waking and dreaming.*

15.2 Drugs and consciousness

The effects of drugs on consciousness provide the most convincing evidence that awareness depends on the brain. This may seem obvious, but I mention it because there are many people who believe that their mind is independent of their brain, and can even survive after its death. This kind of theory becomes very awkward to sustain once you begin to understand the effects of psychoactive drugs. Psychoactive drugs are those that affect mental functioning. They are found in every known culture, and human beings seem to take endless delight in finding ways to change their consciousness. Many psychoactive drugs can be dangerous, or even lethal, if wrongly used, and most cultures have a complex system of rituals, rules, and traditions that limit who can take which drugs, under what circumstances, and with what preparation. An exception is modern Western culture, where prohibition means that such natural protective systems cannot develop, and many of the most powerful psychoactive drugs are bought on the street and taken by young people without any such understanding or protection.

There are several major groups of psychoactive drugs and these have different effects. Anesthetics are those used to abolish consciousness altogether. The first anesthetics were simple gases, such as nitrous oxide or 'laughing gas', which in high doses induce unconsciousness but in low doses are claimed by many to promote mystical states and philosophical insights.

Modern anesthetics usually consist of three separate drugs to reduce pain, induce relaxation, and abolish memory, respectively. One might think that studying anesthetics would be a good way to understand consciousness; we might find out what consciousness is by systematically increasing and decreasing it.

In fact, it is clear that anesthetics work in many different ways, but mostly they affect the entire brain; there is no sign of a 'consciousness center' or a particular process that is switched on and off.

Other psychoactive drugs are used in psychiatry, including antipsychotics, antidepressants, and tranquillizers. Some tranquillizers have become drugs of abuse, as have other depressants that are drugs that depress the central nervous system. These include alcohol (which has both stimulant and depressant effects) and barbiturates. Narcotics include heroin, morphine, codeine, and methadone. These mimic the action of the brain's own endorphins, chemicals involved in stress and reward. They are intensely pleasurable for some people but highly addictive. Stimulants include nicotine, caffeine, cocaine, and amphetamine. Most of these are highly addictive; increasing doses are required to have the same effect, and withdrawal causes unpleasant symptoms and craving for the drug. Cocaine is normally inhaled into the nose but can also be converted to 'crack' and smoked, which means it is faster-acting and therefore more powerful and more addictive.

The amphetamines are a large group including many of the modern designer drugs. An example is *MDMA*, or *ecstasy*, which has a combination of stimulating and hallucinogenic, as well as emotional, effects. The most interesting drugs, from the point of view of understanding consciousness, are the hallucinogens. The term 'hallucinogen' may not be entirely appropriate because some of these drugs do not produce "hallucinations" at all. Indeed, technically, a true hallucination is one that the experiencer confuses with reality – as when a schizophrenic genuinely believes that the voices she hears in her head are coming from the walls of her room. On this definition, most hallucinogenic drugs produce 'pseudo-hallucinations', because the user still knows that none of it is real. For this reason, these drugs are also known as psychedelic, which means mind-manifesting, or psycholytic, which means loosening the mind.

Cannabis is in a class of its own, and is sometimes called a minor psychedelic. It is derived from the beautiful plant Cannabis sativa, or hemp, which has been used for over 5,000 years both medicinally and as a source of tough fibre for ropes and clothes. Many 19th-century artists used cannabis for their work, and Victorians used it for medicine.

The major hallucinogens have far more dramatic effects, are usually much longer lasting, and are harder to control, which perhaps explains why they are less widely used. They include *DMT* (dimethyltryptamine, an ingredient of the South American visionary potion, ayahuasca), psilocybin (found in "magic mushrooms"), mescaline (derived from the peyote cactus), and many synthetic drugs including *LSD* (lysergic acid diethylamide), and various phenethylamines and tryptamines. Most of these drugs resemble one of the four major brain neurotransmitters, acetylcholine, noradrenaline (norepinephrine), dopamine, and serotonin, and interact with their function. They can all be toxic at very high doses, and can exacerbate pre-existing mental illness, but they are not generally addictive. The best known hallucinogen is probably LSD, which became famous in the 1960s when people were urged to 'turn on, tune in, and drop out'. LSD induces a "trip" that lasts about eight to ten hours; so named because the hours seem endless, and often feels like a great journey through life.

Not only are colors dramatically enhanced, but ordinary objects can take on fantastic forms: wallpaper becoming writhing colored snakes, or a passing car turning into a dragon with fifty-foot wings.

Such visions can be delightful and glorious or absolutely terrifying – leading to a 'bad trip'. There is often a sense of the numinous, along with mystical visions, and a loss of the ordinary sense of self. A person can seem to become an animal or another individual, or to merge with the entire universe. An LSD trip is not a journey to be undertaken "lightly." In 1954, Aldous Huxley (1894–1963), author of Brave New World, took mescaline for the first time and described it as an opening of the 'doors of perception'. Ordinary things appeared colorful and fantastic; everything around him became miraculous and the world appeared perfect in its own 'isness'. His descriptions resemble those of mystical experiences, and indeed some people describe such drugs as 'entheogens', or releasers of the God within. This raises the interesting question of whether drugs can induce genuine religious experiences.

In a famous study, American minister and physician Walter Pahnke gave pills to twenty divinity students during the traditional Good Friday service. Half took a placebo and experienced only mild religious feelings, but half had psilocybin, and, of these, eight reported powerful mystical experiences. Critics dismiss these as somehow inferior to 'real' mystical experiences, but this implies that we know what 'real' mystical experiences are.

15.3 Unusual experiences

From out-of-body experiences to fugues and visions, a surprisingly large number of people (perhaps 30–40%) report quite dramatic spontaneous altered states. These are sometimes known as 'exceptional human experiences', especially if they involve a change in the person's sense of self or their relationship to the world. *Out-of-body experiences* (OBEs) are those in which the person seems to have left their body and to be looking at the world from a location outside it. About 20–25% of the population claim to have had at least one such experience. It is usually very brief, although sometimes people report apparently flying great distances or going to other worlds. OBEs are usually pleasant, although they can be frightening, especially if they are combined with sleep paralysis.

Note that this definition does not necessarily imply that anything has left the body – only that the person feels as if it has. Theories differ widely on this point. For example, some people believe that their *spirit*, *soul*, or *consciousness* has left their body and may go on to survive the death of that body. According to the theory of "astral projection", a subtle "astral body" is exteriorized. Many experiments have been tried to test this idea but without success. For example, detectors have been used, including physical instruments, weighing machines, animals, and other people, but no reliable detector of an astral body or soul has ever been found. Alternatively, people having OBEs have been asked to look at targets such as concealed numbers, letters, objects, or scenes. Although many claim to be able to see the targets, their descriptions are generally no more accurate than would be expected by chance. This does not prove that nothing leaves the body, but there is certainly no convincing evidence that it does.

15.4 Mapping Consciousness

Can states of consciousness be mapped? Moving from one state to another can feel like moving in a vast multidimensional space, with some states easy to reach and others far away. Many people have tried to develop such maps, but it is hard to know what the relevant dimensions are.

Psychological theories explain OBEs in terms of how the person's body image and model of reality can change. It has been found that people who tend to dream in bird's eye view or who are good at imagining changed points of view are more likely to have OBEs. OBEs can occur at almost any time, but are most common on the borders of sleep, during deep relaxation, and in moments of fear or stress.

The experience has also been deliberately induced by electrically stimulating part of the brain's right temporal lobe; the region that constructs and controls our body image. When people come close to death, they sometimes report a whole series of strange experiences, collectively known as a near-death experience (NDE).

Although the order varies slightly, and few people experience them all, the most common features are: going down a dark tunnel or through a dark space towards a bright white or golden light; watching one's own body being resuscitated or operated on (an OBE); emotions of joy, acceptance, or deep contentment; flashbacks or a panoramic review of events in one's_life; seeing another world with people who are already dead or a 'being of light'; and finally deciding to return to life rather than enter that other world.

After such experiences people are often changed, claiming to be less selfish or materialistic, and less afraid of death. NDEs have been reported from many different cultures and ages, and seem to be remarkably similar in outline. The main cultural differences are in the details; for example, Christians tend to see Jesus or pearly gates, while Hindus meet ramdoots or see their name written in a great book. Religious believers often claim that the consistency of the experiences proves their own religion's version of life after death. However, the consistency is far better explained by the fact that people of all ages and cultures have similar brains, and those brains react in similar ways to stress, fear, lack of oxygen, or the many other triggers for NDEs. All these triggers can cause the release of pleasure-inducing endorphins, and can set off random neural activity in many parts of the brain. The effects of this random activity depend on the location: activity in visual cortex produces tunnels, spirals, and lights (as do hallucinogenic drugs that have similar neural effects); activity in the temporal lobe induces body image changes and OBEs, and can release floods of memories; and activity in other places can give rise to visions of many kinds, depending on the person's expectation, prior state of mind, and cultural beliefs. There is no doubt that many people really are changed by having an *NDE,* usually for the better, but this may be because of the dramatic brain changes, and because they have had to confront the idea of their own death, rather than because their soul has briefly left their body. One does not need to be near death to have profound experiences, and many quite ordinary people have quite extraordinary experiences in the midst of their everyday lives.

These are usually called 'religious experiences' if they include visions of angels, spiritual beings, or gods, but mystical experiences if they do not. There is no simple way to define or even describe a mystical experience.

They are often said to be ineffable or indescribable, to involve a sense of the numinous, and to convey unexpected knowledge or understanding of the universe which cannot be spoken of. Perhaps most central to the experience is a changed sense of self, whether this is a complete loss of the idea of a separate self or a sense of merging with the universe in oneness.

These experiences usually occur spontaneously and are very brief, but there are methods that can make them more likely, or can gradually bring about similar states of mind.

15.5 Meditation

The common image of meditation is sitting in a *lotus position* (cross-legged) and going into a state of deep relaxation, cut off from the *familiar* world. Some meditation is like this, but there are many different kinds, including walking meditations, and alert and active forms. Meditation used to be practiced primarily within the context of religions, most notably in Hinduism and Buddhism, although the contemplative traditions in Christianity, Sufism, and other religions have similar methods. Today there are many secular forms of meditation, promoted mostly as methods of relaxation and stress reduction; the best known being Transcendental Meditation (*TM*). Most meditation is done sitting down in special postures, such as the full or half lotus in which both feet, or just one foot, rests on the opposite thigh. However, many people meditate in simpler positions, using firm cushions, or sitting on a low bench with their feet tucked underneath. There is nothing magical about these positions. They all have the same aim; that is, to provide a posture that is both relaxed and alert. In meditation there are always two dangers; either becoming drowsy and falling asleep, or becoming agitated by distracting thoughts or discomfort. The special postures provide a firm base, a straight spine, and good breathing; and so help avoid both.

Now what about the mind? Here techniques differ widely, although it is sometimes said that they all have in common the aim of dropping thoughts and training attention – and neither is easy. If you have never tried it, you might like to do the following exercise – just look down and think of nothing for one minute. What happens? The instruction cannot be obeyed.

Thoughts come pouring up from inside, attention is distracted by things happening outside, and there is rarely a moment of silence in the mind. Perhaps this is not surprising. After all, our brains evolved to cope with the world and keep us safe, not to go silent on command. Nevertheless, with extensive training it is possible to calm the mind and let go of all distractions.

Most kinds of meditation entail learning to drop unwanted thoughts. The best advice is not to fight them, nor to engage with them in any way, but just to let them go. This can be used as the entire method, but it is not easy and so various other techniques have been developed. Concentrative meditation uses something else to attend to, giving the mind something to do.

This might be a mantra (a word or phrase repeated silently), as is used in TM, or it might be an object such as a stone, flower, candle, or religious icon. The most common method is to watch the breath. The idea is simply to observe one's natural breathing, feeling the air going in and out, and then count the breaths up to ten. When you get to ten you go back to one and start again. Other kinds of meditation use no support. For example, in Zen it is common to sit with eyes half open and look at a white wall. The aim in this case is 'just sitting' – not something many people can do. Meditation can be done with eyes closed or open. The danger of closing the eyes is either going off into grand fantasies without realizing it, or falling asleep. The danger of open eyes is getting distracted, although it is easier to stay alert.

What is the point of all this? Many people take up meditation because they think it will be relaxing and help them cope with stress. In fact, thousands of experiments have been done on the effects of meditation and the results are rather surprising. When standard measures of relaxation are taken, such as heart rate, breathing, oxygen consumption, skin conductance, or brain activity, meditation is found to be no more relaxing than sitting quietly and reading or listening to music. Indeed, it can even be highly arousing, for example when unwanted thoughts keep coming up and the person struggles to keep a control on their emotions. Certainly in the short term, it seems that meditation is far from a quick fix, and if you want to reduce stress it is probably better to take more exercise than to meditate.

In the long term, however, the effects are more profound. Long-term meditators, that is those who have been practicing for many years or even decades, do enter states of very deep relaxation.

Breathing rate can drop to 3 or 4 breaths a minute, and brain waves slow down from the usual beta (seen in waking activity) or alpha (seen in normal relaxation), to the much slower delta and theta waves. But people who practice for many years are not usually just seeking a method of relaxation. Their reasons for meditating are usually either religious or mystical. That is, they meditate to seek salvation, to help others, or to obtain insight. This is certainly so in Buddhist meditation, and in particular in Zen, which is a form of Buddhism that has few religious trappings and a reputation for using tough methods to reveal direct insight into the nature of the mind.

Some Zen students practice silent illumination, learning to calm the mind so as to look directly into the nature of consciousness. Others use special stories or questions called koans. These are not usually questions that can be answered intellectually, or even understood in any ordinary sense. Rather, they are questions that provoke the questioner into a great state of doubt and perplexity, from which new insights can arise. The ultimate koan is probably 'Who am I?'; a question that turns back on itself and causes the meditator to look deep into immediate experience.

Not finding an obvious 'me' is only the first step of what can be a long journey. In Zen practice people report many enlightenment experiences in which something breaks through and the world is seen in a new way, but these may be only transitory experiences, much like spontaneous mystical experiences. The ultimate aim is said to be complete enlightenment in which the illusions of duality (illusions of a separate self and of someone who acts) are completely gone.

Such practices raise fascinating questions for the science of consciousness. Could we study how the brain changes in such cases and thereby understand what is going on? Are there really progressions through stages or do different people take different routes? Do people really become more compassionate and less selfish after meditation, as is claimed? And, perhaps of most interest, are any of their insights genuine? In both mystical experiences and long-term meditation, people describe seeing through the illusion of separate selfhood, or seeing the world as it truly is. Could they be right? Are these the same illusions that the scientific study of consciousness is struggling with? All we can say for the moment is that the study of consciousness is nowhere near sufficiently well developed to answer such questions, but at least we can begin to ask them.

Ch 16 The Mystery of Consciousness

Ch 16.1 Does consciousness collapse wave functions?

When we described the experimentally demonstrated quantum facts and the quantum theory explaining those facts - as distinct from the theory's several contending interpretations - we presented the undisputed consensus of the physics community. We cannot describe such a consensus in our discussion of consciousness. But there is none. There is, of course, a large amount of undisputed experimental data, but diametrically opposed explanations of that data are strongly held. We have our own take on the matter.

Until the late 1960s, behaviorist-dominated psychology avoided the term "consciousness" in any discussion that presumed to be scientific. There has since been an explosion of interest in consciousness. Some attribute this to the striking developments in brain imaging technology that allow seeing which parts of the brain become active with particular stimuli.

What is Consciousness?

We use the term consciousness as roughly equivalent to "awareness". For most of us, "consciousness" most definitely includes the perception of free choice by the experimenter. This use of "consciousness" is that quite standard in the treatment of the quantum measurement problem. One can know of the existence of consciousness in no other way than through our first-person feeling of awareness, or the second-person reports of others.

The quantum enigma arises from the assumption that experimenters can freely choose between two experiments, two experiments that yield contradictory results. We assume that the experimenters had the "free will" to make that choice. However, we can't evade the quantum enigma by denying the free will of the experimenters, that is, merely by having their choices somehow determined by the electrochemistry of their brains. To evade the quantum enigma the required denial of counterfactual definiteness must be employed. That denial must include the assumption of a "conspiratorial" world.

Ch 17 Quantum Mind Theories

17.1 The quantum mind / quantum consciousness

The *quantum mind* or *quantum consciousness* hypothesis proposes that classical mechanics cannot explain consciousness, while quantum mechanical phenomena, such as "quantum entanglement" and "superposition," may play an important part in the brain's function, and could form the basis of an explanation of *consciousness*.

QUANTUM APPROACHES TO CONSCIOUSNESS

1. Introduction.

Quantum approaches to consciousness are sometimes said to be motivated simply by the idea that quantum theory is a mystery and consciousness is a mystery, so perhaps the two are related. That opinion betrays a profound misunderstanding of the nature of quantum mechanics, which consists fundamentally of a pragmatic scientific solution to the problem of the connection between mind and matter. The key philosophical and scientific achievement of the founders of quantum theory was to forge a rationally coherent and practically useful linkage between the two kinds of descriptions that jointly comprise the foundation of science.

Microtubules

"An affirmative answer can be provided by linking Penrose's rule to Hameroff's belief that consciousness is closely linked to the ***microtubular*** *sub-structure of the neurons.* (Hameroff and Penrose, 1996)"

"It was once thought that the interiors of neurons were basically structureless fluids. That conclusion arose from direct microscopic examinations. But it turns out that in those early studies the internal substructure was wiped out by the fixing agent. It is now known that neurons are filled with an intricate structure of *microtubules*. Each microtubule is a cylindrical structure that can extend over many millimeters."

"The surface of the cylinder is formed by a spiral chain of tubulin molecules, with each circuit formed by thirteen of these molecules. The tubulin molecule has molecular weight of about 110,000 and it exists in two slightly different configurational forms. Each tubulin molecule has a single special electron that can be in one of two relatively stable locations.

Quantum mechanics deals with the observed behaviors of macroscopic systems whenever those behaviors depend sensitively upon the activities of atomic-level entities. Brains are such systems Their behaviors depend strongly upon the effects of, for example, the ions that flow into nerve terminals. Computations show that the quantum uncertainties in the *ioninduced* release of neurotransmitter molecules at the nerve terminals are large (Stapp, 1993, p.133, 152)

The Penrose-Hameroff Theory

Roger Penrose and Stuart Hameroff (Hameroff & Penrose, 1996) have proposed a quantum theory of consciousness that brings together three exciting but controversial ideas. The first pertains to the still-to-be-worked-out quantum theory of gravity. The second involves *the famous incompleteness theorem of Gödel.* The third rests upon the fairly recently discovered microtubular structure of neurons.

Penrose proposes that the abrupt changes of the quantum state that are associated with conscious experiences are generated by the gravitational effects of particles of the brain upon the structure of space-time in the vicinity of the brain. Ordinarily one would think that the effects of gravity *within the brain* would be too minuscule to have any significant effect on the functioning of the brain. But Penrose and Hameroff come up with an estimate of typical times associated with the gravitational effects that are in the tenth of a second range associated with conscious experiences. This fuels the speculation that the abrupt changes in the quantum state that occur in quantum theory are caused not by the entry of thoughts into brain dynamics, but by quantum effects of gravity. But then why should thoughts or consciousness be involved at all? Two reasons are given. Penrose uses Gödel's incompleteness theorem to argue that mental processing cannot be wholly mechanical or algorithmic.

Conclusion

Roger's belief is that each conscious moment is a quantum event. He believes that it makes sense that a joining of unconscious quantum possibilities into definite values emerges as a reality from the collapse of the wave function, or quantum state reduction. The particular values created in each reduction will define the conscious experience and influence behavior.

He goes further by saying that these events are reconfigurations at the most basic level of Einstein's space-time continuum, or the Planck scale. They are much smaller than atoms or quarks, and the Planck scale is quantized and non-random. It has specific geometry, information and logic.

This leads us to:

1) Interconnectedness among living beings can be accounted for by nonlocal quantum entanglement.

2) Interaction with cosmic intelligence may be influence by Penrose noncomputable Platonic wisdom embedded in Planck scale geometry.

3) Existence outside the body: According to Orch OR, consciousness occurs at the fundamental level of Planck scale geometry, normally in and around microtubules between our ears. But when brain coherence is lost, quantum information related to consciousness and the unconscious mind remain in the universe, distributed but still entangled. Science will indeed be able to find proof of the soul through the application of quantum mechanics to neuroscience. Steps are already being taken in this direction."

Today's discussions of free will in psychology or neurophysiology usually focus more narrowly on whether the choices we make are somehow predetermined by the electrochemistry of our brain. This free will issue is therefore peripheral to the quantum enigma. But "free will" constantly comes up in connection with the quantum enigma.

17.2 Free Will – according to medieval religious philosophy

Problems with free will arise in several contexts. Here's an old one: Since God is omnipotent it might seem unfair that we be held responsible for anything we do. God, after all, had control. Medieval theologians resolved this issue by deciding that every train of events starts with a "remote efficient cause' and ends with a "final cause," both in God's hands. Causes in between come about through our free choices, for which we will be held accountable on judgment day.

This medieval concern is not completely remote from that of today's philosophers of morality. Similarly, criminal defense lawyers can make the concern practical by arguing that the defendant's actions were determined by genetics and environment rather that by free will.

Classical physics, Newtonian physics, is completely deterministic. An "all-seeing eye," viewing the situation of the universe at one time, can know its entire future. If classical physics applied to everything, there would be no place for free will.

Modern scientists could dismiss the notion of free will as not their concern and leave it to the philosophers an theologians. However, that dismissal does not come so easily today as scientists study the operation of the brain, its electrochemistry, and its response to stimuli. They deal with the brain as a physical object whose behavior is governed by physical laws. Free will does not fit readily into that picture.

Most neurophysiologists and psychologists tacitly ignore that corner. Some though, taking a physical model to apply broadly, deny that free will exists and claim that our *perception* of free will is only and illusion. The controversy creates what is now called "the hard problem".

How could you demonstrate the existence of free will? Perhaps all we have is our own feeling of free will and the claim of free will that others make. If no demonstration is at all possible, perhaps the existence of free will is meaningless. Here's a counter to that argument: Though you can't demonstrate your feeling of pain to someone else, you know it exists, and it's certainly not meaningless.

A famous free-will experiment has generated fierce argument. In the early 1980s, Benjamin Libet had his subjects flex their wrist at a time of their choice, but without forethought. He determined the order of three critical times: the time of the "readiness potential," a voltage that can be detected with electrodes on the scalp almost a second before any voluntary action actually occurs; the time of the wrist flexing, and the time the subjects reported that they had made their decision to flex by watching a fast-moving clock.

One might expect the order to be (1) decision, (2) readiness potential, (3) action. In fact, the readiness potential preceded the reported decision time. Does this show that some deterministic function in the brain brought about the supposedly free decision? Some, not necessarily Libet, do argue this way. But the times involved are fractions of a second, and the meaning of the reported decision time is hard to evaluate. Moreover, since the wrist action is supposed to be initiated without any "preplanning," the experimental result seems, at best, ambiguous evidence against conscious free will.

Belief in our free will arises from our conscious perception that we make choices between possible alternatives. If free will is just an illusion, and we're all just sophisticated robots controlled by our neurochemistry with perhaps a bit of thermal randomness, is our consciousness then also an illusion?

If you're going to deny free will, stopping at the electrochemistry of the brain is arbitrary. After all, the motivation for suggesting such denial is the Newtonian determinism of classical physics. Being logical, consistent, and thus accepting that reasoning all the way, we come to the completely deterministic world where the all-seeing eye" can know the entire future of everything, including our experimenters' supposedly-free choice leading to the quantum enigma.

Unlike arbitrarily stopping at the electrochemistry of the brain, accepting complete determinism does evade the quantum enigma. For most of us, being "robots" in a completely deterministic world is too much to swallow. However, accepting both free will and the undisputed quantum experiments, we come to the quantum enigma and to quantum theory for an explanation. And quantum theory, unlike classical physics, is not a theory of the physical independent of the experimenters' freely made decisions, their free will.

17.3 According to John Bell:

"It has turned out that quantum mechanics cannot be 'completed' into a locally causal theory, at least as long as one allows ... freely operating experimenters".

Before John Bell's theorem, "free will" - or an explicit assumption of "freely operating experimenters" was not something seen in a book about physics. It was certainly not seen in a serious physics journal. That's of course changed. In December 2010, for example, the prestigious journal Physical Review Letters published a calculation of precisely how much free will would have to be given up to account for the correlations observed by freely operating experimenters performing twin-state photon experiments. It's 14%. Now what that means in human terms is unclear.

In the nineteenth and much of the twentieth century, scientific thinking was generally equated with materialist thinking. Even to psychology departments, consciousness did not warrant serious study. Behaviorism became the dominant view. People were to be studied as "black boxes" that received stimuli as input and provided behaviors as output. Correlating the behaviors with the stimuli was all that science needed to say about what goes on inside. If you knew the behavior corresponding to every stimulus, you would know all there is to know about the mind.

The behaviorist approach had success in revealing how people respond and, in some sense, why they act as they do. But it did not even address the internal state, the feeling of conscious awareness and the making of apparently free choices. According to behaviorism's leading spokesman, B. F. Skinner, the assumption of a conscious free will was unscientific. But with the rise of humanistic psychology in the latter part of the twentieth century, behaviorist ideas seemed sterile.

17.4 The "Hard Problem" of Consciousness

Behaviorism had wandered when, in the early 1990s, David Chalmers, a young Australian philosopher, shook up the study of consciousness by identifying the "hard problem" of consciousness. In a nutshell, the hard problem is that of explaining how the biological brain generates the subjective, inner world of *experience*. Chalmers' "easy problems" include s=such things as the reaction to stimuli and the report ability of mental states, and all the rest of consciousness studies. Chalmers does not imply that his easy problems are easy in any absolute sense.

They are easy only relative to the hard problem. Our present interest in the hard problem of consciousness, or awareness, or experience, arises from its apparent similarity - and connection - to the hard problem of quantum mechanics, the problem of observation.

Chalmers' easy problems often involve the correlation o neural activity with physical aspects of consciousness, the "neural correlates of consciousness." Brain-imaging technology today allows the detailed visualization of metabolic activity inside the thinking, feeling brain and has stimulated fascinating studies of thought processes.

Exploration of what goes on inside the brain is not new. Neurosurgeons have long correlated electrical activity and electrical stimulation with reports of conscious perception by placing electrodes directly on the exposed brain. This is done largely for therapeutic purposes, of course, and scientific experimentation is limited. Electroencephalography (EEG), the detection of electrical potentials on the scalp, is even older. EEG can rapidly detect neuronal activity but can't tell much about where in the brain the activity is taking place.

Positron emission tomography (PET) is better at finding out just where in the brain neurons are firing. Here, radioactive atoms, of oxygen for example, are injected into the blood stream. Radiation detectors and computer analysis can determine where there is an increase in metabolic activity, and can correlate this call for more oxygen with reports of conscious perceptions.

The most spectacular brain imaging technology is functional magnetic resonance imaging (MRI). It is better than PET at localizing activity and involves no radiation. The examined head must, however, be held still in a large, usually noisy magnet. MRI is the medical imaging technology as one of the practical applications of quantum mechanics. fMRI can identify the part of the brain that is using more oxygen during a particular brain function responding to an external stimulus.

fMRI can correlate a brain region with the neural process involved in, say, memory, speech, vision, or reported awareness.

So, is the physical brain that these techniques observe, presumably all there is to the brain, all there is to the mind? Some say that the future goal to achieve is to completely explain consciousness in terms of neuroscience. And make the claim that consciousness would be completely explained because there is nothing to it beyond the neural activity we correlate with the experiences we call "consciousness."

Francis Crick, physicist co-discoverer of the DNA double helix, who turned brain scientist, looked for the "awareness neuron." For him, our subjective experience, our consciousness, is nothing but the activity of such neurons. In his book *The Astonishing Hypothesis* identifies that hypothesis:

"You," your joys and sorrows, your memories and your ambitions, your sense of personal identity an free will, are in fact no more than the behavior of a vast assembly of nerve cells and their associated molecules."

David Chalmer, a principal spokesperson for a point of view, diametrically opposite to Crick's sees explaining consciousness purely in terms of its neural correlates to be *impossible*. At best, Chalmers maintains, such theories tell us something about the physical role consciousness may play, but those physical theories don't tell us how consciousness arises.

So, in conclusion, thus far the jury is out on the phenomenon we call consciousness, which keeps us in the thrones of the quantum enigma.

Ch 18 Quantum Mechanics Mathematical Examples

The following pages contain mathematical explanations and practical examples of quantum mechanics.

Quantum Mechanics can be described in one of two ways. Firstly, through Wave Mechanics, initially developed by Erwin Schrodinger and secondly, by Matrix Mechanics, initially through the work of Werner Heisenberg.

This book uses the Wave Mechanics approach.

Quantum Mechanics

Introduction to Quantum Mechanics

Quantum Mechanics

- Following J.J. Thompson's discovery of the electron, scientists began thinking about the structure of the atom. Niels Bohr proposed a somewhat successful model that did two things:
  - incorporated the idea that energy is quantized.
  - explained the emission and absorption spectrum of hydrogen.
- However, the Bohr model of the atom had many limitations.
  - It could not explain line spectra of more complex atoms
  - It could not explain why certain lines appear brighter
  - It could not explain bonding between atoms in solids & liquids.
  - It did not incorporate the wave-particle duality.
- By the mid-1920s, it became clear that the Bohr theory was not a very satisfactory theory. In 1925, a much more comprehensive theory was developed by two leading physicists — Heisenburg and Schrödinger, that became known as Quantum Mechanics.
- Quantum Mechanics is a comprehensive theory that deals primarly with the atomic and subatomic world.
- Quantum Mechanics has been an extremely successful theory that is accepted by virtually all scientists today. But why? What makes it so successful?
  - It explains the production of line spectra for all atoms and it even explains the finer details, like why certain lines appear brighter.
  - It incorporates the wave-particle duality and quantized energy.
  - It explains many important phenomena, from black-body radiation to bonding of atoms to form molecules.
  - In fact, when Quantum Mechanics is applied to the macroscopic world, it is readily able to produce all laws that deals with classical physics. This is known as the <u>correspondence principle</u>.

Quantum Mechanics

Quantum Mechanics Theory

- According to Quantum mechanics, electrons do not actually reside on defined circular orbits as described by Bohr.

Bohr Model

The electron orbits the nucleus on a well-defined circular pathway that corresponds to a discrete amount of energy.

- However according to Quantum mechanics, electrons can act as waves and are described using the wave function Ψ. Since waves are not localized, we can think of electrons as being spread out over some region around the nucleus as if it were a cloud. The size and shape of these electron clouds can be determined using Schrödinger equations.

- For example, if we solve Schrödinger's equation for the hydrogen atom, we get the following wave function.

$$\Psi(x) = \frac{1}{\sqrt{\pi x_0^3}} e^{-x/x_0}$$

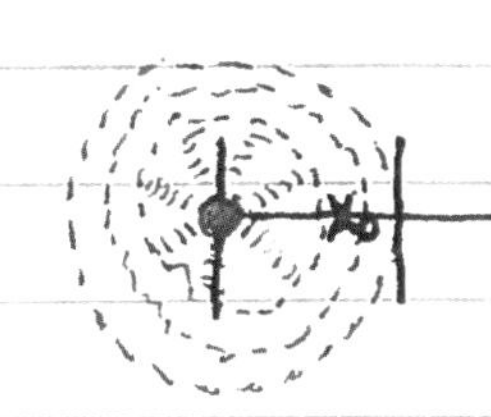

- Taking the square of the absolute value of the wave function gives us the density of the electron cloud. For the hydrogen atom in the ground state, the electron cloud has a spherical shape.

Quantum Mechanics

Statistical vs. Deterministic Theory in Quantum Mechanics

- Classical physics takes on a deterministic approach to studying the behavior of macroscopic objects. According to the deterministic theory, if we know a certain set of quantities as well as all the forces involved, we can predict the future position of the object.

- For example, if we let an object fall from a height of 10m at $t = 0$ seconds, neglecting air resistance, we can determine the exact position of the object at any later time.

10m $\vec{F}_{gravity}$

- Knowing the initial position and initial velocity, we can determine the future pathway the macroscopic object will follow.

- Quantum mechanics however takes a completely different approach. As we already saw in the double-slit experiment with electrons, there is no way of knowing the pathway that any electron will take nor is it possible to know where an electron will end up with extreme certainty. In fact, Heisenberg's uncertainty principle tells us that it is impossible to know the exact velocity and position of an electron at any moment in time.

Quantum Mechanics

Statistical vs. Deterministic Theory in Quantum Mechanics (continued)

- Instead, quantum mechanics uses a statistical approach. Even though we cannot predict where any one electron ends up, we can describe the future position of the electron using probability. This interpretation of nature is known as the Copenhagen interpretation.

- It states that there is an inherent uncertainty in being able to predict the pathways and quantities that describe an object's motion. This uncertainty comes from the fact that ordinary objects do not simply behave as particles but also can behave as waves. According to quantum mechanics, since electrons are not simple particles, they cannot follow any simple pathway that is predicted by the deterministic approach.

Quantum Mechanics

Double-slit Experiment with Electrons (continued)

But what happens if we decrease the number of electrons that pass through the openings to the point where only one electron enters the slit at any given time? At first, the locations of where electrons hit the screen would appear to be completely arbitrary. However, if we allow the experiment to run for a very long time, a pattern will emerge.

CONSTRUCTIVE AND DESTRUCTIVE INTERFERENCE

(1) INITIALLY

(2) AFTER A LONG TIME

Although it would be impossible to predict where each individual electron would strike the screen, we could in fact predict the probability of an electron striking a certain region of screen.

This probability is given by $|\Psi|^2$, the square of the wave function. Inside the dark fringes, the $|\Psi|^2$ is a very small value (~ 0) while in regions where bright fringes are formed, the $|\Psi|^2$ is a large value.

Conclusion

① A diffraction pattern is only produced when wave inference takes place. Since only a single electron travels through any one slit at any given time, this implies that the electron must travel through both openings at the same time to produce wave interference; this only happens if electrons act as a wave.

② For waves Ψ measures displacement; for particles, it measures probability.
(displacement, meaning the amplitude of the constructive wave pattern)
(AMPLITUDE equates to the energy strength of the wave function)

Quantum Mechanics

Schrödinger's Equation (continued)

The given Schrödinger Equation:

$$-\frac{\hbar^2}{2m}\nabla^2\psi + V\psi = i\hbar\frac{\partial\psi}{\partial t}$$

The height of a bowl shaped 3-D curve on a graph at each point in space is represented by the variable (V) in the equation.

The height does not represent a spatial dimension, but the value of the potential energy for our charged particle at each location, which is near the top of the 3D curve.

("m") represents the mass of the particle.

($\hbar$)(h-bar) represents Plank's constant divided by 2π

$\hbar = \frac{h}{2\pi}$ ← the wave

(ψ)(psi) represents the value of the wave-function at each point in space and time.

At each space and time, the wave function is a complex number with a real component and an imaginary component (i)

$\frac{\partial\psi}{\partial x}$: This expression indicates how much the wave function changes as (x) increases by a small amount. This change is a complex number, which is represented graphically as pointed arrow(s).

$\frac{\partial^2\psi}{\partial x^2}$: The amount by which these arrow change as we increase (x) by a small amount is represented by this expression.

Quantum Mechanics

Introduction to the Wave Function

The Wave Function

- Quantum mechanics incorporates the wave-particle duality of nature; that is, it correctly describes particles as waves. But how exactly do you describe a wave? What are the quantities that describe the properties of waves? Well, they are — frequency, wavelength, amplitude (displacement).

- For any electromagnetic wave, the frequency is a measure of the energy of a wave, while the amplitude is the measure of the strength (intensity) of $\vec{E}$ or $\vec{B}$ and is related to the intensity (brightness) as a visual property.

y $\vec{E}$ x z $\vec{B}$

$I \propto A^2 \Rightarrow$
$I \propto \vec{E}^2$

- This means that in order to describe material particles as waves, we must use similar quantities. Louis de Broglie was able to relate the momentum of a particle to the wavelength (or frequency) by the equation:

$$\boxed{\lambda = \frac{h}{p}} \quad \text{since } \lambda = \frac{c}{f} \Rightarrow f = \frac{cp}{h}$$

- But what exactly describes the amplitude (height) of a particle wave?
In EM waves, the $\vec{E}$ and $\vec{B}$ describe the displacement (height), but in quantum mechanics, it is a quantity known as the wave function (Ψ).
This wave function describes the strength of the field produced by the particle wave. Ψ is a function of both time and position (space).

Quantum Mechanics

Introduction to the wave function (continued).

• The wave produced by particles is commonly know as MATTER WAVES.

Interpretation of Matter Waves

• From a wave's point-of-view, $I \propto \vec{E}^2$. However, if we take the view of a particle, the intensity is proportional to the # of incident photons.

– $I \propto N$ and $I \propto \vec{E}^2 \Rightarrow \boxed{N \propto E^2}$ (1) } The greater the # of photons, the stronger the beam of light. Each photon is equal to eachother in strength.

$\vec{E}$ = energy = strength of the electric field

• But for a single photon, this equation can be interpreted in a different way. For a single photon, the E^2 is a measure of the probability of the photon being found at that position.

• Matter waves can be interpreted in a similar way. Equation (1) tells us that the number of particles determine the strength of the matter field; that is, the wave function ψ represents the collection of particles.

ψ^2 := probability density = At any point in space and time, ψ^2 tells us the probability of finding an electron at that point in time and space.

– Amplitude ~ magnitude ~ strength ~ height (of wave)

Quantum Mechanics

Wave Function of Schrödinger's Equation

Time-Dependent Schrödinger Equation

- The Schrödinger equation is the analog of the second law of motion and describes the motion and behavior of systems on the atomic and subatomic levels using the wave function, Ψ.

Classical Mechanics		Quantum Mechanics.
Position (X)	→	Wave Function (Ψ)
Momentum (P)		

- Consider a particle of mass (m) moving along the X-axis. At any position and moment in time, the behavior and motion of the particle is given by the wave function $\Psi(x,t)$.

$\leftarrow \Psi(x,t)$

- $\Psi(x,t)$ could take the form of any continuous function that can be squared and integrated to get a finite answer.

$$i\hbar \frac{\partial \Psi(x,t)}{\partial t} = U(x)\Psi(x,t) - \frac{\hbar^2 k^2}{2m} \cdot \frac{\partial^2 \Psi(x,t)}{\partial x^2}$$

One-dimensional time-dependent Schrödinger equation

- From this equation we see that if we know what the wave function is at some initial time (say $t=0$), we can use that to determine the behavior of that particle at some future time. That is, if we know what $\Psi(x,0)$ looks like, we can predict the future of the motion of the particle.

Quantum Mechanics

Wave Function of Schrodinger's Equation (continued)

- What is the meaning behind the (i) in the time-dependent equation?

  $i = \sqrt{-1}$ • But how can an imaginary number be part of the equation that describes motion in Quantum Mechanics? How can we use a number that does not actually exist to describe the real world?

How is this equation useful?

- It incorporates the great complexity of nature, such as the wave-particle duality of matter.
- It explains a multitude of natural phenomenon, such as the production of unique line spectra by atoms.

Conclusion:

- The Schrödinger equation describes how our universe functions, from the microscopic to the macroscopic worlds.

Example 14 - Quantum Mechanics

The Heisenburg Uncertainty Principle

As we now already know, quantum mechanics deals with atomic and subatomic worlds. Now when we make measurements of objects found on the atomic and subatomic level there always exists an amount of uncertainty that involves those particular measurements. In fact, there is a limit to how precise those measurements can be made. And this is known as the Heisenburg uncertainty principle.

Now as we will see towards the end of this lecture that the Heisenburg Uncertainty Principle is a result of the wave particle duality of matter that exist on the atomic and subatomic level. The HUP absolutely has nothing to do with the actual device or instrument that we use to measure those measurements. So to fully understand what the uncertainty principle is, lets conduct the following thought experiment. Suppose we have an electron and we are asked to determine the position of that electron using some type of hypothetical device. Lets suppose a powerful light microscope. Now, if we want a precise position of that electron at some instant in time, the wavelength of light used by that microscope has to be very small. It has to be on the order of the size of our subatomic particle (our electron).

Now recall, if we decrease the wavelength of the light used by the microscope, we increase the energy of the momentum carried by the photon of light. So, that is given by the equation - the momentum used by the light sent by the microscope (p) is equal to (c) the constant that gives us the speed of light in a vacuum divided by the wavelength of light used by the microscope given by Lambda. So basically if we want a more precise location of the electron in some moment in time we have to decrease the wavelength of light. But if we decrease the wavelength of light, we decrease the denominator and thereby increase (p) the momentum of the photon. 0:02:21

Now why is this relevant in our discussion? So, lets examine the following diagram. This is the eye of the observer, making our observation. Now, this is the scope of our powerful light microscope. So the microscope uses a certain wavelength of light that is shown by the following arrow. Now in order for us to actually see that electron, our light has to actually interact with that electron. So the photon of light basically collides with the electron, reflex, and eventually reaches the eye of the observer. However, what actually takes place between our electron and photon is a collision.

And that collision between the photon and the electron basically transfers some of that energy and momentum to that electron and as a result, the electron moves away with some velocity. So basically, the lower the wavelength is, the higher the momentum of that electron is and the higher the velocity of that electron is after that collision takes place. So, even though we might know the precise position of the electron at that moment in time, because of that collision, our electron will move away with some unknown amount of momentum with some unknown amount of velocity.

And that means we have no way of actually knowing the future position of the electron and we have no way of knowing the momentum the electron has if we know that precise position of that electron at that particular instance in time.

So, once again, if we want a precise measurement of the electron's position, we need a photon with a small wavelength. However, such a photon will carry a large amount of energy as per this equation. And the collision between the photon and the electron will cause that electron to move away with a very high velocity.

Which will change that electron's position. As a result, although we might know what the position of the electron is at that instant in time, we have no way of knowing the momentum of the electron, and therefore, no way of knowing where that electron will be at some future instant in time.

So basically, this uncertainty in knowing the position of the electron and at the same time the momentum of that electron - the velocity of that electron at the same time is the bases of the HUP. Now mathematically, the HUP is given by this equation.

Delta-x multiplied by Delta-y, subscript-x is greater than or equal to H-bar, where H-bar is simply a constant. H-bar is equal to H divides by 2-Pi.

So basically the Delta-x is the change in our position of the electron along the x-axis and delta-p-x is the change in the momentum of our electron, the subatomic particle along that same x-axis.

Now, exactly is the meaning behind this equation. What does this equation actually tell us? So, this equation tells us that the more precisely we know the position of our electron (the smaller the delta-x is), the larger our delta-p is. And the less precisely we know the momentum of our electron.

So, basically, if our delta-x is smaller, to compensate for that smaller delta-x because h-bar is a constant, the delta-p must increase (on the same scale or magnitude as delta-x).

So this basically means that it is impossible to measure precisely the position and at the same time the momentum of our particle at some given moment in time. Now, this does not mean of course that we cannot make precise individual measurements. We can for example determine the precise position of an object (an electron) at some moment in time, but then the momentum would be completely unknown. So that basically translates into this equation. So the smaller our uncertainty there is in our position (the smaller our delta-x is) the greater our uncertainty is in our momentum, which gives us our future position.

So, this basically implies that we have absolutely no way of knowing the future position of our electron if we know exactly where that electron is at that particular moment in time.

Now, once again we have to emphasize the following important point about the HUP. The HUP has absolutely nothing to do with the precision of the measuring instrument that we are using. The HUP is an inherent property of nature and comes from the fact that matter at the subatomic level can act as both a particle and a wave. That is the reason an electron does not have a precise location and momentum simultaneously in any given moment in time because our electron is not actually a particle.

Quantum Mechanics

A simplistic view and analogy of the Heisenberg uncertanity principle

The Heisenberg uncertanity principle states that you can not know both the precise location and precise momentum of a particle at the same time.

The following is a simplistic and analogy of why that is.
The Heisenberg equation is:

$$(\Delta X)\cdot(\Delta y) \geq \frac{\hbar}{2\pi} \; ; \; (h) \text{ is Planck's constant value}$$

Both the ΔX and Δy variables are the values of uncertanity, while the $\frac{\hbar}{2\pi}$ is a constant value.

- Suppose you assign ΔX the value of 2 and Δy also the value of 2. According to the equation $(2)(2)=4$, so the constant $\frac{\hbar}{2\pi}$ will now have the value of 4, and that (→ or less.) can now never change moving forward because $\frac{\hbar}{2\pi}$ is a constant.
- Now suppose you want to know a more precise location of the particle, say the electron. As of now the electron is located somewhere in a space of 2 units along the X-axis; for instance as shown in the illustration.

—2 units of ΔX—

[ooooo|ooooo]
0 1 2

The arbritary position of the electron is radomly given with eight possible electron positions along the X-axis; 8 is arbritrary.

So in order to have better knowledge of where along the X-axis the electron is located you would reduce the ΔX value. Suppose you reduce the ΔX value by half. The new unit value would be 1.

1 unit of ΔX

[oooooooo]
0 1

At this point, to find the electron in this reduced space increases the likelyhood. However, by changing the value of ΔX (increasing the chances of finding the electron by decreasing the uncertanity value), you must increase the Δy value as well in order to balance the equation.

Quantum Mechanics

A simplistic view and analogy of the Heisenburg uncertanity principle (continued).

The new equation is now

$$(\Delta x)(\Delta y) \geq \frac{h}{2\pi} \Longrightarrow (1)(2) \neq 4.$$

So in order to balance the equation you must increase the Δy (momentum) value to 4. This givens $(1)(4) \geq 4$. The equation is now balanced.

The result of increasing the Δy-value increased its uncertanity in momentum. This means we now have even more "less knowledge" about the future of the electon in terms of its velocity, direction, etc. In other words, but modifing the Δx and Δy uncertanity variables, we now know alittle more about the precise location of the electron at the moment but even less about where it will be in the future do to the larger momentum placed upon it.

An example would be a series of strong light photons bombarding our electron. With high frequency photon energy our Δx uncertanity decreases because the electron can be found between the waveform of the photon(s).

ie) our electron can be precisely found between two of our photon's oscillating waves, however, due to the photons strong momentum frequency & amplitude (strength of its energy), the electron could be ejected out from its natural orbit around its atom's necleous into a large quantum 3D space unknown to the observer.

So in conclusion, we can never know both the precise Δx uncertanity of our particle and the precise momentum (where its going) at the same time. This is an inherent property of nature. Nature is probabilistic

QUANTUM MECHANICS

HEISENBERG UNCERTAINTY PRINCIPLE

- Whenever you make a measurement, there is always some amount of uncertainty involved. In fact, there is a limit to how precise a measurement can be.

- Thought Experiment.

Suppose we are asked to find the position of an electron using a powerful light microscope. If we want a precise position of the electron, we must use a wavelength of light that is very small. However, the momentum of a photon with a low λ becomes very large. $\left(P = \frac{c}{\lambda}\right)$.

- If we want a precise measurement of the electron's position, we need a photon with a low wavelength. However, such a photon will carry a large momentum and the collision will cause the electron to move away with a high velocity, which will change the electron's position.

- As a result, although we might know what the position of the electron is at that instant in time, we have no way of knowing the momentum of the electron and therefore no way of knowing where the electron will be at a future time.

- This uncertainty in knowing the position and momentum at the same time is the basis of Heisenberg's uncertainty principle. Mathematically, it can be expressed as:

$$\boxed{(\Delta x)(\Delta p_x) \geq \hbar} \quad \text{where } \hbar = \frac{h}{2\pi} = \frac{6.626 \times 10^{-34}\ J \cdot s}{2\pi} = 1.055 \times 10^{-34}\ J \cdot s$$

- This does not mean of course that we cannot make precise individual measurements. We can for example determine the precise position of an object at some instant in time, but then the momentum would be completely unknown. This means we would have no way of knowing the future position of the object.

QUANTUM MECHANICS

HEISENBERG UNCERTAINTY PRINCIPLE — EXAMPLE #1

AN ELECTRON IS TRAVELING ALONG the X-AXIS with a UNIFORM VELOCITY OF 2×10^6 m/s. IF the velocity has been calculated to a precision of 0.15%, calculate the maximum accuracy with which the position of the electron could be measured at the same instant in time. Assume: $m_e = 9.11\times10^{-31}$ Kg AND $\hbar = 1.055\times10^{-34}$ J.s.

APPLY HEISENBERG UNCERTAINTY PRINCIPLE:

① $(\Delta x)(\Delta p_x) \geq \hbar \rightarrow \Delta x \geq \frac{\hbar}{\Delta p_x} \rightarrow \Delta x \geq \frac{\hbar}{m_e \Delta V_2}$

② $\Delta x \geq \frac{1.055\times10^{-34}\ J\cdot s}{(9.11\times10^{-31}kg)(0.0015)(2\times10^6\ m/s)} \rightarrow \Delta x \geq 3.86\times10^{-8}m \rightarrow 38.6\,nm$

③ SO, 38.6nm; by uncertainty principle, the best (most accurate) simultaneous position will have an uncertainty of 38.6nm (Δx) of the electron.

Note: Recall: $p = mv$ (NON-RELATIVISTIC)

SO, $\Delta p = m\Delta v$ ↳ The uncertainty in the velocity.

NOTE: h = PLANCK'S CONSTANT, which is the smallest space/displacement found in nature.

QUANTUM MECHANICS

HEISENBERG UNCERTAINTY PRINCIPLE — EXAMPLE #2

An electron in the $n=2$ energy level of hydrogen remains there for 80ns before moving down to energy level $n=1$.

(a) Determine the uncertainty in the energy released by the electron in the transition.

$$(\Delta t)(\Delta E) = \hbar \rightarrow \Delta E = \frac{\hbar}{\Delta t} = \frac{(1.055\times10^{-34}\,J.s)}{(80\times10^{-9}s)} = \boxed{1.319\times10^{-27}J}$$

(B) Determine the fraction of energy that this represents with respect to the total energy released by the transition.

$$\frac{1}{\lambda} = R\left(\frac{1}{n_1^2} - \frac{1}{n_2^2}\right) \rightarrow \frac{1}{\lambda} = (1.097\times10^7 m^{-1})\left(1-\frac{1}{4}\right) \rightarrow \lambda = 1.215\times10^{-7} m$$

$$E_{photon} = \frac{hc}{\lambda} = \frac{(6.626\times10^{-34}\,J.s)(3\times10^8\,m/s)}{(1.215\times10^{-7}m)} = 1.636\times10^{-18}\,J$$

$$\text{Fraction in energy} = \frac{1.319\times10^{-27}J}{1.636\times10^{-18}J} = \boxed{8.06\times10^{-10}} \left(\frac{1}{1.24\times10^9}\right)$$

Quantum Mechanics

Heisenburg Uncertainty Principle – Example #3

An electron and a tennis ball are both moving along the x-axis with a velocity of 50m/s. If the mass of the tennis ball is 0.058 Kg and the velocity is measured to a precision of 0.02%, find the uncertainties in position of each one of the objects.

Tennis Ball:

$$(\Delta x)(\Delta p) \approx \hbar \longrightarrow \Delta x \approx \frac{\hbar}{\Delta p} = \frac{\hbar}{m\Delta v} = \frac{(1.055\times10^{-34}\ J.s)}{(0.058\ Kg)(0.0002)(50\ m/s)} = \boxed{1.82\times10^{-31}\ m}$$

↳ small change

Electron:

$$(\Delta x)(\Delta p) \approx \hbar \longrightarrow \Delta x \approx \frac{(1.055\times10^{-34}\ J.s)}{(9.11\times10^{-31}\ Kg)(0.0002)(50\ m/s)} = \boxed{1.16\times10^{-2}\ m}$$

Quantum Mechanics

Wave Packet in Quantum Mechanics

Wave Packets

- Previously we saw that a free particle with zero potential energy is represented by a sinusoidal wave function given by $\Psi(x) = A\sin(kx) + B\cos(kx)$ where k is given by $k = p/\hbar$ and p is the momentum of the free particle.

- By the uncertainty principle, since we know what p is and thus $\Delta p = 0$, this implies that the uncertainty in position of particle is infinitely large ($\Delta x = \infty$). Such a particle with fixed momentum is spread out indefinitely in space.

- Suppose that we want to describe a particle whose position is known to be within some localized region along the horizontal axis.

- If we take the superposition of many plane waves, each with a slightly different wavelength (and thus momentum), we can in fact obtain a single wave that is found within a localized region of space and this is known as a wave packet.

Quantum Mechanics

Wave Packets (continued)

- Basically, a wave packet is a localized wave (found only within a certain region of space) that consists of many superpositioned sinusoidal plane waves with slightly different momenta.

- Since the momentum now has a range of values ($\Delta p \neq 0$), by the uncertainity principle, there is now a finite uncertainity in position (Δx) of the free particle. That is, as the wave packet moves along the x-axis, the particle always remains within the region given by Δx.

Ψ x t_1 t_2 t_3 Δx $\vec{v}$

Quantum Mechanics

Probability Density in Quantum Mechanics

• In the previous discussion, we were able to obtain the time-independent Schrödinger equation from the more general time-dependent equation.

$$i\hbar \frac{\partial \Psi(x,t)}{\partial t} = U(x)\Psi(x,t) - \frac{\hbar^2 K^2}{2m} \cdot \frac{\partial^2 \Psi(x,t)}{\partial x^2}$$

The general form of Schrödinger's equation

⇓

$$E\psi(x) = U(x)\psi(x) - \frac{\hbar^2 K^2}{2m} \cdot \frac{d^2\psi(x)}{dx^2}$$

No time, so t is fixed

Time-independent Schrödinger's equation

• By letting $\Psi(x,t) = \psi(x) \cdot F(t)$ (uppercase (psi), lowercase (psi)), substituting this into the general form and spreading/separating out the variables, we obtained:

$$i\hbar \frac{1}{F(t)} \cdot \frac{\partial F(t)}{\partial t} = E \quad \Big| \quad U(x) - \frac{\hbar^2 K^2}{2m} \frac{1}{\psi(x)} \frac{\partial^2 \psi(x)}{\partial^2} = E$$

(A) — the left-hand side; E: our arbitrary constant

In this lecture, we will solve for $\Psi(x,t)$ (the total wave-function) in terms of $\psi(x)$:

(1) We can rearrange equation (A):

Recall: $i^2 = -1 \Rightarrow i \cdot i = -1 \Rightarrow i = -1/i$

$$\frac{dF(t)}{F(t)} = \frac{E\,dt}{i\hbar}$$

the two infinitely small quantities

$$\Rightarrow \frac{dF(t)}{F(t)} = \frac{-iE\,dt}{\hbar}$$

20%

Quantum Mechanics

Probability Density in Quantum Mechanics

② By integrating both sides : $\Rightarrow \ln f(t) = \frac{-iEt}{\hbar}$

③ If we take the exponential :

$e^{\ln f(t)} = e^{iEt/\hbar} \Rightarrow$ $\boxed{f(t) = e^{-\frac{iEt}{\hbar}}}$

④ This means that the wave function : $\boxed{\Psi(x,t) = \psi(x)e^{-\frac{-iEt}{\hbar}}}$

Interpretation :

$$|\Psi(x,t)|^2 = |\psi(x)|^2 \cdot \underbrace{\left|e^{\frac{-iEt}{\hbar}}\right|^2}_{=1} = |\psi(x)|^2$$

This is the probability density of the particle; notice that the probability density in space does not depend on the time.

Remember : Our original goal in this exercise was to : Take the total wave function uppercase (Psi) Ψ, and the Ψ appears in the general form of the Schrödinger equation, and represent it using our wave-function that app

Quantum Mechanics

Interpretation of Schrödinger's Equation

Schrödinger Equation

- In classical mechanics, Newton's laws of motion and the conservation of energy used to describe the motion and behavior of systems. In quantum mechanics, which incorporates the wave-particle duality of matter, the Schrödinger equation takes the role of describing and predicting the behavior of systems.

- Schrödinger's equation is a differential equation that allows us to determine the wave function Ψ and describes how it changes over time.

$$E \cdot \Psi(x) = U \cdot \Psi(x) - \frac{\hbar^2 K^2}{2m} \cdot \frac{d^2\Psi(x)}{dx^2}$$

one-dimensional time-independent equation

- Ultimately, this differential equation was invented by Erwin Schrödinger in order to describe the "matter waves" produced by particles such as electrons.
In particular, the above equation describes a single free particle moving along a single dimension (X-axis).

- What is a wave function, Ψ (Psi)? In quantum mechanics, a wave function is a quantity that is used to describe the displacement of a matter wave produced by particles (i.e. electrons). By taking the square of the absolute value of the wave function also allows us to determine the probability of finding the particle at some given location.

QUANTUM Mechanics

NORMALIZATION CONSTANT OF PARTICLE IN RIGID BOX

- RECALL THAT THE WAVE FUNCTION THAT REPRESENTS A PARTICLE IN AN INFINITE POTENTIAL WELL (RIGID BOX) IS GIVEN BY:

$\psi(x) = A\sin\left(\frac{n\pi}{\ell}x\right)$, where $A = \sqrt{\frac{2}{\ell}}$ AND where x is the position of the particle along the bottom of the box, along the x-axis.

→ DERIVE THE EQUATION USING NORMALIZATION.

ℓ = WIDTH OF THE RIGID BOX

- The CONSTANT (A) is the AMPLITUDE OF THE WAVE PRODUCED. But what exactly is (A)? For any wave function to be measurable and physically meaningful, it must be NORMALIZED. We can use this to calculate what (A) is. **(A) is the NORMALIZATION CONSTANT!**

① $\int_{-\infty}^{+\infty} |\psi|^2 dx = 1$ (NORMALIZATION): MUST SATISFY THE EQUATION; MUST EQUAL TO 1.

→ PULLING OUT OUR CONSTANT FROM THE INTEGRAL

② $\int_{-\infty}^{+\infty} \left|A\sin\left(\frac{n\pi}{\ell}x\right)\right|^2 dx = A^2 \int_{-\infty}^{+\infty} \sin^2\left(\frac{n\pi}{\ell}x\right) dx = 1$

Now, since the particle moves between $x=0$ AND $x=\ell$, we can replace lower bound with w/0 AND upper bound w/ℓ.

③ $A^2 \int_0^{\ell} \sin^2\left(\frac{n\pi}{\ell}x\right) dx = 1$

$\frac{n\pi}{\ell}x$: θ (THETA)

OUR RIGID BOX

y, 0, ℓ, X

④ Let $\theta = \frac{n\pi}{\ell}x$, then $d\theta = \frac{n\pi}{\ell}dx \Rightarrow dx = \frac{\ell}{n\pi}d\theta$ (solving for dx)

$A^2 \int_0^{n\pi} \sin^2\theta \cdot \left(\frac{\ell}{n\pi}\right) d\theta = \frac{A^2\ell}{n\pi} \int_0^{n\pi} \sin^2\theta \, d\theta = 1$

$\frac{A^2\ell}{n\pi}$ → ALL CONSTANTS

Note: $n\pi = 1$ (RANGE LIMIT)
n = OUR QUANTUM NBR.
→ $\sin(n\pi) = 1$

Quantum Mechanics

Normalization Constant of Particle in Rigid box (continued)

(5) Apply trig. Function: $\sin^2\theta = \frac{1}{2}(1-\cos 2\theta)$

$$\frac{A^2 \ell}{2n\pi}\int_0^{n\pi}(1-\cos 2\theta)\,d\theta = \frac{A^2 \ell}{2n\pi}\left(\theta - \frac{\sin 2\theta}{2}\right)\Bigg|_0^{n\pi} = 1$$

$$\frac{A^2 \ell}{2n\pi}(n\pi) = 1 \Longrightarrow \boxed{A = \sqrt{\frac{2}{\ell}}}$$

amplitude (A) or normalization constant

Solving for A:

$$A^2 = \frac{2}{\ell} \rightarrow A = \sqrt{\frac{2}{\ell}}$$

Example 16 - Quantum Mechanics Examples

Wave Function Constraints and Normalization

In this section we're going to examine some of the constraints that exists when we're dealing with wave functions in quantum mechanics.

In quantum mechanics every single system, for example an electron, has its own wave function associated with that system. Now we know that the way function is obtained by solving Schrodinger's equation. And the wave function itself basically provides us with all the information we need to know about the motion and behavior we need to know about our system. Now our system for example could be an electron. Now if we know what the wave function is which by the way depends on the position X as well as the time t, if we know what the wave function is, this gives us the single value probability amplitude of the wave produced by that system. So basically PSI gives us information about the actual wave itself. It gives us the amplitude of how high or how low our wave goes. Now what if we take the square of the absolute value of our way function. This gives us information about the particle of that system. So basically it gives us the probability distribution, the probability density of that particle that produces that wave. This basically tells us the probability, or how likely it is that we are to find that particle in a certain position x and at certain time t.

Now for this quantity to actually make sense, it has to be between 0 and 1. We can't get a probability that is negative and likewise we can't get a probability that is greater than 1. So basically, in order for a wave function to be useful and be physically meaningful and measurable it must satisfy a certain set of requirements.

Requirement number 1

So the way the function must satisfy, it must be a solution to Schrodinger's equation. To the following equation that's basically the Schrodinger equation. This is the time dependent Schrodinger equation. What exactly do we mean by requirement one. Well to understand what we mean let's recall an analogy. So let's suppose we're dealing in classical mechanics. So we have the following equation. F=ma

Force equals mass times acceleration. (F = ma).

So remember in quantum mechanics, Schrodinger equation is the analogous equation to force equals ma. So in classical mechanics, we use this equation to study the motion of an object. And in quantum mechanics we use the Schrodinger equation. So let's suppose that we know the force on our object being 10 newtons. So if we know the force is 10 newtons that means, lets say if we plug in a value of lets say one kilogram for the mass and two meters /sec squared for our acceleration, we can easily see that one times two dose not equal 10. So one and two are not solutions to this equation. So in the same exact way our wave function must satisfy the solution to the Schrodinger equation. So that's exactly what we mean by step one. So let's suppose our way function doesn't in fact satisfy our requirement number one. So what else must it satisfy.

Requirement number 2

The wave function must in fact be a continuous function and that simply because the probability must always be a continuous value. It must range from 0 to 1. And it must be a value in between. So once again, since the square of the absolute value of the wave function PSI represents the probability of finding our particle at some position X and at some moment time t, then the probability itself must be a continuous value from 0 to 1, and it follows that the wave function itself must also be continuous. So basically the wave function cannot be a discontinuous function. It must be a continuous function.

Requirement number 3

The final requirement states that the wave function must in fact be normalized. The word normalized is a term that basically comes from mathematics, statistics and probability. So what exactly does it mean for a wave function to be normalized for any equation for that matter to be normalized?

So the wave function must correctly give us the probability value. That is, the probability should never exceed one and the probability should never be a negative quantity. More precisely, the probability of finding a particle at anyone location or another should always be one. So what exactly do we mean by that? So lets suppose we are dealing with one dimension. We have some type of particle so let's suppose it's an electron found somewhere along the x-axis. And the x-axis extends two negative infinity and positive infinity in the region along the x-axis on the graph. And the curve above the x-axis represents our wave function. Now let's choose an infinitely small distance or an infinitely small region in space along the x axis, we call dx. Given by the following region in the graph. Now for our wave function to actually be physically measurable or physically meaningful, we must take the square of the absolute value of the wave function (PSI) and multiply it this infinitely small distance given by dx, then this should give us the probability of finding our particle in this infinitely small region given by dx.

Now if we slice our entire x-axis into infinitely small regions. And then we integrate them from negative infinity to positive infinity. If we then take the sum of all these probabilities then the entire probability should give us one. This is exactly what we mean by our wave function being normalized. So basically what exactly does it mean for the integral going from negative to positive infinity of the square of the absolute value PSI being equal to one with respect to dx? So basically the sum of all the probabilities along the entire x-axis, along all of space should always be one. In other words it should be 100%. Now what that basically means is we should be able to find our particle somewhere along the x-axis. And when we integrate all these probabilities or when we sum up all these probabilities, then the probability should in fact be equal to 1. It cannot exceed that and it should not be anything less than 1.

QUANTUM MECHANICS

TIME-INDEPENDENT SCHRÖDINGER EQUATION (EXAMPLE)

- IN QUANTUM MECHANICS, WE DESCRIBE SYSTEMS USING WAVE FUNCTIONS. WHEN WE TREAT A SYSTEM AS A WAVE, THE WAVE FUNCTION REPRESENTS THE DISPLACEMENT OF THE WAVE. IF WE TREAT THE SYSTEM AS A PARTICLE, THEN THE WAVE FUNCTION IS USED TO GIVE US THE PROBABILITY OF FINDING THE PARTICLE AT SOME ($|\psi|^2$).

- IN ORDER TO DESCRIBE ANY SYSTEM IN QUANTUM MECHANICS, WE MUST BE ABLE TO DETERMINE WHAT THE WAVE FUNCTION IS NUMERICALLY. THE SCHRÖDINGER EQUATION IS A DIFFERENTIAL EQUATION THAT WE CAN USE TO SOLVE FOR THE WAVE FUNCTION QUANTITIVELY.

- IN THE SAME WAY THAT ISAAC NEWTON INVENTED THE SECOND LAW OF MOTION ($\Sigma \vec{F} = m\vec{a}$), THE SCHRÖDINGER EQUATION WAS INVENTED (NOT DERIVED) AND CONFIRMED USING EXPERIMENTS. THIS MEANS THAT THERE IS NO WAY TO DERIVE THIS EQUATION.

- WHAT DOES THE SCHRÖDINGER EQUATION ACTUALLY LOOK LIKE?
- IN ORDER TO DETERMINE WHAT FORM THE EQUATION TAKES, WE WILL USE THE CONSERVATION OF ENERGY. WE WILL ALSO ASSUME THAT THE WAVE FUNCTION DOES NOT DEPEND ON TIME AND ONLY DEPENDS ON THE SPATIAL POSITION OF THE SYSTEM. ($\psi(x)$).

---•→------ X-AXIS $p = \frac{h}{\lambda} \longrightarrow \boxed{\lambda = \frac{h}{p}}$ (CONSTANT)

① $\psi(x,t) = A\sin(kx - \omega t) + B\cos(kx - \omega t)$ [CLASSICAL WAVE EQ.]

SINCE WE ARE ASSUMING TIME-INDEPENDENCE, LET $t = 0$.

$\longrightarrow \psi(x) = A\sin(kx) + B\cos(kx)$ ① [WAVE FUNCTION FOR PARTICLE]

$$k = \frac{2\pi}{\lambda} = \frac{2\pi p}{h} = \frac{p}{h/2\pi} = \frac{p}{\hbar}$$

Quantum Mechanics

Time-Independent Schrödinger Equation Example (Contined)

(2) For a particle with mass (m) and velocity (v), the total energy is:

$$E = K + U \rightarrow \frac{1}{2}mv^2 + U \Rightarrow E = \frac{p^2}{2m} + U$$

Since $p = k\hbar \Rightarrow E = \frac{k^2\hbar^2}{2m} + U$ (2)

Conclusion:

- Whatever the Schrödinger equation looks like, it must satisfy equation (2) (conservation of energy) and the solution of the Schrödinger equation must take the form of equation (1).

Quantum Mechanics

Solving for Time-Independent Schrödinger Equation (Non-Relativistic)

Determine the time-independent Schrödinger equation using the more general time-dependent equation. (A)

$$i\hbar \frac{\partial \Psi(x,t)}{\partial t} = U(x)\Psi(x,t) - \frac{\hbar^2 K^2}{2m} \frac{\partial^2 \Psi(x,t)}{\partial x^2}$$

(1) In mathematics, it is sometimes possible to express a function of two variables as the product of two different functions:

$$f(x,y) \rightarrow f(x,y) = g(x) \cdot h(y)$$

$$\boxed{\Psi(x,t) = \psi(x) \cdot f(t)} \quad (B)$$

↳ lower-case psi
↳ upper-case psi

(2) Let us now substitute equation (B) into equation (A)

$$i\hbar \frac{\partial \psi(x) f(t)}{\partial t} = U(x)\psi(x)f(t) - \frac{\hbar^2 K^2}{2m} \frac{\partial^2 \psi(x) f(t)}{\partial x^2}$$

(3) Now we can divide both sides by $\psi(x) \cdot f(t)$ (by seperating our variables)

$$i\hbar \frac{1}{f(t)} \cdot \frac{\partial f(t)}{dt} = U(x) - \frac{\hbar^2 K^2}{2m} \cdot \frac{1}{\psi(x)} \cdot \frac{\partial^2 \psi(x)}{\partial x^2}$$

↳ the potential energy of our particle

In mathematics, this is known as the separation of variables.

(4) Notice that the left-side of the equation is strictly in terms of time, while the right-side of the equation is strictly in terms of position. This implies that both sides are equal to some constant, lets call it (E); an arbitray letter.

$$i\hbar \frac{1}{f(t)} \frac{\partial f(t)}{\partial t} = E \qquad \left| \qquad \underbrace{U(x) - \frac{\hbar^2 K^2}{2m} \cdot \frac{1}{\psi(x)} \cdot \frac{\partial^2 \psi(x)}{\partial x^2}}_{(C)} = E \right.$$

QUANTUM MECHANICS

SOLVING FOR TIME-INDEPENDENT SCHRODINGER EQUATION (CONTINUED)
(NON-RELATIVISTIC)

(5) Now let us take equation (C) and multiply both sides by $\Psi(x)$.

$$U(x)\Psi(x) - \frac{\hbar^2 K^2}{2m} \cdot \frac{\partial^2 \Psi(x)}{\partial x^2} = E\Psi(x)$$

We see that we can readily obtain the time-independent schrödinger equation from the time-dependent equation.

QUANTUM MECHANICS

WAVE FUNCTION CONSTRAINTS AND NORMALIZATION

- In quantum mechanics, every system has a wave function associated with it. This wave function provides us with everything we need to know about the motion and behavior of that system.

$\Psi(x,t)$: = Probability amplitude of the wave

$|\Psi(x,t)|^2$: = Probability distribution of the particle producing the wave.

- In order for a wave function to be useful and physically measurable and meaningful, it must satisfy certain sets of requirements

① It must be a solution to Schrödinger's equation.

$$i\hbar \frac{\partial \Psi(x,t)}{\partial t} = U(x)\Psi(x,t) - \frac{\hbar^2}{2m}\frac{\partial^2 \Psi(x,t)}{\partial x^2}$$

② It must be continuous.

Since $|\Psi|^2$ represents the probability of finding the particle at some x and t, and the probability itself must be a continuous value from 0 to 1, then we see that Ψ must be continuous.

③ It must be normalized.

The wave function must correctly give us the probability value. That is, the probability should never exceed 1. More precisely, the probability of finding a particle at one place or another should always be 1.

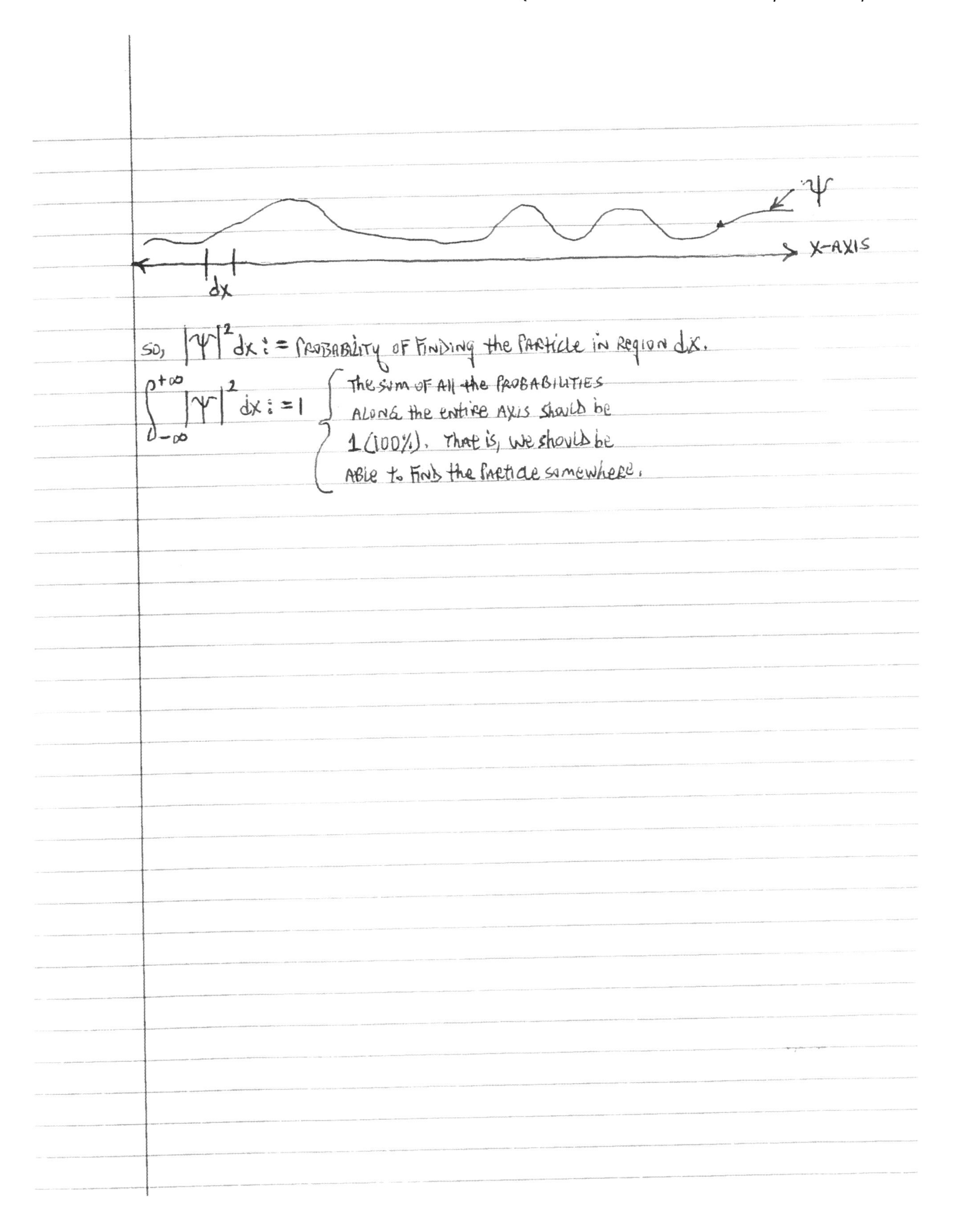
Ψ
X-AXIS
dx
SO, $|\Psi|^2 dx$:= PROBABILITY OF FINDING the PARTICLE in REGION dx.
$\int_{-\infty}^{+\infty} |\Psi|^2 dx := 1$
The sum of All the PROBABILITIES ALONG the entire AXIS should be 1 (100%). That is, we should be Able to Find the Particle somewhere.

Quantum Mechanics

Quantum tunneling

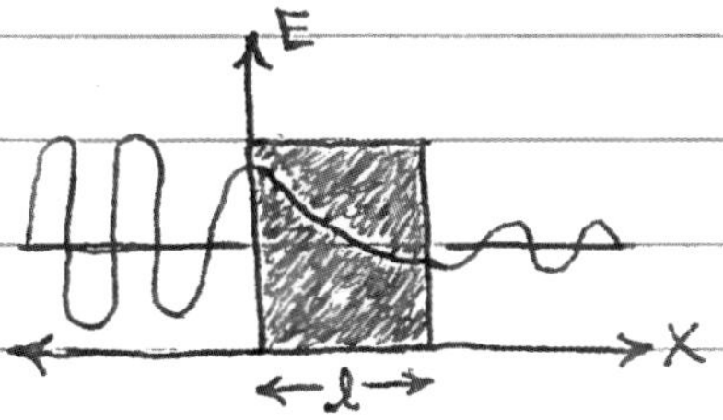

• A free particle has the form of a sinusoidal wave. When the wave enters the potential region, the wave function begins to decay exponentially. But before it decreases to zero, it resurfaces and continues to move, although with a smaller amplitude.

• This ability of the particle to penetrate barriers is known as Quantum Tunneling.

• Transmission Coefficient (T) — Fraction of particles that make it across.
• Reflection Coefficient (R) — Fraction of particle that reflect

① $T + R = 1$
② $T \approx e^{-2\alpha l}$ ↳ proportional, where $\alpha = \sqrt{\dfrac{2m(U_0 - E)}{\hbar}}$.

Conclusion: With Quantum Mechanics, there is always a possibility of the seemingly impossible to occur, all be it extreemly small. DPH

Quantum Mechanics

Quantum Tunneling

- From our discussion of a particle in a finite potential well, we saw that in quantum mechanics, an electron can penetrate a potential barrier and enter a region that is normally forbidden by classical mechanics.

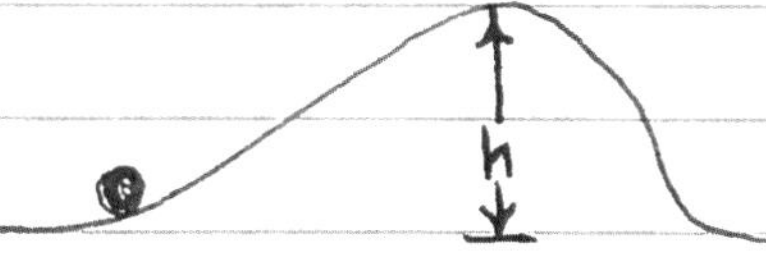

- A ball of mass (m) is on the ground ($U=0$) and is moving with a kinetic energy, K, when it begins ascending a hill of height (h). If the potential of the hill is too high, the ball will never make it to the other side ($K < U = mgh$).

- Suppose now we have a particle of mass (m) and no potential energy ($U=0$) moving with kinetic energy $E_0 = K$ when it encounters a potential barrier such that $K < U_0$. According to classical mechanics, the particle should simply bounce off and travel in reverse.

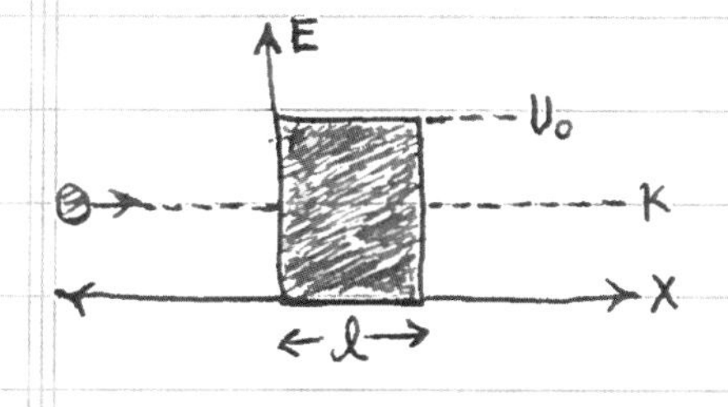

- According to quantum mechanics, there is a non-zero probability that the particle will end up on the other side, even though $K < U_0$.

QUANTUM MECHANICS

QUANTUM TUNNELING EXAMPLE

- The PROBABILITY OF AN ELECTRON PENETRATING a POTENTIAL BARRIER OF 10.0 eV IS EQUAL to 0.8%. IF the width of the POTENTIAL BARRIER is 0.6nm, FIND the energy of the electron. (.8% OF the electrons will cross the barrier)

$T = e^{-2\alpha l}$ where $\alpha = \dfrac{\sqrt{2m(U_0 - E)}}{\hbar}$ (α: Alpha)

ELECTRON-VOLTS CONVERTED INTO JOULES

$10.0\,eV(1.6 \times 10^{-19}\,J/eV) = 1.6 \times 10^{-18}\,J$

ψ, E, $U_0 = 10.0\,eV$, X, l

TAKE the NATURAL log OF BOTH sides:

$\ln T = -2\alpha l \Rightarrow \alpha = \dfrac{-\ln T}{2l} \Rightarrow \dfrac{\sqrt{2m(U_0 - E)}}{\hbar} = \dfrac{-\ln T}{2l}$, NOW REARRANGE the EQUATION to solve FOR E.

$\Rightarrow \sqrt{2m(U_0 - E)} = \dfrac{-\hbar \ln T}{2l} \longrightarrow 2m(U_0 - E) = \left(\dfrac{-\hbar \ln T}{2l}\right)^2$ (T)

$\Rightarrow E = U_0 - \dfrac{\left(\dfrac{-\hbar \ln T}{2l}\right)^2}{2m} = 1.6 \times 10^{-18}\,J - \dfrac{\left(\dfrac{(-1.05 \times 10^{-34}\,\frac{m^2}{s}\,J)\ln(0.008)}{2(0.6 \times 10^{-9}\,m)}\right)^2}{2(9.11 \times 10^{-31}\,kg)}$

$\Rightarrow 1.5 \times 10^{-18}\,J$ (OR 9.4 eV)

↳ The ENERGY OF the ELECTRON AS IT CROSS ALONG the X-AXIS.

NOTE: (T) = TRANSMISSION COEFFICIENT

Quantum Mechanics

Free Particle and Plane Wave Example

- A free particle, say an electron, is moving along the x-axis with an energy of 10.0 eV. If we assume that the potential energy is zero:

(a) Determine the wavelength of the plane wave produced.

$$① \; k = \frac{2\pi}{\lambda} \qquad ② \; k = \sqrt{\frac{2mE}{\hbar^2}}$$

Setting equation ① equal to ②: $\frac{2\pi}{\lambda} = \sqrt{\frac{2mE}{\hbar^2}} \Rightarrow$

$$\lambda = \frac{2\pi}{\sqrt{\frac{2mE}{\hbar^2}}} \Rightarrow$$

$$\lambda = \frac{2\pi}{\left(\frac{2(9.11 \times 10^{-31}\,kg)(10)(1.6 \times 10^{-19}\,J)}{(1.055 \times 10^{-34}\,J \cdot s)^2}\right)^{1/2}} = \boxed{3.89 \times 10^{-10}\,m}$$

(b) Determine the wave function that represents the plane wave (assume: $B = 0$).

$$\psi(x) = A\sin(kx) + B\cos(kx) \Rightarrow \psi(x) = A\sin\left[(1.62 \times 10^{10})x\right]$$

(A → amplitude)

$$k = \frac{2\pi}{\lambda} = \frac{2\pi}{3.89 \times 10^{-10}\,m}$$

$$k = 1.62 \times 10^{-10}$$

QUANTUM Mechanics

WAVE FUNCTION OF PARTICLE IN FINITE POTENTIAL

PARTICLE IN A FINITE POTENTIAL WELL

IF WE SET $X=0$: $\psi_1(0) = \psi_2(0) \Rightarrow$

$$A\sin(0) + B\sin(0) = Ce^0 \Rightarrow \boxed{B=C}$$

$$\frac{d\psi_1(0)}{dx} = \frac{d\psi_2(0)}{dx} \Rightarrow kA\cos(0) - kB\sin(0) = \alpha Ce^0 \Rightarrow \boxed{kA = \alpha C} \quad ②$$

IF WE LET $X=\ell$, WE OBTAIN TWO MORE EQUATIONS THAT RELATE UNKNOWNS C AND D. FINALLY, IF WE NORMALIZE THE WAVE FUNCTION, WE CAN OBTAIN THE FIFTH EQUATION. USING THESE FIVE EQUATIONS, WE CAN SOLVE FOR THE UNKNOWNS A, B, C, D & E.

U_0

$n=1$

I $X=0$ II $X=\ell$ III

- IF WE REARRANGE THIS EQUATION AND MULTIPLY BOTH SIDES BY $-2m/\hbar^2$:

$$\frac{\hbar^2}{2m}\frac{d^2\psi}{dx^2} + E\psi - U_0\psi = 0 \Rightarrow \frac{d^2\psi}{dx^2} - \left[\frac{2m}{\hbar^2}(E-U_0)\right]\psi = 0$$

- LET $\alpha^2 = \frac{2m}{\hbar^2}(E-U_0) \Longrightarrow \frac{d^2\psi}{dx^2} - \alpha^2\psi = 0$ ①

- IF WE SOLVE EQUATION ① FOR ψ: $\psi = Ce^{\alpha x} + De^{-\alpha x}$

Quantum Mechanics

Wave Function of Particle in Finite Potential Well (continued)

- If particle is in Region I, the (x) is always negative. If $X \to -\infty$, (→ approaches) the term, then $e^{-\alpha x} \to \infty$ (→ approaches) and $\psi \to \infty$.
  Since this cannot be true, that means $D = 0$. Therefore, $\psi_1 = Ce^{\alpha x}$, if the particle is in Region I.

- If the particle is in Region III, by a similar argument, $C = 0$.
  Therefore, $\psi_3 = De^{-\alpha x}$, if particle is in Region III.

Recall that for a wave function to be "physically measurable," it must be continuous throughout, even at $X=0$ and $X=\ell$. Also, the slope at these two points must exist.

$$\Longrightarrow \psi_1(0) = \psi_2(0), \quad \frac{d\psi_1}{dx} = \frac{d\psi_2}{dx} \text{ if } x=0$$

$$\psi_2(\ell) = \psi_3(\ell), \quad \frac{d\psi_2}{dx} = \frac{d\psi_3}{dx} \text{ if } x=\ell$$

QUANTUM Mechanics

Finite Potential Well – EXAMPLE

AN ELECTRON with AN ENERGY OF 10.0eV is traveling on the left of the Potential well with a depth of 4.0eV. If the width of the well is 20.0nm. Find the wavelength of the wave formed by the electron when its near the well and when its above the well.

λ, energy, E, K, U_0, X-AXIS, X=0, X=l, 20nm

EQUATIONS:

① $\lambda = \frac{h}{\sqrt{2m(E-U_0)}}$ (LEFT OF WELL) | $\lambda = \frac{h}{\sqrt{2mE}}$ (ABOVE WELL)

② $\lambda = \frac{h}{\sqrt{2mE}} = \frac{6.626\times10^{-34}\ J\cdot s}{\sqrt{2(9.11\times10^{-31}\ kg)(10eV)(1.6\times10^{-19}\ J/eV)}}$

$\lambda = 3.88\times10^{-10}\ m$ (OR 0.388nm) ← wavelength of ELECTRON directly ABOVE the well

③ $\lambda = \frac{h}{\sqrt{2m(E-U_0)}} = \frac{6.626\times10^{-34}\ J\cdot s}{\sqrt{2(9.11\times10^{-31}\ kg)(10eV-4eV)(1.6\times10^{-19}\ J/eV)}}$

$\lambda = 5.01\times10^{-10}\ m$ (OR 0.501nm) ← wavelength of ELECTRON to the left of the WALL.

QUANTUM MECHANICS

WAVE FUNCTION OF PARTICLE IN RIGID BOX

INTERPRETATION OF PARTICLE IN RIGID BOX

$U(x)$

$x=0$ $x=l$ → X-AXIS

- A PARTICLE WITH MASS m IS MOVING ALONG THE BOTTOM OF THE RIGID BOX BETWEEN $x=0$ AND $x=l$.

- THE WAVE FUNCTION $\Psi(x)$ (→ REPRESENTS THE POSITION OF OUR PARTICLE WITHIN THE RIGID BOX.) THAT DESCRIBES THE MOTION OF THE PARTICLE WITH POTENTIAL ENERGY $U(x)=0$ IS GIVEN BY

$$\Psi_n(x) = \sqrt{\frac{2}{l}}\sin\left(\frac{n\pi}{l}x\right)$$

THE WAVEFUNCTION FORMULA THAT DESCRIBES THE BEHAVIOR AND MOVEMENT OF THE PARTICLE.
THE FUNCTION DESCRIBES THE WAVE OF OUR FREE PARTICLE (THE ELECTRON).

n: = QUANTUM NUMBER l: = WIDTH OF BOX

x: = POSITION ALONG THE AXIS $\sqrt{\frac{2}{l}}$ = AMPLITUDE OF THE WAVE, PRODUCED BY THE PARTICLE

$|\Psi|^2$

$n=1$ $n=2$ $n=3$

THE HIGHEST POINT OF THE DISTRIBUTION DENSITY AT ANY GIVEN MOMENT IN TIME, THE PARTICLE IS LIKELY TO BE FOUND AT THE CENTER OF OUR X-AXIS. OUR PARTICLE IS MOST LIKELY TO BE FOUND.

- THESE THREE GRAPHS REPRESENT THE PROBABILITY DENSITY OF THE PARTICLE; THE Y-AXIS GIVES THE VALUE OF $|\Psi|^2$.

NOTE: THE PARTICLE (ELECTRON) MOVES (OSCILLATES) BETWEEN THE TWO CORNERS OF THE BOX

NOTE: IF WE SQUARE THE ENTIRE RIGHT SIDE OF THE EQUATION, THAT WILL GIVE US THE PROBABILITY DENSITY (PROBABILITY DISTRUBUTION) OF OUR PARTICLE AS IT MOVES BETWEEN THE LEFT AND RIGHT SIDES OF THE RIGID BOX.

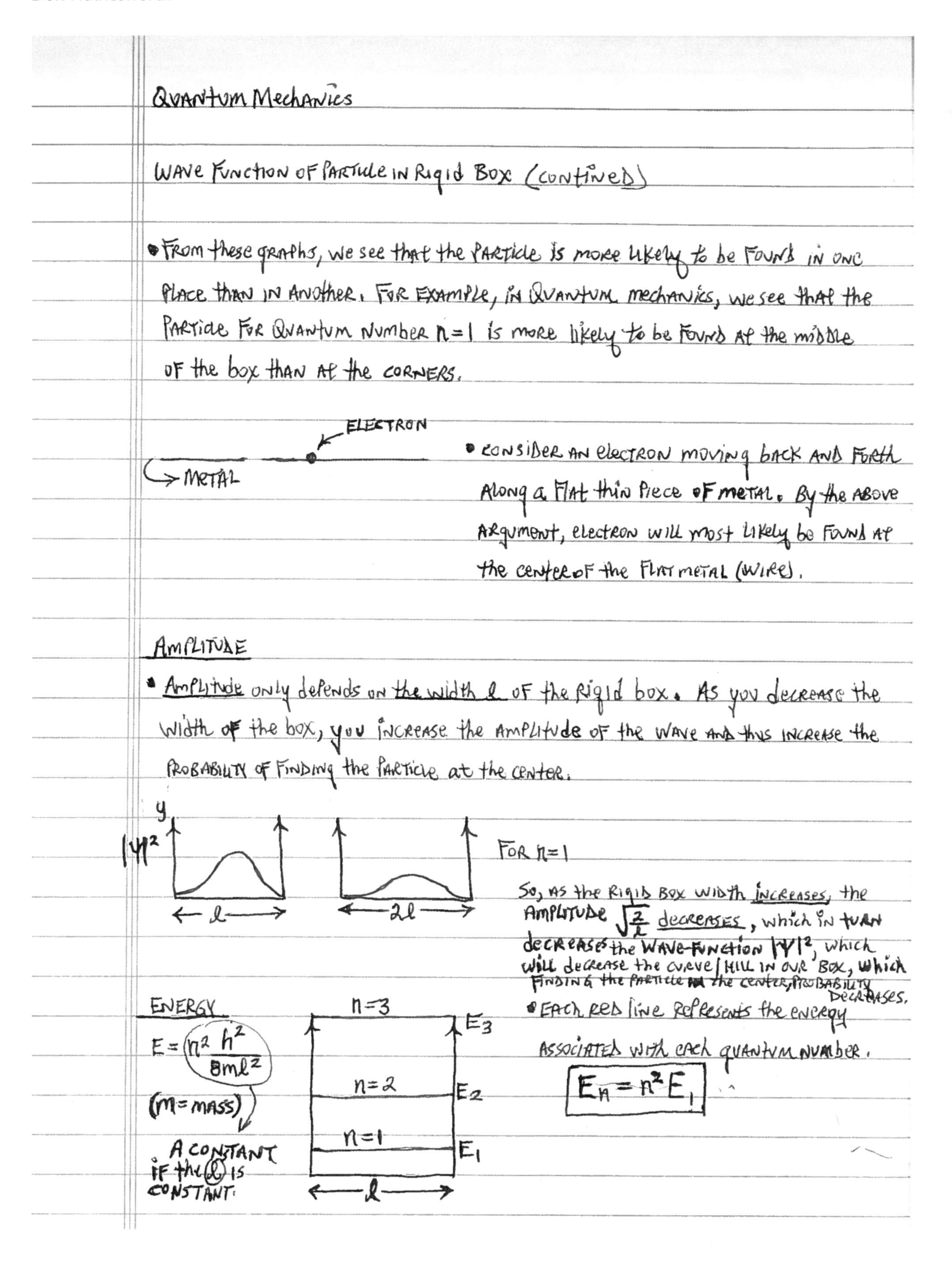

Quantum Mechanics

Wave Function of Particle in Rigid Box (contined)

- From these graphs, we see that the particle is more likely to be found in one place than in another. For example, in quantum mechanics, we see that the particle for quantum number n=1 is more likely to be found at the middle of the box than at the corners.

- Consider an electron moving back and forth along a flat thin piece of metal. By the above argument, electron will most likely be found at the center of the flat metal (wire).

Amplitude

- Amplitude only depends on the width l of the rigid box. As you decrease the width of the box, you increase the amplitude of the wave and thus increase the probability of finding the particle at the center.

For n=1

So, as the rigid box width increases, the amplitude $\sqrt{\frac{2}{l}}$ decreases, which in turn decreases the wave-function $|\psi|^2$, which will decrease the curve/hill in our box, which finding the particle at the center probability decreases.

Energy

$E = n^2 \frac{h^2}{8ml^2}$

(m = mass)

A constant if the l is constant.

- Each red line represents the energy associated with each quantum number.

$$E_n = n^2 E_1$$

QUANTUM MECHANICS

ENERGY OF A PARTICLE IN A RIGID BOX – EXAMPLE

AN ELECTRON is trapped in an infinitely deep square potential well that has a width of 0.2 nm.

(a) Find the zero-point energy of the electron.

$$E_n = n^2 \cdot \frac{h^2}{8ml^2} \rightarrow E_1 = \frac{h^2}{8ml^2} = \frac{(6.626 \times 10^{-34} J.s)^2}{8(9.11 \times 10^{-31} Kg)(0.2 \times 10^{-9} m)^2}$$

$$= \boxed{1.51 \times 10^{-18} J}$$

(B) Find the quantity of energy released when the electron jumps from $n=3$ to $n=1$ quantum number.

n=3 | E_3
n=2 | E_2
n=1 | E_1
E
0.2nm

$$E_3 = n^2 E_1 = (3)^2 (1.51 \times 10^{-18} J) = 1.36 \times 10^{-17} J$$

$$\Delta E = E_3 - E_1 = (1.36 \times 10^{-17} J) - (1.51 \times 10^{-18} J) = \boxed{1.21 \times 10^{-17} J}$$

(C) The energy in part (B) is released as a photon. Find the wavelength of the photon.

$$\Delta E = \frac{hc}{\lambda} \Rightarrow \lambda = \frac{hc}{\Delta E} = \frac{(6.626 \times 10^{-34} J.s)(3 \times 10^{8} m/s)}{1.21 \times 10^{-17} J} = \boxed{1.64 \times 10^{-8} m}$$

Quantum Mechanics

Alpha Decay and Quantum Tunneling

ALPha Decay and Quantum tunneling:

- If certain atoms have naturally unstable nuclei, why do they persist to exist for long periods of time and why don't they decay instantaneously?
  For example, an unstable Uranium-232 atom will undergo an alpha decay but it will take 68.9 years for half of the initial amount to decay.

- The fact that it takes some unstable nuclei such a long time to decay means there must be some type of barrier that is in place and does not allow the alpha particle to escape from the nucleus.

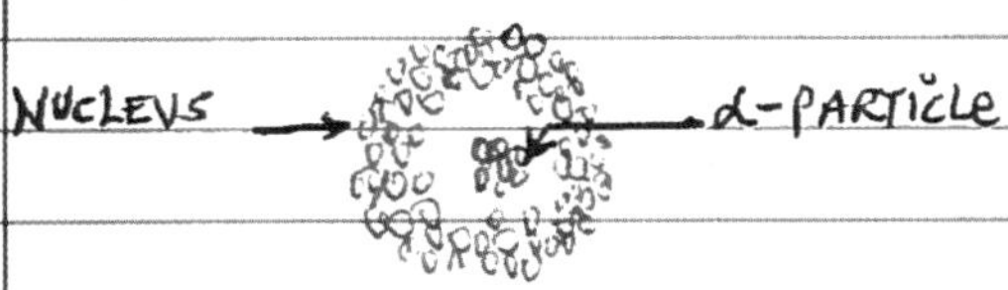

- In order for an alpha particle to escape to the outside of the nucleus, it must overcome the nuclear forces of the other nucleons.

- We can imagine that the alpha particle "moves" about the very small space in the nucleus with some kinetic energy that is less than the energy required to overcome the attraction/attractive nuclear forces. Therefore, the alpha particle is trapped in a finite potential well.

Note: The potential well is equivalent to a rigid box.

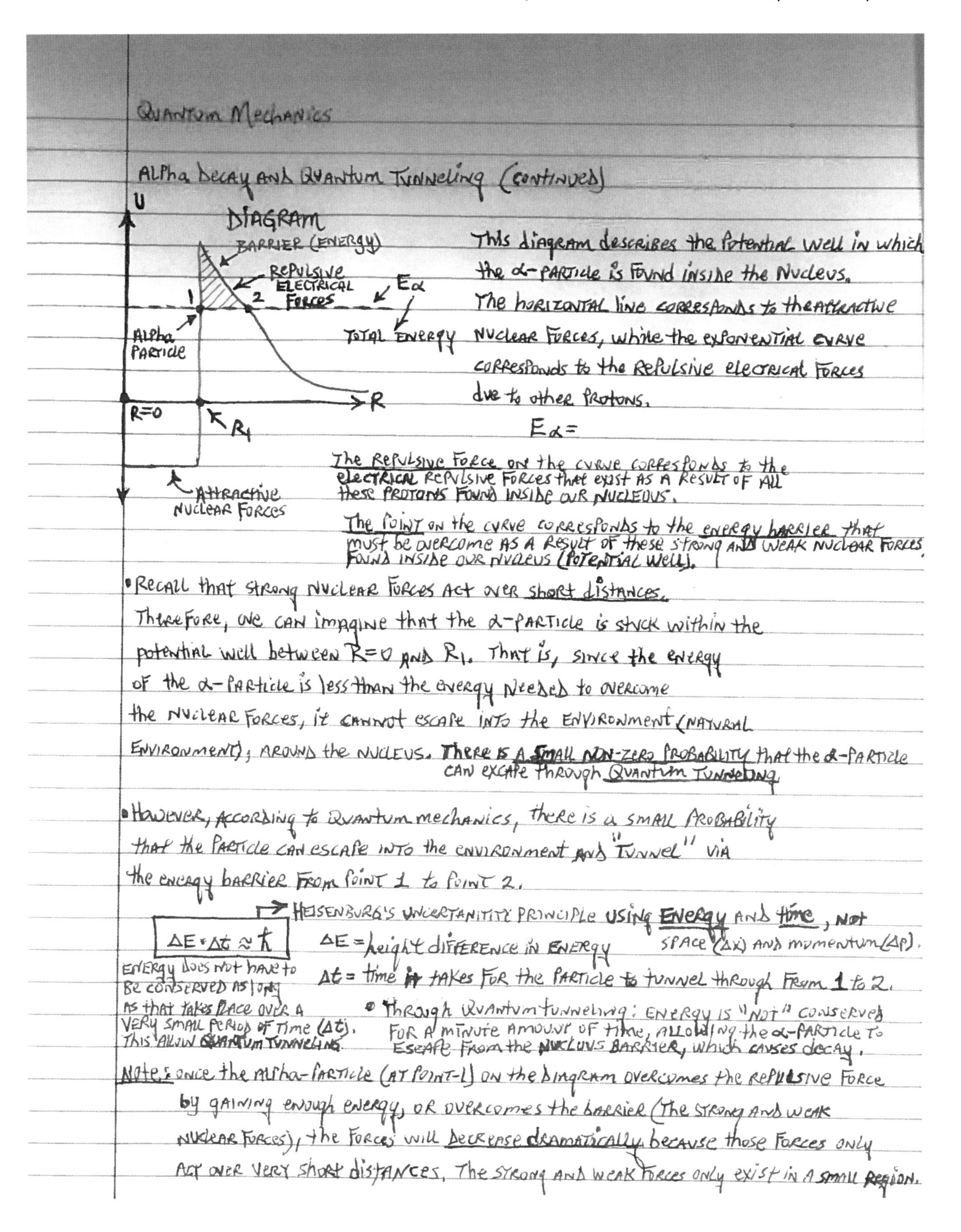
Quantum Mechanics
Alpha Decay and Quantum Tunneling (continued)
U
Diagram
Barrier (Energy)
Repulsive Electrical Forces
1
2
Eα
Total Energy
Alpha Particle
R=0
R
R1
Attractive Nuclear Forces
This diagram describes the potential well in which the α-particle is found inside the nucleus.
The horizontal line corresponds to the attractive nuclear forces, while the exponential curve corresponds to the repulsive electrical forces due to other protons.
Eα =
The repulsive force on the curve corresponds to the electrical repulsive forces that exist as a result of all these protons found inside our nucleous.
The point on the curve corresponds to the energy barrier that must be overcome as a result of these strong and weak nuclear forces found inside our nucleus (potential well).
• Recall that strong nuclear forces act over short distances.
Therefore, one can imagine that the α-particle is stuck within the potential well between R=0 and R1. That is, since the energy of the α-particle is less than the energy needed to overcome the nuclear forces, it cannot escape into the environment (natural environment), around the nucleus. There is a small non-zero probability that the α-particle can escape through quantum tunneling.
• However, according to quantum mechanics, there is a small probability that the particle can escape into the environment and "tunnel" via the energy barrier from point 1 to point 2.
Heisenburg's uncertainitity principle using energy and time, not space (Δx) and momentum (Δp).
ΔE·Δt ≈ ħ
ΔE = height difference in energy
Δt = time it takes for the particle to tunnel through from 1 to 2.
Energy does not have to be conserved as long as that takes place over a very small period of time (Δt). This allow quantum tunneling.
• Through quantum tunneling: energy is "not" conserved for a minute amount of time, allowing the α-particle to escape from the nuclus barrier, which causes decay.
Note: once the alpha-particle (at point-1) on the diagram overcomes the repulsive force by gaining enough energy, or overcomes the barrier (the strong and weak nuclear forces), the forces will decrease dramatically because those forces only act over very short distances. The strong and weak forces only exist in a small region.

QUANTUM MECHANICS

DISINTEGRATION ENERGY OF ALPHA DECAY

Uranium-232, an unstable atom, readily undergoes alpha decay to produce thorium-228. Determine the disintegration energy during this radioactive decay, assuming that the atomic mass of uranium is 232.037156u and that of thorium is 228.028731u (α has a mass of 4.001506u).

$$\boxed{{}^{232}_{92}U \rightarrow {}^{228}_{90}TH + {}^{4}_{2}\alpha}$$

$$M_U c^2 = M_{TH} c^2 + M_\alpha c^2 + Q$$

$$\Rightarrow Q = M_U c^2 - M_{TH} c^2 - M_\alpha c^2$$

$$Q = \left[M_U - (M_{TH} + M_\alpha)\right] c^2$$

$$Q = (232.037156u - (228.028731u + 4.001506u))(1.6606 \times 10^{-27} \tfrac{kg}{u})(3 \times 10^8 m/s)^2$$

$Q = 1.034 \times 10^{-12}$ J OR $\boxed{6.46 \text{ MeV}}$ This energy represents the kinetic energy of the thorium atom and the alpha particle

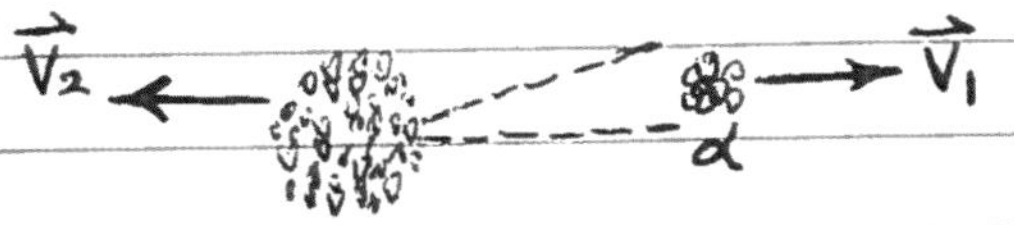

Epilogue

Multi-dimensional universes and Quantum Consciousness theory seems like a natural ending point in our long discussion of Quantum Mechanics, and that perhaps is a sign that the unification of quantum theory and human consciousness may be on the horizon, but there are no certain endings yet. In the last few decades, physics has come a long way in understanding our reality. Quantum field theories such as *QCD* and the electroweak theory, once considered esoteric and treated with suspicion, have now become the Standard Model, routinely tested and confirmed by precision experiments to many decimal places. But precision experiments, both on the ground and in the sky, have also raised questions.

In addition, a consistent and smooth geometry describing *the fabric of space-time* at the quantum level cannot currently be established, due to quantum fluctuations of point particles. Geometry is fundamentally built on dimensionality. For example, a single point in space-time represents a single dimension, two points connected forming a line represents two dimensions. And a surface represents three dimensions or 3D.

Particles located in space-time fluctuate due to the Heisenberg uncertainty principle. So for instance, when a two-particle system (at the sub-atomic level) that creates a Euclidian or Riemann geometric line is formed, the particles' locations become uncertain, thus the line becomes uncertain. Consequently, a smooth geometry cannot to be suitably or efficiently established.

A new kind of geometry must be formulated to accurately define the point, line and surfaces of particles and other sub-atomic *entities* at the sub-atomic level. Because classical and modern geometry is built upon the framework of Euclidian and Riemann (smooth curved) geometry, respectively, Einstein's General theory of Relativity, based upon *Riemann tensor calculus* becomes incompatible to quantum theory when measured or defined at the sub-atomic level.

And, because we work with "classical" and "modern" geometry in the effort in trying to describe quantum effects, we are left with quantum errors that need correction or what is known as (renormalization). This leaves us with only a "good" approximation of the small region of space-time in question. A new formulated new type of geometry that more accurately describes space-time at the sub-atomic level must be employed.

Nobody knows what the future may bring, but the only real surprise would be if there were no surprises. If the business of physics is ever finished, the world will be a much less interesting place in which to live, which is why many physicists and other scientists are happy to leave us with loose ends, and with the prospect of more interesting stories yet to be told.

Glossary

Aberration Small apparent displacement of all fixes stars from their mean locations on the celestial sphere occurs with a period of one year.

Absolute Temperature The temperature measured on the Kelvin scale: zero kelvin = -273.15 degrees Celsius. Absolute temperature is directly related to (kinetic) energy via the equation $E = k^{bxT}$, where k^{b} is Boltzmann's constant. So, a temperature of *K* corresponds to zero energy, and room temperature, 300 K = 27-degrees *C*, corresponds to an energy of 0.025 eV.

Acceleration Is measured in units of centimeters per second per second, or of feet per second or similar units. The acceleration of an automobile might be measured in miles per hour per second. This is the rate at which the velocity of a material particle changes in the course of time. Acceleration involves some change in the speed or in the direction the acceleration involves some change in the speed or the direction of motion, or both. When there is no change in the direction of motion, the acceleration is forward; when there is no change in speed the acceleration is at right angles to the motion itself, as in circular motion.

Action at a Distance According to Newton, bodies affect each other's motions by causing accelerations, whose magnitudes also depend on the masses of the bodies being acted on. The product of acceleration and mass is called the force acting on a body. As this force is exerted across empty space, one speaks of action at a distance, in contrast to Maxwell's theory of electromagnetism, according to which bodies exert forces on each other only through the 'intermediary' of the electromagnetic field. Newton's action-at-a-distance force is characterized by the 'instantaneous' effect of one body on another, regardless of the distance between them. But for electromagnetic fields and for gravitational effects, it is approximately 300,000 km (186,000 miles) per sec. This universal speed of propagation is denoted by the symbol *c – the speed of light.*

A fortiori Reasoning from a premise stronger than the one needed to come to a conclusion at hand.

A priori "Before experience." A priori truths do not require experience or observation to be formulated.

Air Speed The speed relative to the surroundings air mass with which an airplane moves.

Alchemy Medieval and Renaissance proto-chemical practice with goals of turning base metals into gold and concocting the elixir of life (*aqua vitae*)

Alpha Decay The spontaneous emission of alpha particles by certain radioactive nuclei.

Alpha Particle A bundle of two protons and two neutrons. Also, the nucleus of helium.

Alpha Ray An alpha particle.

Alpha Particles These are particles first discovered in radioactive α decay, and later identified as helium nuclei (two protons and two neutrons bound together).

Amplitude See quantum-mechanical amplitude.

Analog A comparison based on functions or structures (e.g., swimming in water is analogous to walking on land).

Analysis Intellectual process of breaking complex ideas down into simpler components; examination of whether an argument is logically valid.

Analytic statements Those whose truth is established by definition (e.g., logic and mathematical statements), rather than by evidence.

Analytic truth A statement that is true wholly because of the meaning of its terms (e.g., "All bachelors are unmarried men.")

Angular Momentum The rotational equivalent of ordinary momentum, being mass *x* velocity *x* orbital radius. It is a vector quantity directed along the axis of rotation. In quantum mechanics, (orbital) angular momentum is quantized in integer multiples of *h-bar (ħ).* This corresponds classically to only certain frequencies of rotation being allowed.

Antiparticles Particles predicted by combining the theories of special relativity and quantum mechanics. For each particle, there must exist an antiparticle with the opposite charge, magnetic moment and other internal quantum numbers (e.g. lepton number, baryon number, strangeness, charm, etc.), but with the same mass, spin and lifetime. Note that certain neutral particles such as the photon and superscript) are their own antiparticles.

A posteriori "After experience." A posteriori truths require experience or observation to be formulated.

Argument A train of thought or sequence of sentences that is meant to prove or be persuasive. Proof requires logical validity; persuasion can be achieved with a probable conclusion or appeal to common sense or intuition.

Artificial intelligence (AI) Idea of higher cognition in machines, proposed by mid-twentieth century cognitive philosophers to both solve the mind-body problem and create models for how the human mind works.

Artificial language Formal language constructed for a precise purpose – such as in logic, computer science, mathematics – or the use of formal language y analytic philosophies.

Assumption A statement believed to be true before proceeding on to another subject.

Asymptotic Freedom A term used to describe the observed decrease in the intrinsic strength of the color force between quarks as they are brought closer together. At asymptotically small separations, the quarks are virtually free. This is in contrast to the electromagnetic force whose intrinsic strength increases as two charges particles approached each other.

Atheism Theological or non-religious position that there is no deity or supernatural entity; belief that God does not exist.

Atomism The theory (first put forward in the fifth century BCE) that all matter is composed of atoms separated by empty space.

Atomism Metaphysical Principle that all physical objects and things are made up of particles that cannot be further divided; scientific principle that some part of reality or language has small parts that are foundational for larger parts or objects and that it is the task of thinkers to discover what those atoms are in a particular domain.

Baryogenesis The process by which the Universe's net baryon number was generated. This explains way the Universe is made predominantly of baryons and not antibaryons.

Baryons The generic term for any strongly interacting particle with half-integer spin in units of h-bar (e.g. the proton, neutron and all their more massive excited resonance states). Hadrons are fermions (half-integral spin particles) such as the nucleons.

Begging the question Fallacy in reasoning whereby one assumes beforehand what remains to be proved.

Beta Particle The spontaneous emission of electrons and neutrinos from certain radioactive nuclei. (ie. an electron.)

Beta (β) Particles Particles first discovered in radioactive β decay - later identified as electrons.

B-Factory Asymmetric electron-positron collider experiments tuned to the γ resonance for the study of bottom quarks.

Big-Bang Theory The most widely accepted theory of the origin of the Universe. It asserts that the Universe began some 10^{10} years ago from a space-time point of infinite energy density (a singularity). The expansion of the Universe since that time is akin to the expansion of the surface of an inflating balloon: every point on the balloon's surface is moving away from every other point.

Black Body An ideal object that absorbs all radiation. It's also a perfect emitter of radiation.

Black Dwarf A burned-out white dwarf star. The Sun will end its life as a black dwarf.

Black Hole An object with such strong gravity that not even light can escape.

Boson Any particle with integer spin: 0, $\hbar$, $2\hbar$, etc.

Bosonic String Theory This is a formulation of string theory that only has bosons. There is no super-symmetry, and since there are no fermions in the theory it cannot describe matter. So it is really just a toy theory. It includes both open and closed strings and it requires 26 space-time dimensions for consistency.

Calculus Newton and German philosopher Gottfried Wilhelm von Leibniz (1646-1716) independently discovered the basic mathematical techniques by which variable quantities are treated. One such quantity is called a *function* of another (the argument), if the value of the former is determined by the value of the latter. For instance, the free-fall of a body is described mathematically in terms of its location as a function of time.

Canon A set of traditional writings that the student is expected to master in a field.

Cartesian Pertaining to or derived from the thought of Rene Descartes, usually in reference to claims that the mind is separate from the body, or that the mind and body are two radically different things or substances.

Cartesian Coordinate System In a Cartesian coordinate system, the coordinate axes is all straight lines. They are at right angles to each other, and they are marked off in terms of the units of length adopted, such as centimeters. Given two points with the respective Cartesian coordinates (x_1, y_1, z_1) and (x_2, y_2, z_2).

Cabbibo Angle (α-Subscript-C) The measure of the probability that one flavor of quark (μ) will change into other flavors (*d* or *s*) under the action of the weak force.

Causal theory of meaning Also known as reference theory of meaning; view that meaning is not mental but in the objects named by words come to name objects based on an original "baptism" linking the word to the object.

Causation The reality and study of how events are connected so that one event or one type of event results in another event or another type of event.

CERN The European Laboratory for particle Physics formerly the Counsel European pour la Recherche Nucleaire), located near Geneva in Switzerland. Here, the resources of the European member nations are pooled to construct the large particle accelerators needed for high-energy experiments. The major project at CERN currently is the LHC.

Chance An occurrence of two events or type of occurrence, with no known causal connections. Chance may be an appearance only, due to lack of information about relevant courses, or chance may be viewed as an effect of the randomness of some events that cannot be determined.

Charm The fourth flavor (i.e. type) of quark, the discovery of which in 1947 contributed both to the acceptance of the reality of quarks and to our understanding of their dynamics. The charmed quark exhibits a property called 'charm' which is conserved in strong interactions.

Chirality The handedness of a relativistic fermion, defined by how it transforms under Lorentz transformations.

Choice A situation in which it is possible to do one or more of several things, or an exercise of autonomy.

Circular argument An argument where the conclusion is the same as its premises.

Cogito Latin, literally meaning "I think"; name for Rene Descartes' conclusion that he exists, in the argument: "I think, therefore I am."

Cognition Mental processes that impart or transmit knowledge, often presumed to be unemotional or not determined by emotion.

Cognitive Science Study of human and primate mental processes, including the processing of perceptual information, and learning.

Coherent Light beams vibrating in step.

Color An attribute which distinguishes otherwise identical quarks of the same flavor. Three colors – red, green and blue – are required to distinguish the three valence quarks of which baryons are composed. It must be stressed that these colors are just labels and have nothing to do with ordinary color. Color is the source of the strong force which binds quarks together inside baryons and mesons, and so the three colors (*r, g, b*) can be thought of as three different color charges analogous to electric charge.

Common Sense What most people believe and are considered correct or justified in believing; received opinion.

Common Sense Philosophy approach to philosophical problems that relies on common sense or ordinary opinions, attributed to Thomas Reid, G.E. Moore and others.

Concept An idea, or more accurately, the meaning of a word.

Conceptual Analysis Philosophical analysis of the meaning of terms.

Conscience Moral intuition that a person has certain obligations, or that some kinds of behavior are morally wrong.

Consciousness Awareness, the human mind in operation, or the human subject as self-aware.

Conservation Law The physics law that specifies that a certain quantity cannot be destroyed.

Conservation of Energy The law that states that the total amount of energy you start a process with has to equal what you end up with.

Contingent Uncertain to happen or something that could be or could have been otherwise.

Contradiction A statement that both asserts and denies the same thing. The law of non-contradiction in logic states that either A or not-A must be true, not both, and not neither.

Contrary Two things or statements are logical contraries if they cannot both be true, but can both be false.

Copernican revolution Named after Nicolaus Copernicus, change in the world view from geocentric to heliocentric theory.

Correspondence theory The theory that the meaning of a word is given by the object to which that word corresponds (problematic if we have no independent knowledge of objects).

Corroboration Scientific standard of confirmation that is less than full proof; the statement that is corroborated could turn out to be false in the future.

Cosmological Argument Argument for the proof of God as the creator of the universe, on the grounds that something must have created the universe; because all events and things have causes, then so must the universe as a whole.

Cosmological Constant A term added by Einstein to the gravitational field equations of his theory of general relativity. Such a term would produce a repulsive antigravity force. There is, at present, some evident for the existence of a cosmological constant.

Cosmological Principle The hypothesis that the Universe is isotropic and homogeneous on very large distance scales.

Cosmopolitanism Idea throughout intellectual history that one should be a "Citizen of the world," opposed to localism and chauvinism.

Counter-factual Hypothetical about the past; e.g., if Aristotle's texts had been lost forever, Western philosophy would have been mire platonic.

Coupling Constant A measure of the intrinsic strength of a force. The coupling constant of a particle force determines how strongly a particle couples to the associated field. For example, alpha = $e^2/\hbar * c = 1/137$ (or equivalently, electric charge *e*) specifies the strength of the coupling of charged particles to the electromagnetic field.

Covering law model Standard for scientific explanation whereby specific events are explained by showing how they are instances of generalizations; e.g., The pond is frozen because it's below 32 degrees Fahrenheit and water freezes at 32 degrees.

Critique Verb or noun referring to criticism that originates from a well-formed intellectual perspective.

Cross-Section (σ) The basic measure of the probability that particles will interact. It corresponds to the effective target area (in, for example, cm-squared) seen by the ingoing particles. It can be derived from the quantum-mechanical interaction probability. A convenient unit for measuring cross-section is the barn (symbol: b), defined as 1 b = 10^{-24} cm-squared. Typical hadronic cross-sections are measured in millibarns; 1 mb = 10-^{-27} cm^2. However, neutrino collision cross-sections are typically much smaller, a^{-39} cm^2.

Cynicism Ancient doctrine of withdrawal for society and return to what is simple and natural, which may be obnoxious to those who are refined.

Dark Matter An unknown invisible substance that can be detected only by its gravitational effects.

Dark Matter and Energy New sources of energy density and matter required to explained cosmological observations. For high energy physics located near the electron-proton machine, *HERA*.

Deduction Method of logical reasoning that is determined by the laws of logic alone.

Deliberate Quality of an action where the agent is aware of what he or she is doing.

Determinism The philosophical view that all things are totally conditioned by antecedent causes.

Determinism Doctrine that all events, including human actions, have causes and that the future can in principle he predicted.

Deuteron The nucleus of deuterium is an isotope of hydrogen. It consists of one proton and one neutron bound together.

Dialectical A progressive process involving what are believed to be opposite, in either conversation or reality, which aims toward truth in conversion and creates change in reality; a philosophical method that posits an initial set of terms or principles and shows how they interact and result in new terms and principles.

Dichotomy A compelling difference in meaning between two things, so that they are of an "either-or" nature.

Differential Calculus Relates the rate at which the function changes to the rate of change of the argument. The ration of these two rates is called the derivative the function. In the instance just cited, the derivative of the displacement with respect to the time is the instantaneous speed of the falling object, which is a function of time in its own right. The derivative of the speed with respect to time, in turn, is the acceleration.

Diffraction A property which distinguishes wave-like motions. When a wave is incident upon a barrier which is broken by a narrow slit (of comparable size to the wavelength), then the slit will act as a new source of secondary waves. (ie.) The bending or spreading of a light beam as it passes through a narrow slit or hole.

Dignity Intrinsic worth that deserves respect. According to Immanuel Kant, rational agents have a dignity that cannot be bought or sold (priced).

Dimensions Physically significant quantities usually have dimensions associated with them. The fundamental dimensions are those of mass *M*, length *L* and time *T*. The dimensions of other quantities can be expressed in terms of these fundamental dimensions. So, for example, momentum has dimensions of mass x velocity (MLT^{-1}) and energy dimensions of force *x* distance (MLT^{-2}). One may define certain quantities in which the dimensions cancel out. These dimensionless quantities are significant in that they are independent of the conventions used to define units of mass, length and time.

Doppler Effect The change in the wavelength perceived by a listener who is in motion relative to a source of sound. It also applies to light and other electromagnetic waves.

Duality The equivalence between different superstring theories.

Duality Separation or split between two things, so they are radically different, for example, the Cartesian duality between mind and body.

Egoism Principle that humans act out of self-interest, egoism can be a description of human behavior or a prescription for it.

Eigenstate, Eigenvalue The eigenvalue of a matrix *M* is a number λ which satisfies the equation Mψ = λψ, with λ not—equal-to. In quantum mechanics, the matrix *M* will correspond to a particular dynamical variable (such as position, energy or momentum) and λ will correspond to the value obtained by measuring that dynamical variable if the system is in the state described the *ψ*. *Ψ* is called an Eigen state of the system.

Elastic Scattering Particle reactions in which the same particles emerge from the reaction as entered it (e.g. π-superscript (minus-sign) p -> π_{-} p). In inelastic scattering, where different and or new particles emerge, energy is used to create new particles.

Eliminative Materialism View in analytic philosophy of mind that references to subject states and attitudes (wanting, willing, intending, feeling. etc.) should be eliminated from scientific and empirical philosophical discourse.

Electric Current The rates at which electric charges move through a conductor or through empty space.

Electrons A fundamental particle that can't be split and carries negative electric charge. As far as we know, electrons have no size. A negatively charged spin-1/2 particle, which interacts via the electromagnetic, weak and gravitational forces. It has a mass of 0.511 MeV/c^2, some 1800 times lighter than the proton.

Elastic Scattering Particle reactions in which the same particles emerge from the reaction as entered it (e.g. π-superscript (minus-sign) p -> π_{-} p). In inelastic scattering, where different and or new particles emerge, energy is used to create new particles.

Empiricism Philosophical position that all knowledge of the world is and should be based on perceptual experience, either directly or indirectly.

Energy The capacity to perform work or the result of doing work.

Entropy A measure of the degree of disorder in a system. A quantitative measure of order (or equivalently information) in a physical system. The units are those of energy/temperature.

Epicycle The path traced be a point on the circumference of a circle as that circle is rolled around the circumference of a large one; used for calculating the orbits of planets up to the seventeenth century.

Epicycles Circles that the planets were supposed to follow in their paths across the heavens, according to Ptolemy's model of the Universe. The centers of those circles moved around the Earth. Copernicus showed that the model was wrong.

Epigenisis Early modern idea that living things develop over time, opposed to preformationism.

Epistemology Theory of knowledge, what counts as knowledge, and how beliefs are justified so as to qualify as knowledge.

Escape Velocity The minimum upward speed that you must give an object so that it leaves the Earth and never falls back. This value is 11 kilometers (7 miles) per hour.

Essence Aristotelian idea of that in a thing which makes it what it is and which is also present in all other things of the same category.

Ether An invisible substance filling all space that physicists of the 19th century invented to explain the motion of light in space. Einstein's theory of relativity did away with it.

Event Horizon The surface of a sphere with a radius equal to the Schwarzschild radius.

Events Occurrences in time, usually distinguished from things or substances.

Evidence Grounds for believing something is true, usually used in empirical context.

Existentialism Philosophical doctrine that truth for humankind begins in concrete human existence instead of from abstractions, and that humans have no pre-constructed nature or essence but must create their characters and lives through actions that they choose to do, and values and meanings that they actively bestow.

Experience Everything or anything that happens to or is encountered by a subject; empirical philosophy, perceptual or sensory occurrences; in pragmatism, the whole of all events, without a subject-object distinction.

Experimental Philosophy Early twenty-first century philosophical method of checking the intuitions philosophers have about widespread beliefs by empirically investigating those beliefs.

External Reality Everything except consciousness r the human mental subject, including the subject's physical body.

Faith Type of belief or attitude that does not require empirical evidence or logical reasoning.

Fallacy Mistake in logic or informal argument.

Falsifiability Standard for the scientific nature of theories and hypotheses, according to Karl Popper; so that if how a theory would be falsified cannot be specified then it is not scientific.

Falsification Process whereby an empirical belief is proved false by a prediction that fails to happen or an event that contradicts an hypothesis.

Falsification Used by Popper as a criterion for genuine scientific claims, namely that they should be capable of being falsified on the basis of contrary evidence.

Fatalism Non-philosophical form of determinism that does not posit causal chains but specific inevitable events.

Fermilab The Fermi National accelerator Laboratory, in Batavia, Illinois, USA. Fermilab is the home of the Tevatron, the world's most powerful accelerator, a pp-bar collider with a maximum collision energy of 1.8 TeV (= 1800 GeV = 1.8×10^{-12} eV).

Fermion Any particle with half-integer spin: $1/2\hbar$, $3/2\hbar$, $5/2\hbar$, etc. All fermions obey Pauli's exclusion principle.

Field The distortion of space due to the presence of a body that exerts a force on other bodies.

Flavor The term used to describe different quark types. There are six quark flavors: up, down, strange, charm, bottom and top.

Forms, Platonic Timeless, ideal entities that enable the appearance of entities in this world and set standards for their excellence.

Foundationalism The attempt to establish a basis for knowledge that requires no further justification (i.e., a point of certainty).

Functionalism Analytic philosophical theory of mind that defines mental processes in terms of computations that are related to brain states.

Galaxy A large island of billions of stars, held together by gravity.

Gamma Rays Electromagnetic radiation emitted by certain radioactive nuclei.

Geocentric Earth-centered. Ptolemy's model of the Universe, with the Sun and the stars revolving around the Earth, was geocentric.

Galaxy A large island of billions of stars, held together by gravity.

Gamma Rays Geocentric Earth-centered. Ptolemy's model of the Universe, with the Sun and the stars revolving around the Earth, was geocentric.

Gamma (Y) Rays Electromagnetic radiation emitted by certain radioactive nuclei. First discovered in radioactive material, and later identified as very high energy photons.

Gauge Theory A theory whose dynamics originate from a symmetry. That is, the formulate describing the theory (in particular, the Lagrangian) are unchanged under certain symmetry transformations called 'gauge' transformations.

Generation Leptons and quarks come in three related sets, called generations or families, consisting of two leptons and two quarks. The first generations consists of (e, v^{-e}; u, d). The second and third generations consist of (μ, v^{μ}, c, s) and (τ, v^{τ}, t, b) respectively.

Gluon Glue ball Gluons are the massless gauge bosons of QCD which mediate the strong color force between quarks. Because of the non-Abelian structure of the theory, gluons can interact with themselves, and may form particles consisting of gluons bound together. The existence of these 'glue balls' has yet to be confirmed.

Gods, gods Transcendent immortal beings with or without high moral qualities, who are more powerful than mortals and are capable of affecting human life as well as creating its material conditions.

Gravitational Acceleration Of all objects on the surface of the Earth is approximately 32 ft. per sec ^{2}.

Graviton A massless spin-2 particle which is the hypothetical quantum of the gravitational field. It mediates the force of gravity in a similar way to that in which the spin-1 gauge boson (i.e. the photon, W+/-, Z^{-0}, and gluons) mediate the other forces.

Hadron The generic name for any particle which experiences the strong nuclear force. (ie.) All particles affected by the strong nuclear force.

Heisenberg's Uncertainty Principle You can't determine with complete accuracy the position of a particle and at the same time measure where and how fast its moving.

Helicity The projection name for any particle which experiences the strong nuclear force.

Higgs Boson A hypothetical, spin-less particle that plays an important role in the Glashow-Weinberg-Salam electroweak theory (and in other theories involving spontaneous symmetry breaking, e.g. GUTs).

Higgs Boson Particles interact with the associated Higgs field, and this interaction gives particles their mass.

Higgs Mechanism A mechanism by which gauge bosons acquire mass through spontaneous symmetry breaking. In the Glashow-Weinberg-Salam electroweak model, for example, Higgs fields are introduced into the theory in a gauge-invariant way. However, the state of minimum energy breaks the local gauge symmetry, generating masses for the W+/- and Z^0 bosons, and giving rise to a real, observable Higgs boson, ϕ.

Holistic Describes an approach, argument or view that considers the operations of the whole of a complex entity (as opposed to its constituent parts).

Hyperon A baryon with non-zero strangeness.

Hypothetico-deductive method The process of devising and testing our explanatory hypotheses against evidence.

Hypothetical Not certain or declarative; a classic hypothetical has the form "If,then."

Idea Something before the mind intellectually, which may or may not represent something outside of the mind.

Idealism Philosophical doctrine that what is ultimately real is mental, rather than physical, sometimes leading to a denial of the existence of an external world.

Identity The nature of a thing whereby it is what it is; in contemporary social philosophy, the social nature, understood as constructed, of different types of human beings in terms of race, ethnicity, or gender.

Identity of Indiscernibles Gottfried Leibniz's principle that if two things are exactly the same then they are the same thing, from which it follows that two things cannot be exactly the same.

Ideology Set of beliefs about how things ought to be or interpretations of events based on ideas of how they ought to be; ideologies are not easily falsified.

Incommensurable Two theories or systems of thought are incommensurable if their key terms cannot be translated into one another.

Individualism Doctrines that value the separate individual, distinct from relationships with others.

Induction Process of reasoning that proceeds from experience to build up knowledge.

Induction (**Inductive inference**) The logical process by which a theory is devised on the basis of cumulative evidence.

Infinite Immeasurably and unthinkably great in magnitude; magnitude without limit.

Inertia The resistance of a body in an attempt to change its motion.

Innate Ideas Ideas or structures present in the mind from birth, which may be literally present in fully developed from, or emerge as the child develops.

Instrumentalism The view that scientific theories are to be assessed by how effective that are at explaining and predicting the phenomena in question.

Integral Calculus The reverses of the procedure of differentiation. It is the technique of adding the increments of a function to reconstitute the original function, such as locating an object from its variable speed. Integral calculus is also used to obtain the areas and volumes of figures whose boundaries are given.

Intentionally The aspect of consciousness that is about something other than itself, such as wanting, thinking, willing, desiring, etc.

Interference The pattern that results from the overlap of two wave motions that run into each other. The overlap can be a reinforcement of the waves or a cancellation.

Intuitionism Doctrine that some things, qualities, or truths, are known directly, with no need for empirical evidence or logical proof.

Isotopic spin or isospin A concept introduced by Heisenberg in 1932 to describe the charge independence of the strong nuclear force. Since the strong force cannot distinguish between a proton and a neutron, Heisenberg proposed that these particles were actually different states of a single particle – the nucleon. He argued that just as the electron comes in two different spin states, so the nucleon comes in two different 'isospin' states. So isospin is a concept analogous to spin which is conserved by the strong interaction. The nucleon is an isospin-1/2 particle, and its third component of isospin determines whether we are talking about a proton (I_3 = +1/2) or a neutron (I_3 = -1/2).

Justice As fairness, justice is treating those who are equal in some respect, the same way, or treating equals equally; distributive justice pertains to how the goods of life are divided among members of a community, nation, or the world.

Kelvin Unit of absolute temperature.

K Meson or Kaon The name of particular spin-0 mesons with non-zero strangeness quantum numbers.

Knowledge The goal of intellectual activity; in classic epistemology, knowledge is defined as true belief that has been arrived at in justified ways; i.e., I do not know something if I believe it and it is true, but I do not know way I believe it or how I have come to believe it. Neither do I know it, if it is true I think it's true because I dreamt it or I heard a voice in my head.

Lagrangian A mathematical expression summarizing the properties and interactions of a physical system. It is essentially the difference between the kinetic energy and potential energy of the system.

Laser Light amplification by stimulated emission of radiation produces a narrow beam of single-wavelength, coherent light.

Laws of Nature Regularity, so that e vents of one type are always followed by another; causal regularity; in religious philosophy. God's laws for human behavior.

Lepton All particles not affected by the strong nuclear force, such as the electron and the muon. The generic name for any spin-1/2 particle which does not feel the strong nuclear force. The six known leptons are the electron, the *muon*, the tau lepton, and their respective neutrinos. The name was originally coined to refer to light particles.

Light Year The distance travelled by light in one year, at a speed of 186,000 miles per second.

Logic Formal systems of rules of inference.

Logical Atomism View that an ideal philosophical language can be constructed in which basic terms will represent fundamental units of reality, usually attributed to Bertrand Russell and Ludwig Wittgenstein (in his early work).

Logical Positivism A school of philosophy from the first half of the twentieth century, which, influenced by the success of science, attempted to equate the meaning of a statement with its method of verification.

Logical Positivism Philosophical doctrine that the physical sciences should set the concerns and subject matter of philosophers; epistemological doctrine that a statement is meaningful if it can be said what in perceptual experience would have some bearing on its truth, or ideally, verify or falsify it.

Mass A subatomic particle that has no mass or a very tiny amount of mass – we don't yet know which. It plays a role in radioactivity.

Materialism Doctrine that what is ultimately real is physical.

Materialism The view that the world consists entirely of physical material (hence opposing any form of mind/body dualism or supernaturalism).

Matter Physical stuff or things, the material world.

Meaning The concept (connotation), or thing(s) in the world (denotation), that a world (denotation), that a word or term symbolizes.

Mechanism Explanation of reality in terms of causes and effects that do not make reference to anything distinctive about living things, but refer only to the movements of inert objects in space.

Mesons Hadrons which are bosons (integral spin particles) such as the pion.

Metaphysics In philosophy, abstract explanations of ordinary things, events, and experience, that refer to entities or processes that are not directly accessible to human perception, but are believed to be foundational for what is perceived.

Mind What is not matter, pertaining to the conscious human subject, a synonym for soul; the complex of perceptions, ideas, thoughts, emotions, memory, feeling, and self-reflection, considered as a whole.

M-Theory Considered to be a class of the 5 superstring theories. An unknown theory consisting in ten space-time dimensions, believed to encompass the five known superstring theories in the ten space-time dimensions. There are different dualities that connect the different types of string theories. They go by the names of S-duality and T-duality.

Monad Self-contained, individual unit of awareness or perception, which is the basic unit of substance; in modern philosophy a monad is a single oneness deriving from Gottfried Leibniz's philosophical system.

Monism Doctrine that there is only one thing in the whole of existence, or that all things are part of the one thing.

Monism Doctrine that there is only one thing in the whole of existence, or that all things are part of the one thing.

Mysterianism Doctrine that we cannot know the ultimate causes or reality of our most important concerns; new mysterianism is the doctrine that we will never know the nature of consciousness or how the mind is related to the body.

Mysticism System of belief that posits knowledge without logical reasoning processes or sensory experience.

Mythology In Western intellectual history, term used to refer to accounts, usually poetic or literary, of the nature and actions of ancient deities; broadly used to refer to narratives within a culture pertaining to beliefs that have no scientific foundation.

Naturalism Analytic and pragmatic philosophical methodology that seeks explanations and solutions o philosophical problems in ways that are compatible with or derived from scientific explanations.

Natural Fallacy G.E. Moore's doctrine that goodness is a non-natural quality so that if one defines it in terms of desired consequences or pleasure, or any other natural property, it can always be asked of something fitting the definition, "Is if good?"

Natural Kind A type of thing that is naturally formed to be what it is and where all members of the kind share certain characteristics.

Natural Philosophy The branch of philosophy which considers the physical world, a term used to include science prior to the eighteenth century.

Natural Philosophy Term for early modern physics, astronomy, and proto-chemistry.

Natural Religion Belief in a deity based on combination of reason and experience, rather than revelation.

Nature The non-human world, or the human idea of the non-human world.

Necessary casual condition An event or thing that is always present if an effect is present (effect need not be present if it is present; e.g., oxygen is a necessary condition for fire).

Necessity Type of connection that is logical in that it cannot be denied without contradiction, or connection between real events such that effects are inevitable given their causes.

Neoplatonism Doctrine from the ancient world, influential throughout philosophical history thereafter, that there exists a transcendental reality in which events determine what happens in this world.

Neutron One of the subatomic particles that make up the nucleus of an atom.

Neutron Star An object only tens of kilometers across but with a mass larger than that of the Sun. Neutron stars are composed mainly of neutrons.

Nominalism Doctrine that all natural kinds are arbitrarily designated as such by human intellectual concerns and activity and that there are no universals in reality, but only in language;

in its modern form, credited to John Locke through Boethius first formulated it, that essences are in the mind and made up by the mind.

Non-Euclidian geometry Coherent geometries with principles other than those laid down by Euclid, allowing, e.g., that parallel lines meet and angles in triangles add up to less than 180 degrees; geometric revolution in the nineteenth century that paved the way for Albert Einstein's theory of general relativity.

Noumena Things in themselves that are not directly perceived or describable by us, contrasted by Immanuel Kant with phenomena which we can perceive.

Nucleons Neutrons and protons.

Numerology Ancient doctrine, attributed to Pythagoras and his followers, that numbers are real entities, present throughout reality, in ways that determine the non-numerical properties of things.

Objective Independent of the mind, as in "objective reality"; in human discourse, a lack of bias; in science, the presumption that the same experiments will yield the same data to different observers.

Objectivism Philosophical system developed by Ayn Rand based n the existence of an external objective world and belief that the Aristotelian law of identity, A *is* A, yields a metaphysical truth about that world.

Observation Perceptual process, with or without the use of manmade instruments (e.g., the thermometers, cameras), for recording what happens.

Occasionalism Causal doctrine attributed to Nicolas Malebranche and others that because we cannot perceive causal connections, there are none in reality, although they do exist in the mind of God.

Ockham's Razor The principle that one should opt for the simplest explanation; generally summarized as 'causes should not be multiplied beyond necessity'.

One, the In Neoplatonism, ultimate ontological, ruling, moral and unified basis of existence that is itself separate from existence and/or may be expressed in it.

Ontological argument Proof for the existence of God, used by Rene Descartes and others, that proceeds from God's qualities, as we think them, this necessary existence.

Ontology The science and study of what exists, pursued as a distinct inquiry, or of what is believed to exist, in a specific domain of inquiry, pursued as a distinct inquiry. Martin Heidegger treated ontology in the first sense, W.V.O. Quine in the second.

Other minds, problem of The problem of how we know that other people have minds, since we cannot directly experience the mind of another as that person experiences it.

Paradigm A theory or complex of theories which together set the parameters of what is accepted as scientifically valid within its particular sphere of study. Kuhn describes how paradigms may eventually be replaced if they prove inadequate.

Paradigm According to Thomas Kuhn, a paradigm is an agreed upon set of beliefs in a mature science that determines the ontology of the field, its experimental methods, and appropriate objects of study. Used more loosely after Kuhn, a paradigm is any dominant worldview, in any area of human activity.

Pauli's Exclusion Principle Two identical fermions cannot have the same charge, spin momentum, quantum numbers etc. within the same region of space.

Perturbation Theory The probability of an event occurring is given by the sum of contributions from a series of increasingly more complex Feynman diagrams.

Phenomena Those things which are known through the senses; in Kant, it is the general term used for sense impressions, as opposed to *noumena*, or things as they are in themselves.

Phenomena The appearances of things or things as they show themselves; evidence; in Heideggerian philosophy, that which shows itself to man.

Phenomenalism Logical atomist or logical positivist view that material objects are made up of sense data.

Phenomenology Philosophical methodology, attributed to Edmund Husserl, in which the structures, processes, and intentional objects of consciousness are observed and analyzed.

Philosophy of Science Study of the principles used for scientific discovery and theory construction, as well as progress in science. Philosophy of science may and has been both descriptive and prescriptive.

Photon A quantum of light.

Photon (γ) The quantum of the electromagnetic field. It is the massless spin-1 gauge boson of QED. Virtual photons mediate the electromagnetic force between charged particles.

Phronesis Practical wisdom; in ancient Greece, both good judgment in ordinary affairs and knowledge of the ultimate goods and ends of life.

Planck Units Fundamental units of length, time, mass, energy, etc. involving Planck's quantum constant, h-bar, Newton's gravitational constant, G, and the speed of light, c. As they incorporate both the quantum and gravitational constants, the Planck units play a key role in theories of quantum gravity. The Planck energy is 10^{-19} GeV.

Platonism Systems of thought or ideas deriving from Plato, according to which there are transcendent entities that support the existence of, and are the ideas or essences of, every kind of thing in the world that humans experience.

Positron The antiparticle of the electron, discovered by Anderson in 1934. It has the same mass and spin as the electron, but opposite charge and magnetic moment.

Possible Not logically contradictory to imagine; what is possible in events need not be probable or likely.

Pragmatism American philosophy, known for an analysis of experience and social relevance; method that analyzes experience as an interactive process between the conscious subject and the world.

Primary qualities A term used by Locke for those qualities thought to inhere in objects, and are therefore independent of the faculties of the observer (e.g. shape).

Prime mover Aristotle's idea of the ultimate cause of the universe, posited because without it causal chains would be infinite; interpreted theologically as an argument for God's existence.

Probability Likelihood of an event happening; standard for prediction that is considered reliable, although it falls short of certainty; theory of how probability is assigned, the logic of likelihood.

Process philosophy Usually attributed to A.N. Whitehead, an ontological perspective that reality and everything in it is made up of events, instead of stable entities; method of analysis whereby what were believed to be things, turn out to be events or happenings over time.

Proof A process involving the manipulation of symbols, which is required to proceed in a certain way in mathematics or logic, for the conclusion to be justified; whenever the conclusion of an argument cannot be false if its premises are true.

Proposition The meaning of a sentence.

Proton One of the subatomic particles that make up the nucleus of an atom. The proton has positive electric charge.

Pulsar A fast-spinning, small, and very dense astronomical body that emits light and radio waves, like a lighthouse.

Qualities In ancient philosophy, qualities were considered "accidents" of substances. In early modern philosophy, a distinction was made between primary and secondary qualities. Primary qualities were mass, size, velocity, number of atoms, etc., whereas secondary qualities were color, odor, sound, etc.; the primary qualities were believed to cause the secondary qualities of perceptions as the result of the effects of the atoms in perceptible objects on sense organs.

Quanta Bundles or packets of energy that can't be split. Light and all electromagnetic radiation are made up of quanta.

Quietism Withdrawal from the world based on intellectual; reasons, such as in ancient skepticism, the impossibility of knowledge.

Quasar An extremely bright astronomical body usually found very far away, that shines with the brightness of trillions of Suns.

Quantum Chromodynamics (QCD) the quantum field theory describing the interactions of quarks through the strong "color" field (whose quanta are gluons). QCD is a gauge theory with the non-Abelian gauge symmetry group SU(3)-sub-c.

Quantum Electrodynamics (QED) The quantum field theory describing the interactions between electrically charged particles through the electromagnetic field (whose quantum is the photon). QED is the gauge theory with the Abelian gauge symmetry group U(1).

Quantum Field Theory The theory used to describe the physics of elementary particles. According to this theory, particles are localized quanta of these fields.

Quantum-Mechanical Amplitude A mathematical quantity in quantum mechanics whose absolute square determines the probability of a particular process occurring.

Quantum Theory The theory used to describe physical systems which are very small, having atomic dimensions or less. A feature of the theory is that certain quantities (e.g. energy, angular momentum, light) can only exist in certain discrete amounts, called quanta.

Radioactive (α) Alpha Decay Occurs when an element is not big enough to split into two, but would still like to shed some weight to move up the binding energy curve to a region of greater stability.

Radioactivity The spontaneous emission of particles or radiation from certain atoms.

Rationalism Doctrine opposed to empiricism, according to which knowledge about the world can be present or developed by means or reason, without prior experience.

Rationality Good sense, following the rules of logic and accepting available evidence in forming beliefs and making decisions about action.

Realism The view that scientific theories are descriptions of independent and actual, if unobservable, phenomena.

Realism Naïve realism is the philosophical version of the ordinary belief in the existence of an external physical world, which common sense philosophers think requires no special prof; in medieval philosophy, the belief that universals exist apart from particular objects that are similar, or exist in those objects.

Red Giant A large bright star with a low surface temperature. The Sun will spend its late years as a red giant, eventually ending as a burned-out star called a black dwarf.

Reductionism Doctrine that some things are "nothing but" other more fundamental or perceptible things, as in reducing material objects to atoms or sense data; methodological principle of explanation, whereby one statement or theory is reduced to another if it can be logically derived from it; e.g., the reduction of statements about chemical interactions in chemistry to statements about atoms in physics.

Reductionist The view that all statements about the world are made relative to some particular viewpoint and thus cannot be objective or final.

Reference Frame A reference point for a moving object.

Relativism Descriptive moral doctrine that different circumstances, agents, and cultures have different and often conflicting rules of behavior or value; prescriptive moral doctrine that there are no universal rules of behavior or human values.

Relativism The view that all statements about the world are made relative to some particular viewpoint and thus cannot be objective or final.

Rhetoric The art or skill or speaking or writing to persuade or impress listeners or readings.

Science Precise, rigorous, and formal system of thought and study of the world, including human beings, which in Aristotelian and Cartesian thought was believed to yield certain knowledge, but by the modern period was accepted as most probable knowledge. Since the nineteenth century the sciences have been divided into the physical sciences (e.g., psychology, sociology, anthropology, history), with more theoretical agreement and precision about data attributed to the physical sciences.

Scientific revolution The beginning of modern empirical science, in practice and theory, during the sixteenth and seventeenth centuries, following the Copernican revolution and epitomized by Isaac Newton; term used by Thomas Kuhn to refer to radical change in perspective within a scientific field.

Scientism The view that science gives the only valid interpretation of reality.

Schwarzschild Radius The largest radius of a star below which light will be trapped, as in a black hole.

Semantics Meanings or theory of meaning.

Sense data Sensory impressions of different senses (e.g, greenness, hardness, coldness) directly experienced in the present; believed by logical positivists to be the foundation of empirical knowledge.

Singularity A point in space-time at which the space-time curvature and other physical quantities become infinite and the laws of physics break down. Places where the laws of physics break down or don't apply.

Skepticism Doubt about otherwise plausible claims, as in skepticism about the existence of the external world or other minds. Before the modern period, skepticism was often used to show that knowledge was impossible so that other mental attitudes, such as quietude or faith, could be pursued. Academic skepticism was the view that no knowledge is possible, *pyrrhonic* skepticism the view that we cannot know whether any knowledge is possible.

Social Darwinism Late-nineteenth century application of principles of Darwinian evolution to human society, stressing competition and "survival of the fittest," often used or misused to support social inequality and advocate eugenics programs.

Solipsism Doctrine that I cannot know anything except my own mind and its contents; doctrine that I cannot know that anything exists beyond myself and my mind.

Sophism Form of rhetoric in ancient Greece whereby either side of an argument could be taken up; also associated with cultural relativism and cosmopolitanism during that time.

Soul Immaterial part of the self, considered paramount for morality and identity, and which may or not be believed to survive death.

Secondary Qualities A term used by Locke for those qualities used in the description of an object that are determined by the sensory organs of the perceiver (e.g. color).

Space According to Isaac Newton, space is an objective reality; according to Immanuel Kant it is a condition for human experience.

Spacetime singularity A theoretical point of infinite density and no extension, from which the present universe, including space and time themselves, is thought to have evolved.

Space-Time The combination of the three dimensions of space and one of time. It's needed in relativity because time and space are linked together.

Spectrum The array of the components of the emission of electromagnetic radiation arranged according to wavelength.

Synthetic Statements Those whose truth depends upon evidence (see analytic statements).

Standard Model This refers collectively to the successful theories of QCD and the Glashow-Weinberg-Salam electroweak model.

String Theory A theory in which the fundamental constituents of matter are not point-particles but tiny one-dimensional objects, which we can think of as strings. These strings are so minute (only 10^{-33} cm long) that, even at current experimental energies, they seem to behave just like particles. So, according to string theory, what we call 'elementary particles' are actually tiny strings, each of which is vibrating in a way characteristic of the particular 'elementary particle.'

Substance According to Aristotle and medieval philosophers, a living thing or other being that can exist independently; ultimate substratum or reality, as in Rene Descartes' material and immaterial substances; category of thought according to Immanuel Kant. The idea of substance as underlying substratum was rejected by empirical philosophers beginning with John Locke.

Sufficient cause Something that if present always has a certain effect, although it need not be present whenever the effect is.

Synthetic truth A statement that is true of the world.

Superstring Theory A new theory of the structure of matter that says that all elementary particles are represented by tiny vibrating strings. The theory promises to unify all the forces of nature.

Super-Symmetry An extension of Lorentz symmetry, relating fermions and bosons. If super-symmetry is a true symmetry of nature, then every 'ordinary' particle has a corresponding 'super-partner' which differs in spin by half a unit.

Tachyons Hypothetical particles that would always travel faster than light - and "faster than light" can mean "backwards in time."

Teleological Determined by a future end, goal, or purpose.

The final cause The theory that the meaning of a word is given by the object to which that word corresponds (problematic if we have no independent knowledge of objects).

Theism Belief in transcendental or non-natural beings or god(s).

Theory A linguistic system that can be used to explain experience, although everything asserted in the theory may not have a foundation in experience.

Thermodynamics The study of heat and thermal effects.

Time Aristotle defined time as a measurement of events. Since then, distinctions have been made between objective time as measured by clocks, time as a condition of perceptual experience (Immanuel Kant), and time as subjective experience (Henri Bergson), and time as constructed by the human apprehension of past, present, and future (Martin Heidegger). Bertrand Russell said that it was a contingent matter that we remember the past instead of the future.

TOEA 'theory of everything ', the attempt to find a single theory too account for the four fundamental forces (or interactive forces) of nature (gravity; electromagnetic; strong and weak nuclear) which describe the way in which elementary particles interact.

Transcendental argument Philosophical method attributed to Immanuel Kant of determining what must be true for human beings to be able to have the kind of experience they do; this process of "transcendental deduction" is rigorous in that it does not posit more than is necessary to account for experience.

Transcendentalism Positing entities that exist separately from experience; New England transcendentalism was a nineteenth century philosophical and literary movement that combined romantic ideas of the individual in natural environments with philosophical ideas from both Plato and Immanuel Kant.

Truth In modern analytic philosophy, a quality of statements or propositions. A statement is true according to the correspondence theory of truth if it accurately represents reality; in the coherence theory, true statements are compatible or consistent with other accepted knowledge.

Type I String Theory This version of string theory includes both bosons and fermions. Particle interactions include super-symmetry and a gauge group SO(32). This theory and all that follow require 10 space-time dimensions for consistency.

Type II-A String Theory This version of string theory also includes super-symmetry, and open and closed strings. Open strings in type II-A string theory have their ends attached to higher-dimensional objects called D-branes. Fermions in this theory are not chiral.

Type II-B String Theory Like type II-A string theory, but it has chiral fermions. Heterotic string theory includes super-symmetry and only allows closed strings. It has a gauge group called E_8 * E_8. The left- and right- moving modes on the string actually require different numbers of space-time dimensions (10 and 26). There are actually two heterotic string theories.

Underdeterminism The situation in which we do not have enough evidential data to choose between competing theses.

Utilitarianism Theory by which an action is judged according to its expected results.

Uniform Motion Motion at a constant speed along a straight line.

Universal grammar Innate grammar present in all human beings, enabling them to learn a finite number of natural languages, as posited in different formulations by Noam Chomsky.

Universals General terms like "cat" and "dog." From ancient Greek through early modern philosophy there was a debate about whether universals themselves were real or only particulars were real, or universals were real insofar as they existed "in" particulars.

Vacuum The state of minimum energy (or ground state) of a quantum theory. It is the quantum state in which no real particles are present. However, because of Heisenberg's uncertainty principle, the vacuum is actually seething with virtual particles which constantly materialize, propagate a short distance and then disappear.

Validity Characteristic of an argument that proceeds according to rules of logic.

Value Something worth having, striving for, or retaining, which has intrinsic, usually non-monetary worth, or imparts such worth to other things.

Velocity It is characterized by both magnitude (the speed of the particle) and direction.

Verificationism Logical positivist doctrine that the meaning of a sentence is how it would be verified or falsified in perceptual experience and that only sentences that can be verified or falsified by perception are meaningful.

Virtual Particle A particle that exists only for the brief moment allowed by Heisenberg's uncertainty principle.

Virtue Trait of character considered excellent or morally good, or a disposition to behave in such ways.

Virtual Processes A quantum-mechanical processes which do not conserve energy and momentum over microscopic timescales, in accordance with Heisenberg's uncertainty principle. These processes cannot be observed.

Vitalism Scientifically outdated view of a life force accounting for what is distinct about living things and their abilities to reproduce themselves, which was largely put to rest by James Watson and Francis Crick's discovery of the model of DNA.

Wave A mechanism for the transmission of energy.

Wavefunction A mathematical function describing the behavior of a particle according to quantum mechanics. The wavefunction satisfies Schrodinger's wave equation. Furthermore, the probability of finding the particle at a particular point in space is given by the absolute square of the wavefunction.

Wavelength The length of an oscillation; the distance between two crests or two valleys in a wave.

White Dwarf A small, dense star. One of the final stages in the life of the Sun.

Weltanschauung Terms used for an overall view of the world, through which experience is interpreted.

Worm Hole A tunnel through space that connects two black holes at different places in the Universe. We don't yet know if worm holes really exist.

'J. A. Wheeler, Geometrodynamics 1962).

(Academic, New York, t tF. J. Tipler, Phys. Rev. D 17, 2521 (1978); T. A. Roman, Phys. Rev. D 33, 3526 (1986), and 37, 546 (1988).

'Deutsch and Candelas, in Ref. 10.

'These renormalized stresses and energy density are the same as for flat plates in flat spacetime. The wormhole's spacetime geometry will modify them by fractional amounts +s/ro, which we ignore. The nonzero extrinsic curvature of the plates, which is of order s/r$, will give rise to renormalized 2C. W. Misner, K. S. Thorne, and J. A. Wheeler, Gravita- tion (Freeman, San Francisco, 1973).

M. D. Kruskal, Phys. Rev. 119, 1743 (1960).4M. S. Morris and K. S. Thorne, Am. J. Phys. (to be pub-lished).STheorem 2 of R. P. Geroch, J. Math. Phys. 8, 782 (1967) forfurtherdiscussion,seeR.P.Geroch,Ph.D.thesis,Princeton stressesandenergydensityoforder(hs/ro) (l~s/2)'which University, 1967 (unpublished).sF. J. Tipler, Phys. Rev. Lett. 37, 879 (1976); Sec. 5 of F. J.

Tipler, Ann. Phys. (N.Y.) 108, 1 (1977).7The method for mathematical construction of such space-times is given by P. Yodzis, Commun. Math. Phys. 26, 39 (1972). John Friedman and Zhang Huai (private communica- tion) have given a lovely explicit example in which two wormholes are created in a compact region of spacetime.

SS. W. Hawking, Phys. Rev. D 37, 904 (1988).H. Kandrup and P. Mazur, "Particle Creation and Topolo-gy Change in Quantum Cosmology" (unpublished).'Proposition 9.2.8 of S. %. Hawking and G. F. R. Ellis, The Large Scale Structure of Space Time (-Cambridge Univ.

Press, Cambridge, 1973), as adapted to the wormhole topology and as modified by replacement of the weak energy condition by the AWEC with an argument from D. Deutsch and P. Can- delas, Phys. Rev. D 20, 3063 (1980), and references therein. We thank Don Page for pointing out a variant of this argument.

diverge at the plates; but these can be made negligible by giv- ing the plates a finite skin depth 8—(s /ro) 't «ro, see Ref.

12.'4F. J. Tipler, C. J. S. Clarke, and G. F. R. Ellis, in Einstein Centenary Volume, edited by A. Held and P. Bergman (Ple- num, New York, 1979).

'C. W. Misner, in Relativity Theory and Astrophysics I: Relativity and Cosmologyedit, ed by J. Ehlers (American Mathematical Society, Providence, RI, 1967), p. 160; also Sec. 5.8 of Ref. 10.

'C. W. Misner and A. H. Taub, Zh. Eksp. Teor. Fiz. 55, 233 (1968) [Sov. Phys. JETP 28, 122 (1969)].

'For a discussion of Cauchy horizons with structures similar to this, see F. J. Tipler, Ph. D. thesis, University of Maryland, 1976 (unpublished), Chap. 4.

'For a discussion of uniqueness of evolution of nonglobally hyperbolic spacetimes from a somewhat diA'erent viewpoint, see R. M. Wald, J. Math. Phys. 21, 2802 (1980).

'Parallel Worlds' – Michio Kaku – Anchor Books pp. 140-162

'The Fabric of the Cosmos – Space,Tme, and the Texture of Reality –Brian Green-Vintage Books, pages 120-132

'COSMOS' -Carl Sagan pp. 1-18

'In Search of Schrodinger's Cat' – John Gribbin, pp. 175-182, 205-226

'Consciousness' – A very short Introduction – Susan Blackmore pp 324-362

'Einstein for Dummies – E-MC^2' – Carlos L. Calle, pp 245-248. 282-320

'Black Holes and Baby Universes' Steven Hawking, pp. 110-125, 140-142

'Warped Passages – Unraveling the Mysteries of the Universe's Hidden Dimensions' – Lisa Randall, pp. 262-268, 282-294

'In Search of the Multiverse – Parallel World, Hidden Dimensions, and the Ultimate Quest for the Frontiers of Reality –John Gribbin – John Wiley & Sons, Inc. pp. 18-33, 114-23, 33-108

'A Briefer History of Time' – Steven Hawking – pp 128-146

'Einstein's Theory of Relativity' – Max Born – pp. 166-172, 214-224

'The Big Bang' – pp.320-355

'Understanding Einstein's Theories of Relativity – Man's New Perspective on the Cosmos – Stan Gibilisco – pp. 172-178, 210-234

'The Theory of Everything – The origin and fate of the universe' – Steven W. Hawking – pp. 151-156, 172-185

'A Briefer History of Time' – Steven Hawking – pp. 156-162, 174-187

'The God Effect' – Brian Clegg – pp. 420-432, 510-518

'The Riddle of Gravitation'- Peter G. Bergmann, pp.138-140, 156-168